Carbon Nanostructures

Carbon is intimately connected to almost everything we deal with in a daily basis. Due to its outstanding properties, such as high stability at environmental conditions, different hybridizations, strong covalent bond formation and easy of compounds formation, carbon has been a topic of scientific interest in several areas. Indeed, starting in the 19th century, chemists have devoted a whole field to study carbon-based compounds, which is, nowadays, known as Organic Chemistry. Remarkably, the last 30 years have been witnessing an exponential advance in the science involving carbon and carbon structures. Since the discovery of Fullerenes in 1985, which was awarded the Nobel Prize in Chemistry in 1996, carbon nanostructures have been attracting a great deal of attention from the research community. This public interest dramatically increased after the publications by the Iijima and Bethune groups on single-wall carbon nanotubes in 1993 and found a "new research era" with the isolation of a monolayer of carbon atoms, also called graphene, which conducted groundbreaking experiments demonstrating outstanding phenomena such as the Klein-Tunneling and the fractional quantum hall effect. No wonder, graphene was the object of the 2010 Nobel Prize in Physics.

The "Carbon Nanostructures" book series covers the state-of-art in the research of nanocarbons and their applications. Topics related to carbon allotropes such as diamond, graphite, graphene, fullerenes, metallofullerenes, solid C60, bucky onions, foams, nanotubes and nanocones, including history, theory, synthesis, chemistry & physics, Biophysics & engineering, characterization methods, properties and applications are welcome. Within the "Carbon Nanostructures" book series, the reader will find valuable, up-to-date account of both the newer and traditional forms of carbon. This book series supports the rapid expansion of this field and is a valuable resource for students and professionals in several different areas.

Springer welcomes new book ideas from potential authors. If you are interested in publishing your book in this series, please contact us via mayra.castro@springer.com

**** Indexed by Scopus (2018) ****

J. Tharini · Sabu Thomas
Editors

Carbon Nanomaterials and their Composites as Adsorbents

Springer

Editors
J. Tharini
Department of Science and Humanities
Solamalai College of Engineering
Madurai, India

Sabu Thomas
Mahatma Gandhi University
Kottayam, India

ISSN 2191-3005 ISSN 2191-3013 (electronic)
Carbon Nanostructures
ISBN 978-3-031-48718-7 ISBN 978-3-031-48719-4 (eBook)
https://doi.org/10.1007/978-3-031-48719-4

This Springer imprint is published by the registered company Springer Nature Switzerland AG
The registered company address is: Gewerbestrasse 11, 6330 Cham, Switzerland

Paper in this product is recyclable.

To Richard Smalley, Robert Curl and Harry Kroto

Founders of Buckminsterfullerene

Foreword

It is true that in the past few decades, discovery of new allotrope, namely, carbon nanotubes, Buckminsterfullerene, graphene oxide has created the landmark in the history of adsorption process. The rapid development of nanoscience and nanotechnology has significantly improved the adsorption mechanism. The discovery of carbon nanomaterials has been the promising adsorbents for the metal ion adsorption, effluents from the battery industries, chemical industry, pharmaceutical industries, dyes, and pesticides. A versatile and highly efficient carbon nanomaterial-based composite aids in the removal of toxic pollutants from the water. Carbon nanomaterials could replace traditional adsorbents including zeolites, activated carbon from solid wastes, silica gel, molecular sieves, activated alumina, ion-exchange resins, clays, agricultural wastes, biosorbents and miscellaneous adsorbents because of the high sorption potential, malleable surface charges, enormous pore size, large surface area, and rapid adsorption kinetics.

First part of the book deals with the fundamentals of nanomaterials, classification of carbon nanomaterials, and synthesis of carbon nanomaterials. It helps the reader to keep update the latest success and failures of nanomaterials. The literature in this field is exploding in such a manner that is extremely useful to have a concise overview. The well-documented explanation for the synthesis of carbon nanomaterials will benefit the chemist, physicist, biochemist, engineers, biologist, and another interdisciplinary scientist around the globe. The second part of the book gives the recent literatures for the adsorption and desorption of organic dyes, pesticides, micro-pollutants, heavy metals, and aromatic compounds. These studies will help the environmental scientist to learn latest literatures on the adsorption studies. The final chapter deals with the theoretical modeling of the carbon nanotubes. Finally, I recommend this book for the undergraduate, postgraduate student, experimental scientist, and computational scientist. This book is recommended for both academic and industrial researchers.

Prof. Daniel Pasquini
Instituto de Química
Universidade Federal de Uberlândia
Uberlândia, Brazil
daniel.pasquini@ufu.br

Preface

Adsorption is one of the most important fields of research for water purification. It is also the subject of surface phenomenon which creates a layer of the adsorbate on the surface of the adsorbent. It is the important topic of physical chemistry that every student enjoys from school level to the graduate level. The target audience of this book is, namely, undergraduate students, postgraduate students of environmental engineering, chemistry, environmental science, and Ph.D. students who are active in the area of environmental remediation. This book is mainly designed for both academics and industrial researchers. The goal of the book is to explain the basic concepts of nanomaterial adsorption. We focus on basic concepts of nanomaterials and their uses as the nanoadsorbents for the water purification. The basic concepts of adsorption are well explained with suitable examples. The principal aim of the book is to reach the audience with the latest trends on adsorption for the removal of metal, micro-pollutants, food adulterants, aromatic compounds, pesticides, dyes, and oil particles. This book motivates the student to study absorption at the introductory university level to the advanced level of research.

This book gives the overview of the interdisciplinary aspects of adsorption and practical applications for the removal of hazardous chemicals from the water. We focus on the different types of adsorption isotherms, namely, Langmuir isotherm, Freundlich isotherm, BET adsorption isotherm, and others for water purification. Adsorption kinetics studies are well illustrated with different models, namely, first-order, second-order, and third-order kinetics. The readers will be able to understand the basic mathematics concepts of kinetic models and further optimize these for the complex system. The required background is the basic knowledge of chemistry, biology, mathematics, and nanoscience at the first year level of the university. This book is divided into 16 chapters. Chapters "Introduction to Nanomaterials"–"Application and Research Progress of Nanomaterials as Adsorbents in Environment Field" explain the introduction to nanomaterials, carbon nanoadsorbent, their synthesis, and surface modification. Chapters "Adsorption Isotherms and Kinetic Models" and "Adsorption and Desorption of Adulterants in the Food Industry" explain the fundamental adsorption isotherm, kinetic models, thermodynamic parameters, adsorption mechanism, and experimental adsorption techniques. Chapters "Adsorption and Desorption of Micropollutants"–"Purification of Water Using Carbon Nanomaterials" describe the adsorption and desorption of aromatic compounds, pesticides, heavy metal, micro-pollutants, food adulterant, oil particles, and dye molecules. The last two chapters deal with the theoretical modeling of carbon nanomaterials for the adsorption studies.

Madurai, India J. Tharini
Kottayam, India Sabu Thomas

Acknowledgements

First we thank the almighty for giving this wonderful opportunity to complete this book. We express our sincere thanks for family members and friends for their support and cooperation. We thank all the authors for their wonderful contributions towards the publication of the book. We are very much indebted to all the review experts who have made very critical comments for each chapter. We express our sincere thanks to the publishers for their patience. We express our deep sense of gratitude to our co-workers and assistants. We are indebted to our mentor Dr. R. Saraswathi, Retd. Professor, Department of Materials Science, School of Chemistry, Madurai Kamaraj University for her inspiration and support.

Contents

Introduction to Nanomaterials 1
K. K. Wang, P. V. Chai, and W. L. Ang

Carbon Nanoadsorbents 17
Aruna Yadav and Surender Kumar

Carbon Nanomaterials and Their Composites as Adsorbents 39
Khlood A. Alrefaey, A. B. Abdel-Aziz, Lobna A. Said,
Irene Samy Fahim, and Ahmed G. Radwan

Biological Properties of Carbon Based Materials for Biomedical Applications 63
A. Joel Amalan, M. Kanagalakshmi, K. Hemkumar, and Anitha Pius

Synthesis Strategies of Various Carbon Materials 75
Seema Panicker, Javad B. M. Parambath, and Ahmed A. Mohamed

Surface Modification and Engineering of Nanoscale Absorbent and Their Composite 89
D. Aravind, P. Diwahar, M. Bharathi, K. Prakalathan, M. S. Prasanth,
and Sridhar Jayavel

Application and Research Progress of Nanomaterials as Adsorbents in Environment Field 105
Hailan Qin, Yunkang Liu, Huan Liu, Siyuan Di, and Shukui Zhu

Adsorption Isotherms and Kinetic Models 135
M. Kanagalakshmi, S. Gopika Devi, P. Ananthi, and Anitha Pius

Adsorption and Desorption of Adulterants in the Food Industry 155
Manisha Manimuthu, Sridhar Jayavel, Johni Rexliene Michael,
Balaji Viswanathan, and Ashwini Karuppaswamy

Adsorption and Desorption of Micropollutants 175
K. Hemkumar, P. Ananthi, M. Kanagalakshmi, and Anitha Pius

Recent Trends in the Adsorption and Desorption of Heavy Metals 187
P. Ananthi, K. Hemkumar, S. Subasini, and Anitha Pius

Carbon Nanomaterials for Adsorption and Desorption of Pesticides 203
Astha Tripathi, Seema Lal, and Pratibha Kumari

Adsorption and Desorption of Dyes 229
S. Gopika Devi, A. Joel Amalan, S. Subasini, and Anitha Pius

Two-Dimensional Carbon-Based Materials for Sorption of Selected Aromatic Compounds in Water 247
Adedapo O. Adeola and Patricia B. C. Forbes

Purification of Water Using Carbon Nanomaterials 261
S. Subasini, M. Kanagalakshmi, and Anitha Pius

Mathematical Modeling of Nonlinear Vibrations of Single-Walled Carbon Nanotubes 275
Sekar Rekha and Lakshmanan Rajendran

Editors and Contributors

About the Editors

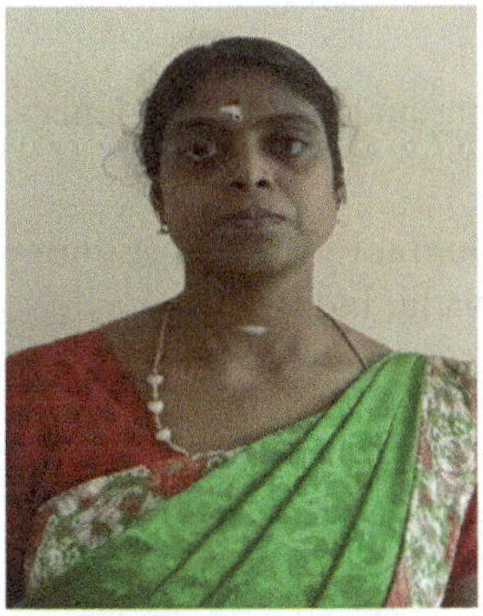

Dr. J. Tharini is working as the Assistant Professor, Department of Chemistry, Solamalai College of Engineering affiliated to the Anna University, Chennai, India. Her area of interest includes adsorption, chemical kinetics, nanoscience, and electrochemistry. She has both teaching and research experience for more than a decades. She has more than 20 research publications in the international referred journals and has published 1 book on popular science. She has supervised more than 8 master graduate projects in chemistry.

Prof. Sabu Thomas is the Director of the International and Interuniversity Centre for Nanoscience and Nanotechnology, Mahatma Gandhi University, Kottayam, Kerala, India. He is also currently the Chairman of the Trivandrum Engineering Science and Technology Research Park (TrESPARK), Trivandrum, Kerala, India. He was the former Vice Chancellor of Mahatma Gandhi University, Kottayam, Kerala, India. He is a highly committed teacher and a remarkably active researcher well-known nationally and internationally for his outstanding contributions in polymer science and nanotechnology. He has published over

1400 research articles in international refereed journals and has also edited and written 210 books. His H-index is 137 and total citations are more than 89,000. He has supervised 125 Ph.D. theses. He has received many international and national awards and recognitions. Under the leadership of Prof. Thomas, Mahatma Gandhi University has been transformed into a top university in India.

Contributors

A. B. Abdel-Aziz October High Institute for Engineering & Technology, Giza, Egypt

Adedapo O. Adeola Department of Chemical Sciences, Adekunle Ajasin University, Ondo State, Akungba-Akoko, Nigeria

Khlood A. Alrefaey SESC Center, School of Engineering and Applied Sciences, Nile University, Giza, Egypt

A. Joel Amalan Department of Chemistry, The Gandhigram Rural Institute—Deemed to be University, Gandhigram, Dindigul, Tamil Nadu, India

P. Ananthi Department of Chemistry, The Gandhigram Rural Institute - Deemed to be University, Tamil Nadu, Dindigul, India

W. L. Ang Department of Chemical and Process Engineering, Faculty of Engineering and Built Environment, Universiti Kebangsaan Malaysia, UKM, Bangi, Selangor, Malaysia;
Centre for Sustainable Process Technology, Faculty of Engineering and Built Environment, Universiti Kebangsaan Malaysia, UKM, Bangi, Selangor, Malaysia

D. Aravind University Science Instrumentation Centre, Madurai Kamaraj University, Madurai, Tamil Nadu, India

M. Bharathi Department of Mechanical Engineering, KIT, Coimbatore, India

P. V. Chai Department of Chemical and Petroleum Engineering, Faculty of Engineering, Technology and Built Environment, UCSI University, Cheras, Kuala Lumpur, Malaysia

S. Gopika Devi Department of Chemistry, The Gandhigram Rural Institute—Deemed to be University, Gandhigram, Dindigul, Tamil Nadu, India

Siyuan Di State Key Laboratory of Biogeology and Environmental Geology, China University of Geosciences, Wuhan, China

P. Diwahar Central Institute of Plastics Engineering and Technology, Chennai, India

Irene Samy Fahim SESC Center, School of Engineering and Applied Sciences, Nile University, Giza, Egypt

Patricia B. C. Forbes Department of Chemistry, Faculty of Natural and Agricultural Sciences, University of Pretoria, Hatfield, Pretoria, South Africa

K. Hemkumar Department of Chemistry, The Gandhigram Rural Institute—Deemed to be University, Gandhigram, Dindigul, Tamil Nadu, India

Sridhar Jayavel Department of Biotechnology (DDE), Madurai Kamaraj University, Madurai, India

M. Kanagalakshmi Department of Chemistry, The Gandhigram Rural Institute—Deemed to be University, Gandhigram, Dindigul, Tamil Nadu, India

Ashwini Karuppaswamy Department of Biotechnology (DDE), Madurai Kamaraj University, Madurai, India

Surender Kumar Department of Chemistry, Chaudhary Bansi Lal University, Bhiwani, Haryana, India

Pratibha Kumari Bioorganic Material Research Laboratory, Deshbandhu College, University of Delhi, Kalkaji, New Delhi, India

Seema Lal Bioorganic Material Research Laboratory, Deshbandhu College, University of Delhi, Kalkaji, New Delhi, India

Huan Liu State Key Laboratory of Biogeology and Environmental Geology, China University of Geosciences, Wuhan, China

Yunkang Liu State Key Laboratory of Biogeology and Environmental Geology, China University of Geosciences, Wuhan, China

Manisha Manimuthu Department of Biotechnology (DDE), Madurai Kamaraj University, Madurai, India

Johni Rexliene Michael Department of Biotechnology (DDE), Madurai Kamaraj University, Madurai, India

Ahmed A. Mohamed Department of Chemistry, College of Sciences, Sharjah, UAE

Seema Panicker Fatima College of Health Sciences, Ajman, UAE

Javad B. M. Parambath Department of Chemistry, College of Sciences, Sharjah, UAE

Anitha Pius Department of Chemistry, The Gandhigram Rural Institute—Deemed to be University, Gandhigram, Dindigul, Tamil Nadu, India

K. Prakalathan Department of Plastic Technology, CIPET, Chennai, India

M. S. Prasanth Central Institute of Plastics Engineering and Technology, Chennai, India

Hailan Qin State Key Laboratory of Biogeology and Environmental Geology, China University of Geosciences, Wuhan, China

Ahmed G. Radwan Nanoelectronics Integrated Systems Center (NISC), Nile University, Giza, Egypt;
Engineering Mathematics and Physics Dept, Faculty of Engineering, Cairo University, Giza, Egypt

Lakshmanan Rajendran Department of Mathematics, AMET Deemed to be University, Chennai, India

Sekar Rekha Department of Mathematics, St. Joseph College of Engineering, Sriperumbudur, Chennai, India

Lobna A. Said Nanoelectronics Integrated Systems Center (NISC), Nile University, Giza, Egypt

S. Subasini Department of Chemistry, The Gandhigram Rural Institute—Deemed to be University, Gandhigram, Dindigul, Tamil Nadu, India

Astha Tripathi Bioorganic Material Research Laboratory, Deshbandhu College, University of Delhi, Kalkaji, New Delhi, India

Balaji Viswanathan Department of Biotechnology (DDE), Madurai Kamaraj University, Madurai, India

K. K. Wang Department of Chemical and Petroleum Engineering, Faculty of Engineering, Technology and Built Environment, UCSI University, Cheras, Kuala Lumpur, Malaysia

Aruna Yadav Department of Chemistry, Chaudhary Bansi Lal University, Bhiwani, Haryana, India

Shukui Zhu State Key Laboratory of Biogeology and Environmental Geology, China University of Geosciences, Wuhan, China

Abbreviations

β-CD	β-Cyclodextrin
ADM	Adomian Decomposition Method
AFM	Atomic Force Microscope
ANN	Artificial Neural Network
AOP	Advanced Oxidation Process
ARB	Accumulative Roll Bonding
BET	Brunauer–Emmett–Teller
BJH	Barrett–Joyner–Halenda
CN	Carbon Nanoparticles
CNs	Carbon-based Nanoadsorbents
CNTs	Carbon Nanotubes
CQDs	Carbon Quantum Dots
CVD	Chemical Vapor Deposition
DEO	Differential Evolution Optimization
DND	Detonation Nanodiamond
DWCNTs	Double-Walled Carbon Nanotubes
ECAP	Equal Channel Angular Pressing
ED	Electrodialysis
FETs	Field Effect Transistors
$G\text{-}C_3N_4$	Graphitic-Carbon Nitride
GA	Genetic Algorithm
GF	Graphite Flakes
GO	Graphene Oxide
GW	Graphene Wool
HPHT	High-Pressure High-Temperature
HPT	High-Pressure Torsion
LUMO	Lowest Unoccupied Molecular Orbital
MAHs	Monocyclic Aromatic Hydrocarbons
MO	Methyl Orange
MOFs	Metal–Organic Frameworks
MP	Micropollutant

MS	Mass Spectrometry
MSA	Moth Search Algorithm
MWCNT	Multiwalled Carbon Nanotubes
NDs	Nanodiamonds
NPs	Nanoparticles
NTCs	Nontubular Carbons
PAHs	Polycyclic Aromatic Hydrocarbons
PANI	Polyaniline
PET	Polyethylene Terephthalate
PPCPs	Pharmaceuticals and Personal Care Products
PS-MS	Paper Spray Mass Spectrometry
RGO	Reduced Graphene Oxide
RO	Reverse Osmosis
SEM	Scanning Electron Microscope
STM	Scanning Tunneling Microscope
SWCNT	Single-walled Carbon Nanotube
TCFs	Transparent Conductive Films
TEM	Tunneling Electron Microscope
TG	Thermogravimetric
TNT	Trinitrotoluene
VIM	Variational Iteration Technique
XRD	X-ray Diffraction

Introduction to Nanomaterials

K. K. Wang, P. V. Chai, and W. L. Ang

Abstract Nanomaterials have grown over the years globally associated to their advancement in the past decades and remarkable impact to the various applications. This chapter discussed the properties of the nanomaterials namely electrical, magnetic, mechanical, and antibacterial properties. Subsequently, the types of nanomaterials selected to be discussed in this book chapter are carbon, inorganic, organic, and composite based nanomaterials. The latter part of the book chapter shall focus on the classification of nanomaterials from zero dimensional to three dimensional and the synthesis of nanomaterials typically on bottom-up and top-downapproach.

Keywords Nanomaterials · Properties of nanomaterials · Classification of nanomaterials · Bottom-up and top-down approach

1 Introduction to Nanomaterials

The word nanomaterials can be separated into two different words which are nano and materials. The word nano in Greek means dwarf. In the current International System of Units, nano is described as a prefix whereby one nanometer equals to billionth of a meter (10^{-9}). Therefore, nanomaterials can be described as a class of materials that have singular or multiple dimensions in nanometric scale. The size of nanoparticles is set to be ranging from 1 to 100 nm.

The idea of nanotechnology is first inspired and proposed in the year of 1959 by the famous physicist Richard Feynman. During a talk with the title "There's

K. K. Wang · P. V. Chai (✉)
Department of Chemical and Petroleum Engineering, Faculty of Engineering, Technology and Built Environment, UCSI University, Cheras, 56000 Kuala Lumpur, Malaysia
e-mail: chaipv@ucsiuniversity.edu.my

W. L. Ang
Department of Chemical and Process Engineering, Faculty of Engineering and Built Environment, Universiti Kebangsaan Malaysia, UKM, 43600 Bangi, Selangor, Malaysia

Centre for Sustainable Process Technology, Faculty of Engineering and Built Environment, Universiti Kebangsaan Malaysia, UKM, 43600 Bangi, Selangor, Malaysia

J. Tharini and S. Thomas (eds.), *Carbon Nanomaterials and their Composites as Adsorbents*, Carbon Nanostructures, https://doi.org/10.1007/978-3-031-48719-4_1

Plenty of Room at the Bottom", Feynman sparked the possibilities of controlling and manipulating individual atoms or molecules. The idea came to fruition after the development of scanning tunneling microscope (STM). The STM technology allowed the researchers to catch sight of individual atoms.

1.1 Properties of Nanomaterials

Different nanomaterials exhibit different physicochemical properties. These properties have made the nanomaterials special in various applications. Examples of different properties are listed down such as electrical properties, magnetic properties, mechanical properties, and anti-bacterial properties.

1.1.1 Electrical Properties

Recently, polyaniline (PANI) based nanomaterials have attracted the interest of many researchers due to their good electrical properties despite their polymer nature. The PANI is also easy to produce and has high environmental stability [1]. The PANI is now very popular in the manufacture of various sensors, cells, and conductors [2–5]. For example, a study from Wen et al., 2018 found that the addition of 0.5 weight percentage of PANI into 60 weight percentage of silver-filled flexible conductors had greatly brought down the electrical resistivity by one-thirtieth of original. Besides, the addition of PANI had also increased the bending stability by a large margin [5]. Other than that, reduced graphene oxide (RGO) particles are also one of the nanomaterials that attract the attention of many researchers. For example, a study from Yang et al. (2017) showed that the 23 μm reduced graphene oxide with cellulose nanofiber film has an ultra-high electrical conductivity [6].

1.1.2 Magnetic Properties

Some of the nanomaterials exhibit strong magnetic properties. This unique property can allow various applications in the science and engineering industries. For example, iron oxide (Fe_2O_3) nanoparticles have been used as pollutants cleanup agent or biomedical materials due to their magnetic properties [7–9]. For example, iron oxide has been used for magnetic drug delivery in the human body [9]. On the other hand, iron oxide nanoparticles have also been used for the cleaning of marine oil spills. The cleaning of oil spills can be done using the iron oxide based nano adsorbent as the oil can be adsorbed onto the surface of the nano adsorbent. After that, the oil can be recollected using a magnet as the iron oxide based nano adsorbent has magnetic properties [7, 8]. Figure 1 shows the magnetic properties of Fe_2O_3 nanoparticles in oil remediation.

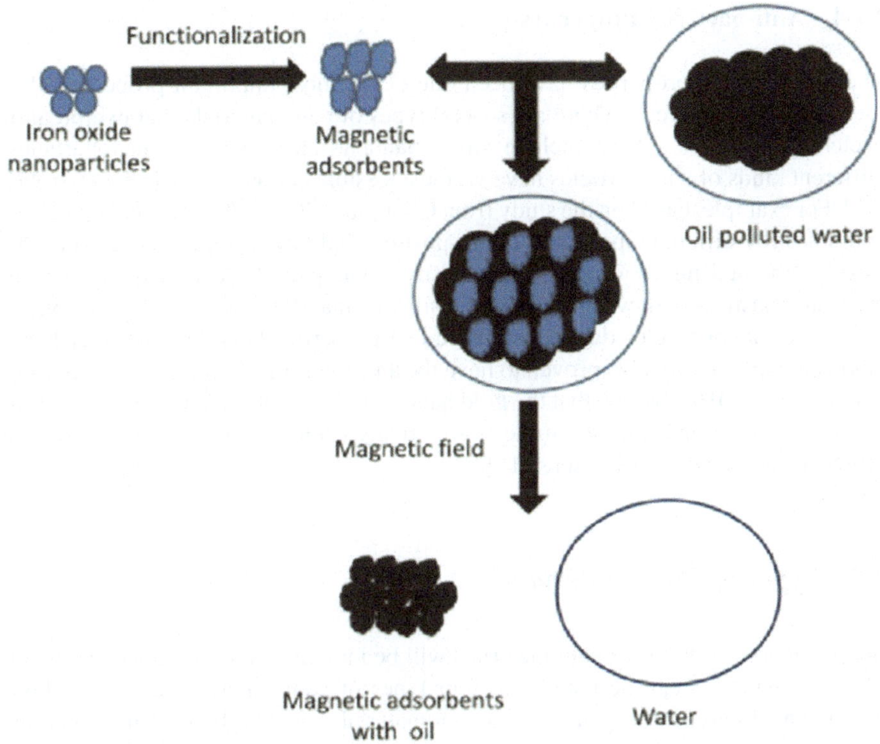

Fig. 1 Magnetic adsorbent for oil remediation [7]

1.1.3 Mechanical Properties

Carbon nanotubes (CNTs) are a type of nanomaterial that exhibits higher tensile strength than normal steel with lower density. Based on the study from Osmani et al. (2014), the tensile strength of multiwall CNT is 150.000 GPa compared to the 0.400 GPa from steel. For the density, the multiwall CNT is 2.600 g/cm^3 compared to the 7.800 g/cm^3 from steel. Lower density equals to lighter weight per unit volume [10]. This concludes that the CNTs are much stronger but lighter compared to conventional steel. Besides, the study from Hassan et al. (2022) also proved that the application of CNTs can improve the mechanical properties of concrete pavement. Based on the study, increasing weight percentage of CNTs can increase the compressive strength, tensile strength, and flexure strength of the concrete specimens after 28 days [11]. This shows that CNTs not only have high strength but are also able to increase the strength of other products such as nanofillers.

1.1.4 Anti-bacterial Properties

The anti-bacterial properties have been one of the most important properties after the COVID-19 outbreak. There are several types of nanomaterials that exhibit anti-bacterial properties. These include silver nanoparticles, and gold nanoparticles. Different kinds of nanoparticles have various possible actions towards the microbes [12]. For example, based on the study from Choi et al. (2008), the silver nanoparticles could destroy different species of Gram-positive and Gram-negative bacteria [13]. Besides, the anti-microbial properties of silver nanoparticles can also be found in medical industries as they can be found in silicone maxilla facial prostheses covered with silver nanoparticles due to the antifungal properties [14]. On the other hand, gold nanoparticles are also proven to have the anti-microbial ability. The study from Kumar et al. (2016) showed that the gold nanoparticles synthesized using auric chloride as precursor and sodium citrate as stabilizing agent are good inhibitors for the growth of water borne pathogens [15].

1.2 Types of Nanomaterials

In this section, all types of nanomaterials will be identified and introduced in detail. The nanomaterials can be listed into four types of nanomaterials namely carbon-based nanomaterials, inorganic-based nanomaterials, organic-based nanomaterials, and composite-based nanomaterials [16].

1.2.1 Carbon-Based Nanomaterials

Carbon is one of the most abundant elements on the Earth. Therefore, the attraction towards nano carbon studies has grown recently. Based on current research, carbon can be used in the production of engineering materials such as fullerenes, carbon nanotubes, and graphene [17]. Different engineering materials exhibit different strengths and applications. For example, fullerenes exhibit the lowest unoccupied molecular orbital (LUMO). This can cause the C_{60} fullerenes to reduce up to six electrons therefore stabilizes negative charges [18, 19]. Fullerenes are suitable to be used in manufacturing solar cells [18]. Figure 2 illustrates the structure of C_{60} fullerene.

Besides, the carbon nanotubes (CNTs), same as fullerenes, are formed by carbon only but have a cylindrical shape which is different compared to fullerenes. The CNTs have sparked research interests due to their high electrical conductivity, large surface area, light, and high mechanical strength characteristics. CNTs are very useful in manufacturing supercapacitors [20–22]. Furthermore, due to high mechanical strength, the CNTs can also be used in manufacturing aircraft fuselage and bullet proof vest [23, 24]. Figure 3 shows the structure of single-walled CNT.

Fig. 2 Structure of C_{60} fullerene [17]

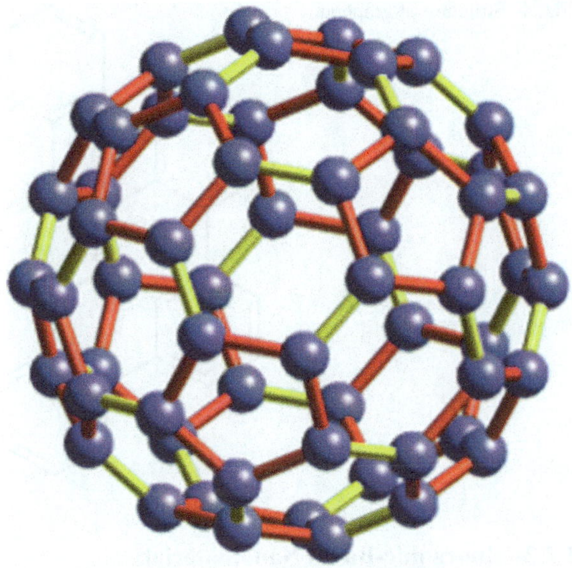

Fig. 3 Structure of single-walled CNT [25]

Graphene is arranged in two dimensions in hexagonal lattice structure formed by carbon atoms only. Like CNTs and fullerenes, graphene also exhibits high levels of electrical conductivity. It can be used in manufacturing lithium-ion batteries. Graphene can be used to tackle the issue of low power density of the current lithium-ion batteries [26]. Figure 4 shows the structure of graphene.

Fig. 4 Structure of graphene

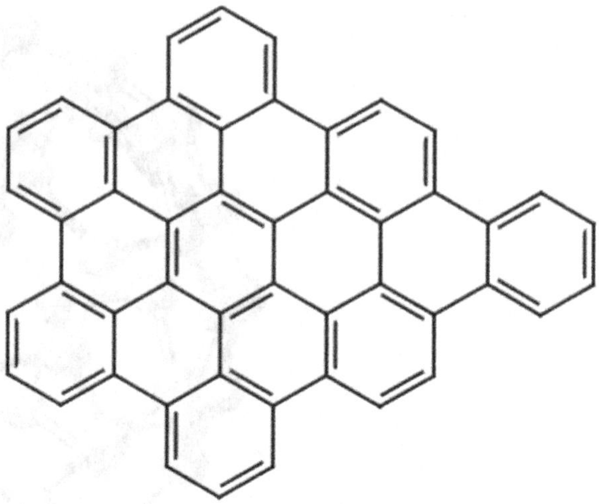

1.2.2 Inorganic-Based Nanomaterials

Inorganic-based nanomaterials can be divided into two groups which are metal-based nanomaterials and metal oxide-based nanomaterials. Some examples of metal-based nanomaterials include silver, gold, and copper. On the other hand, some examples of metal oxide-based nanomaterials are copper oxide, iron oxide, and silicon dioxide. Each of them can be applied in different ways due to their unique properties. Back to Sect. 1.1, silver and gold nanoparticles can be used for antimicrobial purposes whereas iron oxide that has magnetic properties can be used as pollutants clean up agents [7, 8, 12].

1.2.3 Organic-Based Nanomaterials

Organic-based nanomaterials are the nanomaterials that are formed from organic materials excluding carbon materials. Examples of organic-based nanomaterials include dendrimers and micelle [16]. Dendrimers are synthetic polymers that are highly ordered with large number of repeating unit branches. The terminal functionalities on the surface can be anionic, neutral, or cationic [27]. Dendrimers such as glucose-modified carbosilane dendrimers can be utilized as drug delivery carriers for cancer therapy [28].

1.2.4 Composite-Based Nanomaterials

Composite-based nanomaterials are the nanomaterials that are formed from the combination of carbon-based with metal-based, metal oxide-based with organic-based, metal-based with organic-based and etcetera [16]. Nanocomposites have

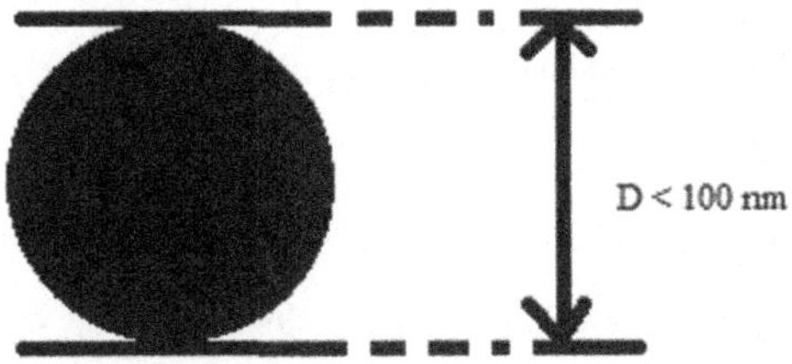

Fig. 5 Structure of nanosphere

attracted much attention as they can be employed in musculoskeletal engineering [29]. A study from Ahadian et al. (2014) utilized composites of carbon nanotubes (CNTs) with methacrylated gelatin polymer in fabricating muscle myofiber [30]. Besides, studies from Jayakumar et al. (2011) used nanocomposites of chitosan polymer coupled with TiO_2 nanoneedles or ZrO_2 in the fabrication of bone tissues [31, 32].

1.3 Classification of Nanomaterials

The nanomaterials can be categorized into several categories according to their different number of dimensions that are not in the nanometric scale which is less than 100 nm. The categories include zero-dimensional nanomaterials, one-dimensional nanomaterials, two-dimensional nanomaterials, and three-dimensional nanomaterials.

1.3.1 Zero-Dimensional Nanomaterials

The zero-dimensional nanomaterial is set to have all three dimensions including length, width, and height, within nanometric range (less than 100 nm). For example, quantum dots and nanospheres are categorized as zero-dimensional nanomaterials. Figure 5 shows an example of the structure and sizing of a nanosphere. The transmission electron microscopy (TEM) image of nanosphere can refer to the following journal [33].

1.3.2 One-Dimensional Nanomaterials

The one-dimensional nanomaterials are defined as the nanomaterials that have two dimensions in the nanometric scale whereas the other dimension is not in the nanometric scale. Examples of one-dimensional nanomaterials include nanotubes and nanowires. Figure 6 shows the structure of nanotubes and nanowires with given dimensions. Both the nanotubes and nanowires have lengths of more than 100 nm whereas the diameters are less than 100 nm.

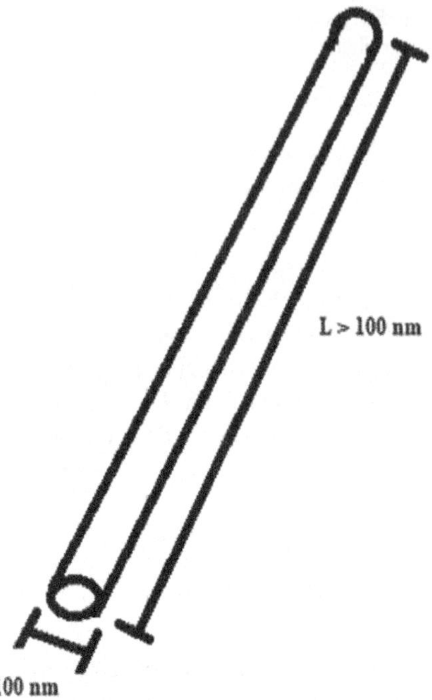

Fig. 6 Structure of nanowires and nanotubes

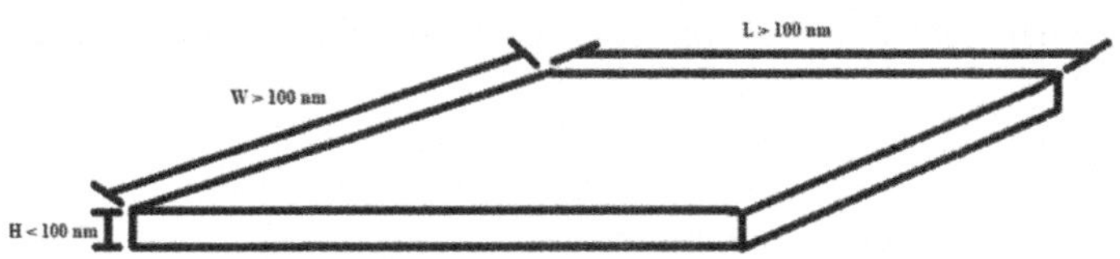

Fig. 7 Structure of nanofilm and nanoplate

1.3.3 Two-Dimensional Nanomaterials

The two-dimensional nanomaterials are a category of nanomaterials that has only one dimension in nanometric scale whereas the other two dimensions are not. The example of two-dimensional nanomaterials includes nanofilms and nanoplates. Figure 7 shows the structure of nanoplate and nanofilm with dimensions. It was seen that nanoplates have lengths and widths of more than 100 nm whereas the heights are less than 100 nm.

1.3.4 Three-Dimensional Nanomaterials

Three-dimensional nanomaterials are defined as a category of nanomaterials with no dimension either length, width, or height, is within the nanometric scale. Normally,

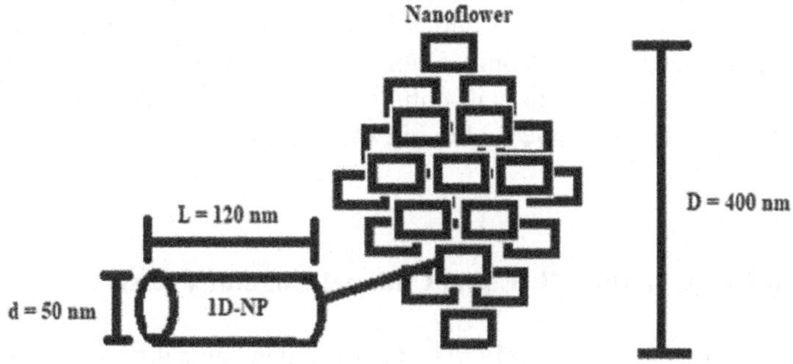

Fig. 8 Structure of nanoflower formed using one-dimensional nanotubes

this category of nanomaterial is assembled or built by zero-dimensional, one-dimensional, and two-dimensional nanomaterials. One of the examples of three-dimensional nanomaterials is nanoflowers. Figure 8 shows a drawn example of the structure of nanoflowers prepared from one-dimensional nanomaterials. From the image, it is noticeable that the nanoflower is formed from one-dimensional nanotubes with the diameters of each nanotube equal to 50 nm (less than 100 nm) and the length of each nanotube equal to 120 nm (more than 100 nm). Whereas for the whole nanoflower, all dimensions exceeded 100 nm which diameters equal to 400 nm.

1.4 Synthesis of Nanomaterials

There are two main approaches in synthesizing nanomaterials which are top-down approach and bottom-up approach. The concept behind the top-down approach is when a bulk material is broken down into smaller nano pieces. There are several methods that are under the top-down approach. These include mechanical alloying, equal channel angular pressing, high pressure torsion, and accumulative roll bonding. Besides, for bottom-up approach, the concept behind is when individual atoms are being brought together to form nanomaterials. The bottom-up approach starts building nanoparticles from either liquid or solid precursor. There are also several methods under the bottom-up approach which include co-precipitation method, sol–gel method, spray-conversion method, physical vapor deposition, and chemical vapor deposition.

1.4.1 Top-Down Method I (Mechanical Alloying)

Mechanical alloying or milling is the easiest method in producing nanomaterials. There are several types of milling methods in producing nanomaterials such as ball milling, vibratory milling, and planetary milling [34–36]. During the ball milling process, the bulk materials will be inserted into a hollow cylindrical chamber together with small balls made from zirconia, alumina, and steel. The chamber rotates horizontally, and the bulk materials will collide with the small balls using centrifugal forces. The figure of mechanical alloying of graphene can refer to the following journal [37].

1.4.2 Top-Down Method II (Equal Channel Angular Pressing)

Equal channel angular pressing (ECAP) is also a mechanical process in producing nanomaterials. During production, the nanomaterial samples are produced by pressing the needed materials through a die. When the pressing process starts, the sample passes through the die with shear deformation. The shear induces recrystallization to fine grain sizes [38]. The die contains two channels with an angle called die channel angle [39]. The pressing of the materials is done by a plunger as shown in the journal [38].

1.4.3 Top-Down Method III (High Pressure Torsion)

High pressure torsion (HPT), like ECAP, is also a severe plastic deformation method [40]. This method allows the generation of nanograins and ultrafine grains in metallic materials [41]. During the process, high pressure is exerted onto the materials by two plungers. One of the plungers is fixed and the other is rotating at a fixed speed. This exerts intense shear stress onto the material [42]. The working mechanism is demonstrated in the journal [42].

1.4.4 Top-Down Method IV (Accumulative Roll Bonding)

One of the newest severe plastic deformation methods is accumulative roll bonding (ARB). The ARB is a method to form ultrafine-grain structures in a single-phase metal [43]. The ARB process follows as a 4 mm thick sheet is rolled into 2 mm sheet with twice the original length. Then, the 2 mm sheet with twice the length is cut into two 2 mm sheets with original length. After that, the two 2 mm sheets are degreased and stacked together to make original thickness and length. Finally, the stack is then rolled again, and the cycle goes on. This method is able to allow large amount of plastic strain to the sheet without changing the dimensions. The tensile strengths can also be increased [44]. The figure of working mechanism of the ARB process can refer to the journal [45].

1.4.5 Bottom-Up Method I (Co-Precipitation Method)

Co-precipitation method is one of the easiest ways in synthesizing nanoparticles. This is done in several steps such as chemical precursor preparation, mixing of precursors, addition of precipitators (reducing agent), and purification of precipitated solids (nanoparticles) [46]. For example, the method from Massart utilized precursor's ferric chloride and ferrous chloride with the addition of ammonia solutions as precipitators, the iron nanoparticles can be obtained by centrifugation and decantation. A study from Sheng et al. (2012) also proved that the co-precipitation method can prepare Mn-Ce/TiO_2 nanocatalyst by following the similar steps [47]. The synthesis of α-Fe_2O_3 via the co-precipitation method can refer in the following journal [48].

1.4.6 Bottom-Up Method II (Sol–gel Method)

The Sol–gel method is slightly different compared to the co-precipitation method. In this method, the precursor used is molecular and usually is metal alkoxide. The precursor selected will then be dissolved in solvent of either water or alcohol followed by continuous heating and stirring to obtain gel. After that, the wet gel will further undergo drying process. Then, the final dried gel will be powdered and calcined [49]. Many researchers have been using the sol–gel methods in catalyst synthesis [50]. For example, a study from Singh et al. (2014) proved that the addition of solvent into silica source and further undergo hydrolysis, aging, and drying, the product of silica powder will be obtained as shown in the journal [51].

1.4.7 Bottom-Up Method III (Spray Conversion Method)

Spray conversion method, as known as spray pyrolysis, is a simple aerosol decomposition technique [52]. In this process, a selected precursor solution is prepared. The solution is then atomized into small droplets. These droplets are then moved towards the heated substrate surface and a thin film is generated [53]. For example, a study from Hong et al. (2008) synthesized nano sized Co_3O_4 powder by using spray conversion method. The experiment dissolved cobalt nitrate in distilled water and the solution was fed into a heated nozzle. The solution was spray dried in hot air of 250 °C [54]. The spray pyrolysis deposition is demonstrated in the following journal [52].

1.4.8 Bottom-Up Method IV (Physical Vapor Deposition)

Physical vapor deposition is a process whereby solid material is evaporated in a vacuum. This technique has been widely used in the coating industry due to the formation of thin layers of nanomaterials on a surface [55]. There are several types of physical vapor deposition methods such as evaporation, sputtering, and laser ablation. Evaporation is a process whereby the atoms or molecules reach the substrates due to thermal vaporization. This happens without the collisions from residual gas molecules [56]. Besides, sputtering is a process when the vaporization process of atoms occurs due to the bombardment of energetic ions towards the sputtering target [57]. Different from the other two, laser ablation is a method whereby an intense pulse of laser beam irradiates the source of interest causing the vaporization of atoms from source [58]. The processes and working mechanisms of the physical vapor deposition, sputtering process and laser ablation respectively can be shown in the following journals [55, 57, 59].

1.4.9 Bottom-Up Method V (Chemical Vapor Deposition)

The chemical vapor deposition process is similar to the physical vapor deposition process. But both processes have some differences between them. For example, in physical vapor deposition, it uses only physical forces in depositing the thin layer while the chemical vapor deposition uses chemical processes. Therefore, chemical vapor deposition is more complex compared to the physical vapor deposition process. There are several steps in the process. Firstly, the precursors gas will be transported towards the surface of substrate. Then, the gas molecules will be adsorbed on the surface and the catalyzed heterogeneous surface reaction will be carried out. After that, the surface diffusion of molecules to the growth sites occurs and nucleation starts. Lastly, the by-products from the chemical reaction will be removed [60]. The figure of the chemical vapor deposition process can refer in the following journal [61].

References

1. Beygisangchin, M., Rashid, S.A., Shafie, S., Sadrolhosseini, A.R., Lim, H.N.: Preparations, properties, and applications of polyaniline and polyaniline thin films—a review. Polymers (Basel) **13**(12) (2021). https://doi.org/10.3390/polym13122003
2. Ahilfi, D.N., Alkabbi, A.S., Mohammed, K.A., Ziadan, K.M.: Fabrication and characterization of polyaniline/CdSe device for applications in nano structured solar cells. In: IOP Conference Series: Materials Science and Engineering, IOP Publishing Ltd (2020).https://doi.org/10.1088/1757-899X/928/7/072069
3. Song, Y., Xu, Y., Guo, Q., Hua, Z., Yin, F., Yuan, W.: MXene-derived TiO_2 nanoparticles intercalating between RGO nanosheets: an assembly for highly sensitive gas detection. ACS Appl. Mater. Interfaces **13**(33), 39772–39780 (2021). https://doi.org/10.1021/acsami.1c12154
4. Krishnapriya, R., Nizamudeen, C., Saini, B., Mozumder, M.S., Sharma, R.K., Mourad, A.H.I.: MOF-derived Co2+-doped TiO2 nanoparticles as photoanodes for dye-sensitized solar cells. Sci. Rep. **11**(1) (2021). https://doi.org/10.1038/s41598-021-95844-4
5. Wen, J. et al.: Fabrication of high performance printed flexible conductors by doping of polyaniline nanomaterials into silver paste. J. Mater. Chem. C Mater **7** (2018). https://doi.org/10.1039/C8TC05391J

6. Yang, W., et al.: Ultrathin flexible reduced graphene oxide/cellulose nanofiber composite films with strongly anisotropic thermal conductivity and efficient electromagnetic interference shielding. J. Mater. Chem. C **5**(15), 3748–3756 (2017). https://doi.org/10.1039/C7TC00400A
7. Qiao, K. et al.: Application of magnetic adsorbents based on iron oxide nanoparticles for oil spill remediation: a review. J. Taiwan Inst. Chem. Eng. **97** (2019). https://doi.org/10.1016/j.jtice.2019.01.029
8. Singh, H., Bhardwaj, N., Arya, S., Khatri, M.: Environmental impacts of oil spills and their remediation by magnetic nanomaterials. Environ. Nanotechnol. Monit. Manag. **14**, 100305 (2020). https://doi.org/10.1016/j.enmm.2020.100305
9. Socoliucet, V. al.: Magnetic nanoparticle systems for nanomedicine—a materials science perspective. Magnetochemistry **6**(1). MDPI AG, pp. 1–36 (2020). https://doi.org/10.3390/magnetochemistry6010002
10. Osmani, R., Kulkarni, A., A.H., Bhosale, R., Ghodake, P., Harkare, B.: Carbon nanotubes: an impending carter in therapeutics. Int. J. Pharmaceutical Clin. Res. **6**, 84–96 (2014)
11. Hassan, A., Galal, S., Hassan, A., Salman, A.: Utilization of carbon nanotubes and steel fibers to improve the mechanical properties of concrete pavement. Beni. Suef. Univ. J. Basic Appl. Sci. **11**(1) (2022). https://doi.org/10.1186/s43088-022-00300-5
12. Staroń, A., Długosz, O.: Antimicrobial properties of nanoparticles in the context of advantages and potential risks of their use. J. Environ. Sci. Health Part A **56**, 1–14 (2021). https://doi.org/10.1080/10934529.2021.1917936
13. Choi, O., Deng, K.K., Kim, N.J., Ross, L., Surampalli, R.Y., Hu, Z.: The inhibitory effects of silver nanoparticles, silver ions, and silver chloride colloids on microbial growth. Water Res. **42**(12), 3066–3074 (2008). https://doi.org/10.1016/J.WATRES.2008.02.021
14. Meran, Z., Besinis, A., Peralta, T., Handy, R.: Antifungal properties and biocompatibility of silver nanoparticle coatings on silicone maxillofacial prostheses in vitro. J. Biomed. Mater. Res. B Appl. Biomater **106** (2017). https://doi.org/10.1002/jbm.b.33917
15. Kumar, R. et al.: Synthesis and antimicrobial effects of colloidal gold nanoparticles against prevalent water borne bacterial pathogens. Cogent. Chem. **2** (2016). https://doi.org/10.1080/23312009.2016.1192522
16. Majhi, K., Yadav, M.: Synthesis of inorganic nanomaterials using carbohydrates, 109–135 (2021). https://doi.org/10.1016/B978-0-12-821887-7.00003-3
17. Siqueira Jr., J.R., Oliveira, O.: Carbon-Based Nanomaterials, pp. 233–249 (2017). https://doi.org/10.1016/B978-0-323-49782-4.00009-7
18. Thompson, B., Frechet, J.: Polymer–fullerene composite solar cells. Angew. Chem. Int. Ed. Engl. **47**, 58–77 (2008). https://doi.org/10.1002/anie.200702506
19. Günes, S., Neugebauer, H., Sariciftci, N.S.: Conjugated polymer-based organic solar cells. Chem. Rev. **107**(4), 1324–1338 (2007). https://doi.org/10.1021/cr050149z
20. Li, X., Wei, B.: Supercapacitors based on nanostructured carbon. Nano Energy **2**(2), 159–173 (2013). https://doi.org/10.1016/j.nanoen.2012.09.008
21. Yu, G., Xie, X., Pan, L.J., Bao, Z., Cui, Y.: Hybrid nanostructured materials for high-performance electrochemical capacitors. Nano Energy **2**, 213–234 (2013). https://doi.org/10.1016/j.nanoen.2012.10.006
22. Lee, S.W., Kim, J., Chen, S., Hammond, P.T., Shao-Horn, Y.: Carbon nanotube/manganese oxide ultrathin film electrodes for electrochemical capacitors. ACS Nano **4**(7), 3889–3896 (2010). https://doi.org/10.1021/nn100681d
23. Gohardani, O., Elola, M.C., Elizetxea, C.: Potential and prospective implementation of carbon nanotubes on next generation aircraft and space vehicles: a review of current and expected applications in aerospace sciences. Progress Aerospace Sci. **70**. Elsevier Ltd, pp. 42–68 (2014). https://doi.org/10.1016/j.paerosci.2014.05.002
24. Obradović, V., Stojanović, D.B., Živković, I., Radojević, V., Uskoković, P.S., Aleksić, R.: Dynamic mechanical and impact properties of composites reinforced with carbon nanotubes. Fibers Polymers **16**(1), 138–145 (2015). https://doi.org/10.1007/s12221-015-0138-2

25. Gooding, J.: Nanostructuring electrodes with carbon nanotubes: a review on electrochemistry and applications for sensing. Electrochim. Acta **50**, 3049–3060 (2005). https://doi.org/10.1016/j.electacta.2004.08.052
26. Li, Y., Wu, J., Chopra, N.: Nano-carbon-based hybrids and heterostructures: progress in growth and application for lithium-ion batteries. J. Mater. Sci. **50**, 7843–7865 (2015). https://doi.org/10.1007/s10853-015-9429-7
27. Lyu, Z., Ding, L., Huang, A.Y.-T., Kao, C.-L., Peng, L.: Poly(amidoamine) dendrimers: covalent and supramolecular synthesis. Mater. Today Chem. **13**, 34–48 (2019). https://doi.org/10.1016/j.mtchem.2019.04.004
28. Wrobel, D., et al.: Glucose-modified carbosilane dendrimers: Interaction with model membranes and human serum albumin. Int. J. Pharm. **579**, 119138 (2020). https://doi.org/10.1016/j.ijpharm.2020.119138
29. Narayanan, N., Kuang, L., Ponte, M., Chain, C., Deng, M.: Design and fabrication of nanocomposites for musculoskeletal tissue regeneration, 3–29 (2016). https://doi.org/10.1016/B978-1-78242-452-9.00001-7
30. Ahadian, S. et al.: Hybrid hydrogels containing vertically aligned carbon nanotubes with anisotropic electrical conductivity for muscle myofiber fabrication. Sci. Rep. **4** (2014). https://doi.org/10.1038/srep04271
31. Jayakumar, R., Ramachandran, R., Divyarani, V.V., Chennazhi, K.P., Tamura, H., Nair, S.V.: Fabrication of chitin–chitosan/nano TiO_2-composite scaffolds for tissue engineering applications. Int. J. Biol. Macromol. **48**(2), 336–344 (2011). https://doi.org/10.1016/j.ijbiomac.2010.12.010
32. Jayakumar, R., et al.: Fabrication of chitin–chitosan/nano ZrO_2 composite scaffolds for tissue engineering applications. Int. J. Biol. Macromol. **49**(3), 274–280 (2011). https://doi.org/10.1016/j.ijbiomac.2011.04.020
33. Verma, M.S., Chen, P.Z., Jones, L., Gu, F.X.: Controlling 'chemical nose' biosensor characteristics by modulating gold nanoparticle shape and concentration. Sens. Biosensing Res. **5**, 13–18 (2015). https://doi.org/10.1016/j.sbsr.2015.04.007
34. Neikov, O.D., Yefimov, N.A., Stepanovich, S.N.: Production of Noble Metal Powders, 2nd ed.Elsevier, pp. 831–846 (2018)
35. Yadav, T., Yadav, R.M., Singh, D.: Mechanical milling: a top down approach for the synthesis of nanomaterials and nanocomposites. Nanosci. Nanotechnol. **2**, 22–48 (2012). https://doi.org/10.5923/j.nn.20120203.01
36. Suryanarayana, C.: Mechanical milling. In: Powder Metal Technologies and Applications. ASM Handbook, ASM International, pp. 80–90 (1998)
37. Kumar, C., Pattammattel, A.: Synthetic routes to graphene preparation from the perspectives of possible biological applications. In Introduction to Graphene, pp. 17–44 (2017). https://doi.org/10.1016/B978-0-12-813182-4.00002-7
38. Milligan, W.: Mechanical behavior of bulk nanocrystalline and ultrafine-grain metals. Comprehensive Struct. Integrity **8**, 529–550 (2007). https://doi.org/10.1016/B0-08-043749-4/08145-3
39. Ravisankar, B.: Equal-channel angular pressing (ECAP). In: Handbook of Mechanical Nanostructuring, pp. 277–297 (2015). https://doi.org/10.1002/9783527674947.ch13
40. Valiev, R., Estrin, Y., Horita, Z., Langdon, T., Zehetbauer, M., Zhu, Y.: Producing bulk ultrafine-grained materials by severe plastic deformation [J]. JOM **58**, 33–39 (2006). https://doi.org/10.1007/s11837-006-0213-7
41. Zhilyaev, A., Langdon, T.: Using high-pressure torsion for metal processing: fundamentals and applications. Progress Mater. Sci. Prog. Mater Sci. **53**, 893–979 (2008). https://doi.org/10.1016/j.pmatsci.2008.03.002
42. Edalati, K., Horita, Z.: A review on high-pressure torsion (HPT) from 1935 to 1988. Mater. Sci. Eng. A **652**, 325–352 (2016). https://doi.org/10.1016/j.msea.2015.11.074
43. Tsuji, N.: Bulk nanostructured metals and alloys produced by accumulative roll-bonding. In: Nanostructured Metals and Alloys: Processing, Microstructure, Mechanical Properties and Applications, pp. 40–58 (2011). https://doi.org/10.1533/9780857091123.1.40

44. Lenard, J.G.: Severe plastic deformation - accumulative roll bonding. In: Primer on Flat Rolling, pp. 303–322 (2013)
45. Rao, G.N.M., Kumar, V.R.M.: A review on recent advances in accumulative roll bonding of similar, dissimilar and metal matrix composites. Mater Today Proc. **56**, A13–A18 (2022). https://doi.org/10.1016/j.matpr.2021.11.608
46. Massart, R.: Preparation of aqueous magnetic liquids in alkaline and acidic media. IEEE Trans. Magn. **17**(2), 1247–1248 (1981). https://doi.org/10.1109/TMAG.1981.1061188
47. Sheng, Z., Hu, Y., Xue, J., Wang, X., Liao, W.: A novel co-precipitation method for preparation of Mn-Ce/TiO_2 composites for NOx reduction with NH3 at low temperature. Environ. Technol. (United Kingdom) **33**(21), 2421–2428 (2012). https://doi.org/10.1080/09593330.2012.671370
48. Kushwaha, P., Chauhan, P.: Synthesis of spherical and Rod-Like EDTA assisted α-Fe_2O_3 nanoparticles via Co-precipitation method. Mater Today Proc. **44** (2021). https://doi.org/10.1016/j.matpr.2021.02.450
49. Bokov, D., et al.: Nanomaterial by Sol-Gel method: synthesis and application. Adv. Mater. Sci. Eng. **2021**, 5102014 (2021). https://doi.org/10.1155/2021/5102014
50. Cauqui, M.A., Rodríguez-Izquierdo, J.M.: Application of the sol-gel methods to catalyst preparation. J. Non Cryst. Solids **147–148**, 724–738 (1992). https://doi.org/10.1016/S0022-3093(05)80707-0
51. Singh, L.P. et al.: Sol-Gel processing of silica nanoparticles and their applications. Adv. Colloidal Interface Sci. **214**, 17–37 (2014)
52. Farooq, S., Ngaini, Z., Farooq, S.: Manufacturing and design of smart polymer composites, pp. 27–84 (2021). https://doi.org/10.1016/B978-0-12-819961-9.00003-7
53. Tahir, M., Rafique, M., Rafique, M., Nawaz, T., Rizwan, M., Tanveer, M.: Photocatalytic Nanomaterials for degradation of organic pollutant and heavy metals (2020). https://doi.org/10.1016/B978-0-12-821192-2.00008-5
54. Hong, S.-H., Bae, J.-S., Ahn, H.-J.: Synthesis of nano-sized Co3O4 powder by spray conversion method for anode material of lithium battery. Met. Mater. Int. **14**(2), 229–232 (2008). https://doi.org/10.3365/met.mat.2008.04.229
55. Parvej, M.S., Khan, M.I., Hossain, M.K.: Preparation of nanoparticle-based polymer composites, pp. 55–94 (2022). https://doi.org/10.1016/B978-0-12-824272-8.00013-0
56. Mattox, D.M.: Vacuum evaporation and vacuum deposition. In: Mattox, D.M. (Ed.), Handbook of Physical Vapor Deposition (PVD) Processing (Second Edition), pp. 195–235. William Andrew Publishing, Boston (2010). https://doi.org/10.1016/B978-0-8155-2037-5.00006-X
57. Pujahari, R.: Solar cell technology, pp. 27–60 (2021). https://doi.org/10.1016/B978-0-12-823710-6.00007-8.
58. Liu, Z.: Laser applied coatings. Shreir's Corrosion **4**, 2622–2635 (2010). https://doi.org/10.1016/B978-044452787-5.00141-4
59. Nair, G.B., Dhoble, S.J.: Phosphor-converted LEDs. In: Nair, G.B., Dhoble, S.J. (Eds.), The Fundamentals and Applications of Light-Emitting Diodes, pp. 87–126. Woodhead Publishing (2021). https://doi.org/10.1016/B978-0-12-819605-2.00004-5
60. Saeed, M., Alshammari, Y., Majeed, S.A., Alnasrallah, E.: Chemical vapour deposition of graphene—synthesis, characterisation, and applications: a review. Molecules **25**, 3856 (2020). https://doi.org/10.3390/molecules25173856
61. Choy, K.-L.: Chemical vapour deposition of coatings. Prog. Mater. Sci. **48**, 57–170 (2003). https://doi.org/10.1016/S0079-6425(01)00009-3

Carbon Nanoadsorbents

Aruna Yadav and Surender Kumar

Abstract Carbon is one of the highly explored materials because of its exceptional properties such as high chemical stability, superior mechanical strength, catenation property, anisotropy and conductivity. Researchers have investigated numerous morphologies of carbon, specially at the nanoscale for their utilization in the advancements of various operational areas. This chapter deals with the overview of various carbon based nanoadsorbents (CNs): Carbon nanotubes (CNTs), Graphene and its derivatives, fullerenes and carbon quantum dots (CQDs). The structure, exceptional properties and synthetic strategies of these nanoadsorbents have been discussed and compared. The shifts and advancements in these strategies were observed towards a greener perspective, so that CNs can be adopted as a suitable and efficient alternative for conventional techniques of water purification.

Keywords Adsorption · Carbon nanotubes · Graphene · Fullerenes · Carbon quantum dots

1 Introduction

The discharge of untreated or inadequately treated wastewater from diverse sources including contaminants such as pesticides, dyes, pharmaceutical and personal care products, heavy metals, bacteria, microplastics and so on has resulted in the fast depletion of freshwater resources [1]. Traditional wastewater treatment methods include floatation, coagulation-flocculation, filtration, sedimentation, membrane separation, ion exchange, reverse osmosis, oxidation and adsorption [2]. These conventional approaches are incapable of achieving the requisite level of decontamination in order to achieve accurate and cost-effective discharge requirements. Some modern technologies for wastewater cleanup include advanced oxidation processes, photocatalysis under UV/visible light irradiation, ozonation, activated carbon adsorption and

A. Yadav · S. Kumar (✉)
Department of Chemistry, Chaudhary Bansi Lal University, Bhiwani 127031, Haryana, India
e-mail: skaushik@cblu.ac.in

J. Tharini and S. Thomas (eds.), *Carbon Nanomaterials and their Composites as Adsorbents*, Carbon Nanostructures, https://doi.org/10.1007/978-3-031-48719-4_2

electro-precipitation, although they are costly and energy-intensive [3, 4]. Adsorption methods have largely superseded these approaches in the last ten years because of their inexpensive production costs, good selectivity, applicability at low concentrations, and ability to be used in both batch and continuous processes. Additionally, these adsorbents can be regenerated with various eluents and reused for subsequent adsorptive cycles. Carbon-based nanomaterials have properties such as high mechanical strength, high surface area and low overall production costs that make them suitable for adsorption-based wastewater treatment. Recently, the application of the adsorbents, produced through an overall green process, for the removal of pollutants, like dyes, heavy metals, pharmaceutical and personal care products, pesticides etc., from the environment is highly needed in the current scenario to eliminatethe toxic effects caused by the use of chemicals in the conventional synthetic procedures [1, 5]. In this context, this chapter aims to give an overview of CNs (CNTs, Graphenes and fullerenes) with the main focus on their structure, exceptional properties, method of synthesis and applications.

2 Carbon Nanoadsorbents (CNs)

According to the IUPAC (2012) recommendations, nanomaterials are the substances whose dimensions of particles in any form vary from 1 to 100 nm [6]. The most prevalent types of nanomaterials are: nanopowders, nanofilaments, nanowires, nanotubes, nanoblocks, nanofilms and nanocables [7]. Nanomaterials can be prepared using a variety of physical, chemical, or mechanical processes, depending on the required properties. One of the aspects responsible for a particular nanomaterial's properties is the relationship between surface area and volume. Nanomaterials are distinguishable from one another based on shape and dimension and can be classified under four categories [7]:

- Zero-dimensional (0D)-fullerenes, quantum dots, nanoparticles, molecules, clusters, grains, nanopowders etc.
- One-dimensional (1D)-nanotubes (CNT), nanowires, nanofilaments, springs, needles etc.
- Two-dimensional (2D)-nanofilms (Graphene), nanolayers
- Three-dimensional (3D)-nanocrystalline structure, fiber skeletons, powder skeletons etc.

Carbon is one of the most investigated materials for a variety of reasons, including high chemical stability, exceptional mechanical strength, catenation property, anisotropy, and conductivity. Different morphologies of carbon have been studied by researchers, including spheres (fullerenes), sheets (graphene and graphene oxide), hollow tubes (carbon nanotubes) and dots (CQDs) [8]. The high surface area to volume ratio of the carbon-based nanoadsorbents, make them an excellent material for the wastewater purification applications. Moreover, when these nanomaterials are mixed with biomass, magnetic nanoparticles, polymers, metal oxides, and many other substances, unique properties are frequently observed [9]. These materials are quite competent and have the potential to be utilized in the advancement of many operational areas such as environment remediation, energy generation and storage, catalysis [10, 11]. There are varieties of methods to synthesize these nanomaterials as discussed subsequently under each section. Further, CNs may be functionalized through covalent or noncovalent interactions to increase the dispersion in aqueous solutions as well as increase their adsorption properties [12, 13].

2.1 *Carbon Nanotubes (CNTs)*

CNTs are 1D CNswhere the graphite sheet gets rolled to form cylindrical tubes with diameter in nanoscale and can have an aspect ratio more than 1000 [13]. CNTs can be classified as: single-walled CNTs (SWCNTs) and multi-walled CNTs (MWCNTs) depending upon the number of graphene sheets that get rolled [13, 14]. SWCNTs consist of a single graphene layer rolled up into cylindrical shape whereas MWCNTs consist of two or more graphene layers that coaxially roll up into concentric cylindrical shells with van der Waals forces between adjacent layers. CNTs can have three chiralitics: armchair, zigzag and chiral one on the basis of the rolling angle of graphene layer. The chirality of CNT is defined by the chiral vector (C_h) as per Eq. 1.

$$C_h = na_1 + ma_2 \tag{1}$$

where n and m are the integers showing the number of steps along the unit vectors, a_1 and a_2, of the hexagonal lattice. The three different sorts of carbon atom orientations around the nanotube circumference are identified using this approach. The nanotube is arm-chair, if $n = m$, whereas, the nanotube is zig-zag, if $m = 0$. Otherwise, the nanotube is chiral [15]. Further, the armchair and zigzag CNTs can be of varying diameters on the basis of the number of carbon atoms in the cross section of CNT as shown in Fig. 1. Interestingly, MWCNT has multiple concentric graphene layers where each layer can have dissimilar chirality as well. However, this also makes the study of its physical properties is more complex than that of SWCNT.

2.1.1 Synthesis of CNTs

Iijima performed the first CNT synthesis unintentionally using arc discharge [18]. However, the current scenario is entirely different as there are numerous methods for the synthesis of CNTs, resulting in the formation of different types of CNTs with specific characteristics such as size and purity as shown in Fig. 2. Typically, arc discharge, laser ablation, high pressure CO and chemical vapordeposition (CVD) are the major reported techniques for synthesis of CNTs [19, 20].

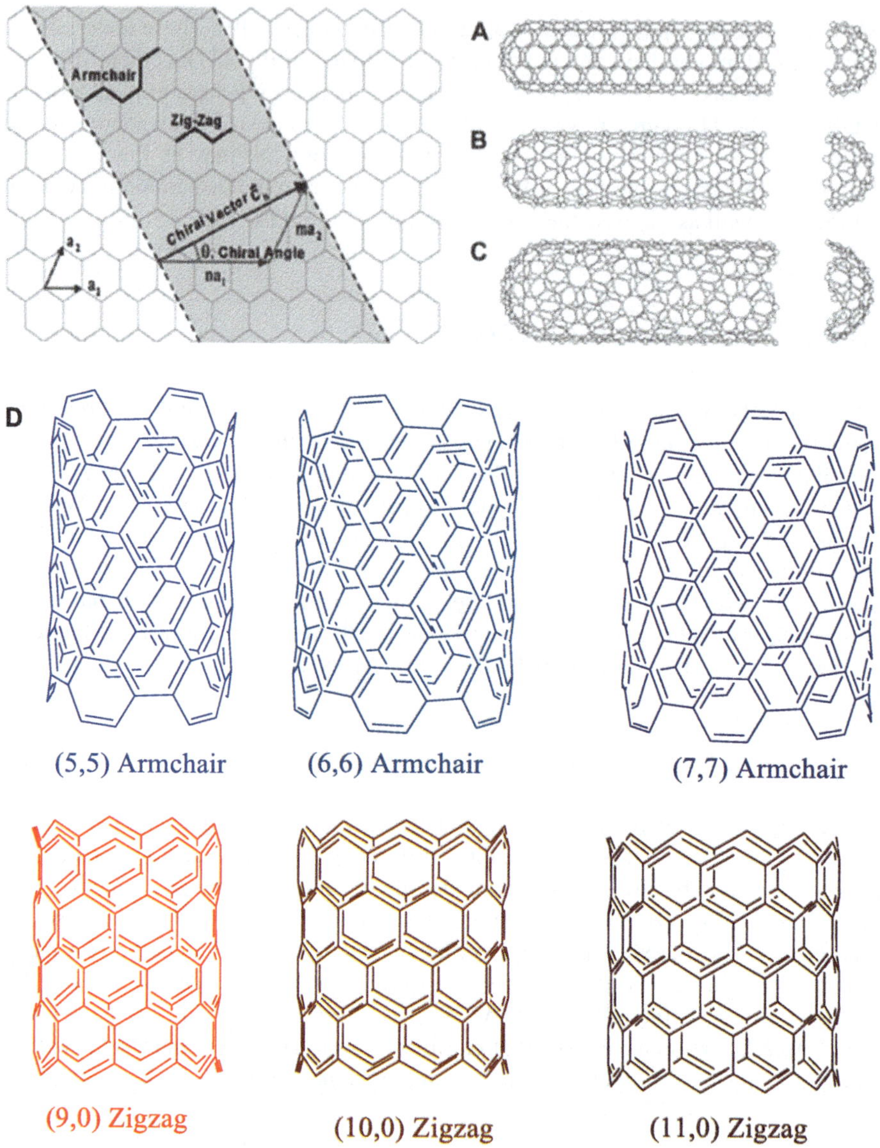

Fig. 1 Chiralities of CNT **a** armchair; **b** zigzag; **c** chiral [15–17] (Reprinted from [16] © 1995 with permission from Elsevier); **d** Different types of arm chairs and zigzag CNTs

Arc discharge technique involves the use of a highly pure graphite anode and cathode under an inert atmosphere of helium gas. In the laser ablation technique, vaporization of a graphite target is carried out with the aid of laser, often in a quartz tubular furnace under an inert atmosphere of argon/helium gas at about five hundered

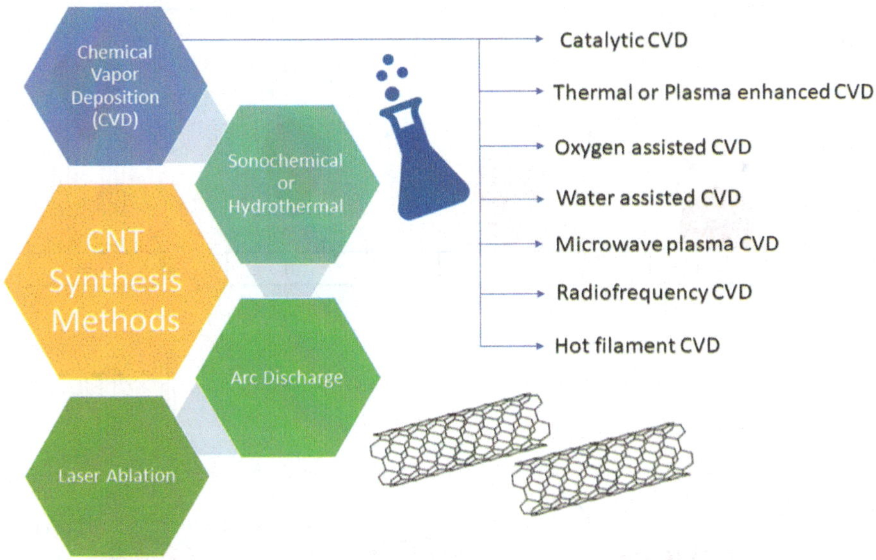

Fig. 2 Method of synthesis of CNTs

torr of pressure. The as-synthesized CNTs are then further collected on a water-cooled metal target [21, 22]. SWCNTs as well as MWCNTs can be synthesized by varying the reaction conditions. Hydrothermal synthesis is a solution reaction-based approach that can take place in a wide range of temperature commonly used for the preparation of CNs. Ultrasonic treatment prior to hydrothermal condition can produce considerable amount of CNTs [23]. The CVD technique is utilized most commonly in the current scenario to synthesize both MWCNTs and SWCNTs with significant yield and purity [24]. There are numerous different types of CVD techniques like water assisted CVD, catalytic chemical vapor deposition, oxygen assisted CVD, thermal or plasma enhanced CVD, microwave plasma, hot-filament CVD or radiofrequency CVD [22]. Among these, the catalytic CVD has gained much more research interest as it is a facile technique which can produce CNTs in a considerable amount. Importantly, the particle size, morphology, alignment, length and density can be controlled using this technique. It involves the decomposition of a carbon containing solvent or hydrocarbon gas in an inert atmosphere which results in the deposition of CNTs on a substrate/support/reactor [24]. One of the limitations related to CVD technique is purification requirements after their synthesis. The CNTs produced with CVD are generally contaminated with impurities such as graphene flakes, metal catalyst and amorphous carbon, which needs to be removed prior to their application [24].

The growth of the CNTs occurs either by tip growth mechanism or root/base growth mechanism on the basis of interactions existing between metal-support [22]. Tip growth mechanism is observed in case of a weak interaction, where the hydrocarbon decomposes on the metal top, while carbon diffuses through the metal [25].

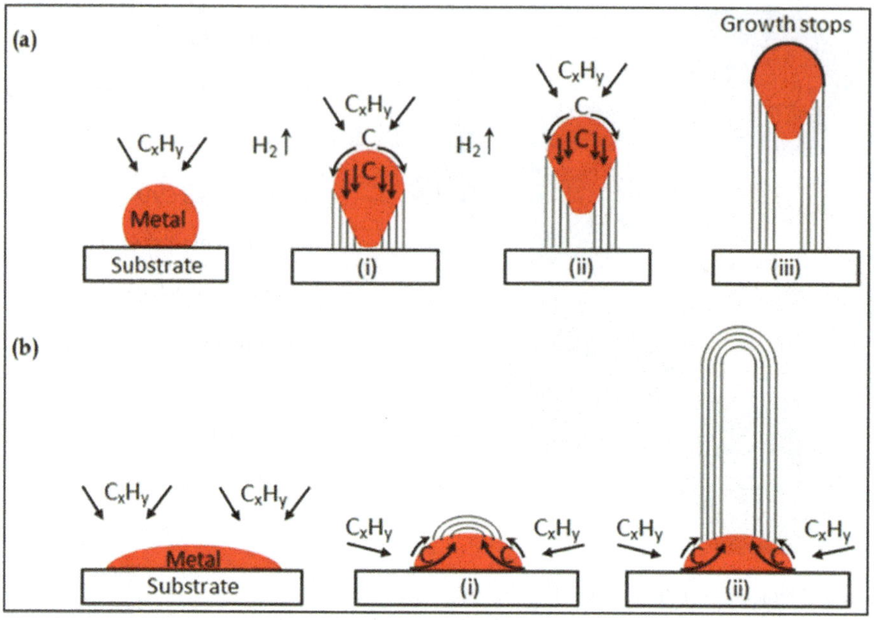

Fig. 3 CNT growth mechanism in CVD **a** tip-growth mechanism **b** root growth mechanism [24, 26] (Reprinted from [26] © 2011 with permission from IntechOpen)

Root/Base growth mechanism is observed in case of strong interaction among catalyst and the support (Fig. 3). Similar to the tip-growth mechanism, here also the initiation occurs in a same fashion, hydrocarbon decomposes on the metal top while the carbon initiates to diffuse through the metal. But, in this case, the metal particle is unable to attain more height owing to the strong catalyst-substrate interaction. Thus, the CNT grows on top of the metal and the catalytic activity stops subsequently upon metal coverage by the excess of carbon, resulting in the CNT growth termination [24].

Furthermore, the growth of SWCNTs or MWCNTs is controlled by various factors such as type of catalyst used, catalyst concentration, particle size of catalyst, thickness of catalyst layer, catalyst preparation conditions, growth temperature, pressure, growth time and gas flow rate [22, 27–29]. For example, CNTs of smaller diameters are promoted using Co catalyst, while, Ni produces the best quality CNT with higher yield using the CVD process [28]. Similarly, rice husk derived SiO_2 was developed as a green and economical catalyst support where SWCNT of larger diameter (>2 nm) were selectively produced [29].

Manufacturing of CNTs at a larger scale requires utilization of greener precursors and development of techniques to diminish the overall processing costs and the negative effects on the environment. The rising demand for CNTs puts a pressure on the nonrenewable resources needed in their manufacture and raises the carbon footprint. Therefore, green synthesis techniques must be employed to ensure an environmentally benign and sustainable manufacturing process. Some of the method for green synthesis of CNTs are shown in Table 1.

Recently, researchers have shown much more scientific attention towards the fabrication of CNT-based composites for functional, environmental and structural applications [15, 44–48]. However, the use of CNTs at practical scale remains limited because of the problems associated with dispersion of entangled-CNT during processing, which results in the poor interactions between the interface of CNTs and the polymer matrix or other composite material [13]. It has been demonstrated that the mechanical, electrical and adsorptive potential of composites containing these entangled or agglomerated CNTs is significantly affected in comparison to the theoretical expectations for well-dispersed CNTs. To overcome this challenge various methods have been employed for the dispersion of CNTs such as ultrasonication, calendaring process, ball milling, stirring [15].

The most popular method for dispersing nanoparticles is ultrasonication. According to this method's basic premise, attenuated waves are created in the molecules of the solvent through which ultrasound travels as it undergoes a sequence of compressions [49]. These shock waves induce separation of individual nanoparticle from the bundled form. Various low viscosity solvents such as water, ethanol or acetone can be used as medium for the ultrasonic dispersion of CNTs. However, care must be taken as the structure of CNTs can be seriously damaged if the ultrasonic treatment is too aggressive (using higher frequencies or probe sonicator) and/or too long [15]. Another method of dispersion/mixing/homogenization of viscous materials such as CNT suspension, is the use of a machine with three roll mills, known as calender. The shear force produced by the rollers is typically used to effectively disperse CNTs in a polymer matrix [15]. Materials can be ground into an incredibly fine powder using the ball milling technique for a variety of applications. The milling process generates a high pressure locally owing to the frequent impact among the tiny, rigid balls placed in a concealed container, which results in the dispersion of CNT particles [15]. Stirring is a typical method for dispersing particles in liquid systems, and it can also be used to mix CNTs into a matrix of polymer or other composites. The mixing speed, shape and size of the propeller, influence the dispersion of nanomaterials [15]. Recent research frequently uses a combination of the aforesaid procedures, such as ultrasonication-ball milling and ultrasonication-stirring.

2.2 *Graphene and Its Derivatives*

The use of graphene and its derivatives as adsorbents to eliminate toxic substances from the environment has gained popularity in recent years. Graphene-based nanomaterials have proved to be an ideal material due to its unique physicochemical properties, such as high thermal conductivity, specific surface area, electron mobility and mechanical strength [7]. In graphene-based materials, sp^2-hybridized carbon atoms are arranged in a 2D layer to form a hexagonal structure that resembles an orderly honeycomb [50, 51]. Graphene has attracted scientific efforts since 2004, when it was

Table 1 Some method of synthesis of CNTs using green precursors

Green Precursor	Method of preparation	Catalyst	Type of CNT produced	References
Coconut oil	CVD	Fe	MWCNT (80–100 nm) (3–4 μm) 58% purity	[30]
Palm oil	CVD	Ferrocene/ silicon	SWCNT and MWCNT (0.6–1.2 nm) (110 μm) 90% purity	[31]
Rice Straw	Pyrolysis	Fe–Ni/Al_2O_3	MWCNT (15–40 nm)	[32]
Walnut extract	CVD		MWNT (8–15 nm) (3600 μm)	[33]
Camphor extracted from latex of *Cinnamomum camphora*	CVD	Fe–Co/Zeolite	MWCNT (10 nm) 88%	[34]
Camphor	CVD	Ferrocene	SWCNT (1.2–1.3 nm) MWCNT (20–40 nm) ~90%	[35]
Eucalyptus oil	Spray pyrolysis	Silica-zeolite support impregnated with Fe/Co	SWCNT (0.79–1.71 nm)	[36]
Wood saw dust mixed with the reducing agent (commercial zinc), calcite (bed material)	Pyrolysis	Ferrocene or Fe/Mo/MgO	CNTs (thinner than 50 nm)	[37]
Neem oil	Spray pyrolysis-assisted CVD method	Ferrocene	MWCNT (15 and 30 nm) (20 to 40 μm)	[38]
Turpentine oil	Spray pyrolysis	Fe-Co/Zeolite	SWCNT (7–20 nm) 58%	[39]
Turpentine oil	Spray pyrolysis	Ferrocene	MWCNT (15–40 nm) (70–130 μm)	[40]
Castor oil	Spray pyrolysis	Ferrocene	MWCNT (20–60 nm) (5–10 μm)	[41]

(continued)

Table 1 (continued)

Green Precursor	Method of preparation	Catalyst	Type of CNT produced	References
Olive oil	Spray pyrolysis	$NiCl_2$	SWCNT (27–31 nm)	[42]
Lignin	CVD	Lignin micelles as soft template	MWCNT (10–110 nm) (1–6 μm)	[43]

initially extracted from graphite using the mechanical exfoliation process [52]. Since then, its properties have been researched with the intention of using it in a variety of processes, including adsorption, separation and photocatalysis [53, 54]. They also have significant contribution in the segregation of toxic pollutants occurring in the petroleum industry effluents.

Graphene has typically been employed as an efficient support material and co-catalyst to create photocatalysts because of its conjugated system along with the free π electrons resulting in high electrical conductivity and huge surface area. The principal challenges encountered in actual applications, however, are its low dispersion, ease of restacking, and challenging recovery from aqueous solution after treatment. As a result, many methods are being investigated for the tailoring of graphene by adding various functional groups (e.g., O, N, P, B, chalcogens and halogens) [55]. Graphene oxide/ Graphite oxide (GO) and reduced graphene oxide (rGO), are the most common graphene derivatives with oxygen-containing functional groups (i.e., hydroxyl, carboxyl, and epoxy groups), whose compositions cannot be properly defined because of various levels of oxidation as illustrated in Fig. 4. These modified groups present at the interface of adsorbent and adsorbate serve as reactive centers promoting photodegradation in addition to making GO and rGO hydrophilic [55, 56]. Furthermore, the O atom being highly electronegative can attract the positively charged contaminants via electrostatic interactions, leading to their adsorption on GO and rGO surface [9, 50].

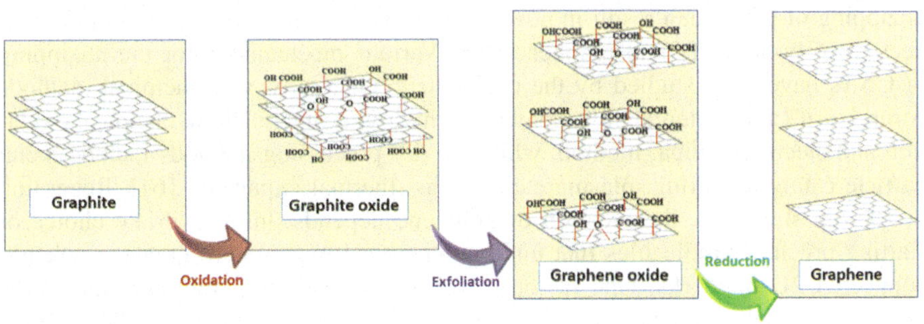

Fig. 4 Graphene and its derivatives

2.2.1 Synthesis of Graphene and Its Derivatives

As shown in Fig. 5, two separate strategies: the top-down approach and bottom-up approach, have been established for synthesizing graphene species to fulfil a variety of applications. The top-down approach involves the use of external forces which can overcome the weak forces of attraction (e.g., electrostatic interactions, hydrogen bonding, aromatic stacking and van der Waal forces) between adjacent layers of graphene leading to delamination of graphite into graphene. In contrast, the latter method relies on organic synthesis and 2D anisotropic assembly of tiny molecular modules [54, 57]. In top-down techniques, bulk carbonaceous materials (graphite) are peeled into flakes to produce graphene derivatives. The conventional mechanical exfoliation involves the operation of two opposite directional routes [58]. Graphite is well known for its lubricative nature, and its layers can undergo relative motion in sidewise direction if subjected to a shear force-based process through ball milling. The inevitable destruction and fragmentation of large flakes into smaller particles occurs during rolling movement along with constant collisions. In contrast, scotch tape method, a type of mechanical exfoliation, can simply exfoliate graphite multiple layers into thin layers of graphene by overcoming significant interlayer attractive forces when subjected to a normal force [57]. However, neither of the two uncontrolled and low-yield techniques can satisfy actual application requirements. This limitation can be overcome by the ultrasound assisted liquid phase exfoliation that can segregate individual layers of graphite easily with a significant yield [59]. The shear forces and the cavitation effect generated during the sonication process works on the graphite during exfoliation. To purify graphene and its derivatives, ideal solvents must be used, and they must also reduce interfacial tension for stable dispersion [54]. Recently, water, a green solvent, serves as an alternative for conventional toxic solvent for the exfoliation of graphite because of its low cost, non-toxicity and low boiling point. However, the process requires the addition of chemicals such as sodium lauryl sulfate surfactant, Perylene tetracarboxylate surfactant, vinyl imidazole-based polymer, which helps in production of stabilized suspensions [60–62]. Apart from this, unzipping of CNTs can also produce graphene at a bulk level [63]. CNTs can be completely unzipped at the nanoscale to form graphene nanoribbons, but partial unzipping of CNTs can result in novel carbon-based nanostructures with interconnection of both CNTs and graphene [64]. Various mechanisms for the unzipping of CNTs have been studied by the researchers. Generally, wet chemical methods comprise of Li ion intercalation, sonication, hydrothermal reaction, chemical oxidation and electrochemical method, while the dry processing methods include metal particle catalytic cutting, plasma etching, fast thermal expansion [64]. Regarding bottom-up strategies, the creation of graphene depends on appropriate choice of starting precursor molecules that undergo chemical reactions to create covalently connected 2D networks. One of the effective approach for producing large-area monolayer graphene or its derivatives with higher structural quality is CVD, which is dependent on type of growth precursors, temperatures, pressures, and the reaction environment as in case of CNTs [57]. Here also, C-atoms generated from the breakdown of hydrocarbons either in gaseous phase or in liquid phase, progressively grow

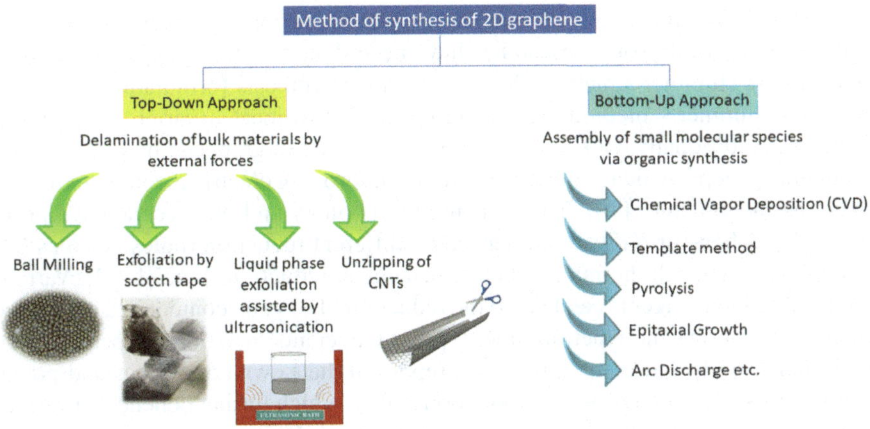

Fig. 5 Method of synthesis of 2D graphene

into substantial domains on metal catalyst throughout this procedure, and graphene is then removed from the substrate surface for additional usage.

However, it is critically necessary to have graphene sheets with a desirable surface area, minimal flaws, and strong conductivity for theoretical research and real-world applications. More notably, due to their abundance of O-containing surface functional groups, structural defects, heteroatomic irregularity and other distinctive properties, graphene derivatives (such as GO and rGO) have steadily evolved and provide potential opportunities for advanced applications for sustainable future [65].

GO is one of the oxidized form of graphene, with C atoms (sp^2- hybridized) in conductive π states and some C atoms (sp^3-hybridized bonded to O), in nonconductive σ states [54]. The surface of GO is hydrophilic in nature as it possesses abundant functional groups like carboxyl (−COOH), hydroxyl (−OH), keto (C=O) and epoxy (C–O–C) [66]. Similar to graphene, synthesis strategies of GO also include the top-down and bottom-up approaches. GO can be prepared by the exfoliation of graphite oxide as shown in Fig. 4. The first attempt at GO preparation was made by Brodie in 1859, who treated Ceylon graphite with an oxidative solution of potassium chlorate and fuming nitric acid [67]. In 1898, Staudenmaier modified the Brodie's method by using a combination of concentrated H_2SO_4, fuming HNO_3 and $KClO_3$ for the oxidation of graphite to GO [68]. Following the Brodie study, alternative approaches for the formulation of GO were proposed, the majority of which used mixture containing powerful oxidizers [69]. Hummer's technique, which involved utilizing graphite powder (100 g), $NaNO_3$ (50 g), H_2SO_4 (2.3 L), and $KMnO_4$ (300 g) to oxidize graphene, is one of the most popular techniques used by researchers to produce GO. In comparison to Brodie's method, Hummer's method has some advantages like the oxidation process is relatively fast and takes a few hours to produce a greater amount of GO, as well as oxygen. However, the emission of explosive gases (ClO_2) and toxic gases (NO_2, N_2O_4) is the main drawback of this approach. Additionally, the residual ions, Na^+ and NO_3^-, are hard to eliminate from the reaction mixture. In order to reduce the hazardous consequences related to the traditional approach, some researches have been done that try to suggest improvements to the Hummer's method. Many innovative synthesis formulations based on modified Hummer's methods are being developed to meet a variety of practical demands. Additionally, GO can be partially converted to graphene by removing O-containing groups, which is notably useful for improving electrical conductivity and changing the structure [70]. The structure morphology and physicochemical properties of graphene or rGO are influenced by different reduction routes (such as UV irradiation, chemical, thermal, and electrochemical reduction) [56, 71, 72]. Various chemical reducing agents can lead to the reduction of oxygen content in a controlled manner by altering the reaction time, type of reductants, reaction temperature and environment [73]. Different approaches represent their own advantages and potentials in terms of their large-scale production ability, which further depends on various factors such as energy usage, overall cost, toxicity analysis etc.

GO sheets has been reduced chemically with the help of various reducing agents including borohydrides [74], hydrazine [75], gaseous hydrogen (after thermal expansion) [76], hydroquinone [77], strong alkaline solutions [78], hydrohalic acids [79], metal–acid mixtures [80], S/N/O- containing reducing agents [73] etc. Thermal reduction has also been employed for the reduction of GO to rGO that involves the heat treatment to eliminate the oxide functional groups from the GO surfaces [72]. However, recently, some green techniques have been emerged as an alternative to above reduction methods [81]. For example, carbohydrates (glucose, sucrose and fructose) have been used for the reduction of exfoliation of graphite oxide into graphene nanosheets through a green and facile approach, where, the reducing agents as well as the oxidized products were environmentally benign [82]. Similarly, naturally occurring anti-oxidants such as carotenoids extracted from the vegetable (carrot, sweet potato etc.) have been explored for the green, facile and cost effective approach towards graphene synthesis [83]. As per another report, graphene nanosheets were synthesized with pomegranate juice, where it performed functions of reductant as well as a capping agent. Initially, a modified Hummer method was adopted to oxidize graphite, which was then converted to GO and lastly, GO was reduced to graphene nanosheet by pomegranate juice [84]. *Verbena officinalis* extract, lemon juice, carrot, bamboo shoot extract are also utilized as a green reducing agent for converting GO to RGO [85–87]. Interestingly, Polyethylene terephthalate (PET) bottle waste has been used as a source material in the synthesis of graphene. The process involves pyrolysis of PET waste in a closed system under autogenic pressure. It was reported that the BET surface area of the as-synthesized graphene was 721.7 m^2/g, while that of PET was less than 2 m^2/g, which made graphene a promising candidate for the adsorption of methylene blue and acid blue 25 dyes [88].

2.3 *Fullerenes*

The fullerenes are the 0D CNs, having sp^2-like hybridization owing to its curvature effect. Many investigations have been conducted on fullerenes since C60 was discovered in 1985 [89]. Fullerenes have been extensively explored for its applications in solar cells, drug delivery systems, antioxidants, nano-sensors, wastewater purification and many more, because of their chemical and electronic properties [90]. C60, the smallest fullerene, has the diameter of 7 Å (approx.) and contains twelve pentagonal rings and twenty hexagonal rings of carbon atoms as shown in Fig. 6a [12]. Interestingly, the number of carbon atoms of a fullerene is always even varying from 20, 60, 70, 82, 100, 180 to 960. C60 fullerene is highly susceptible to chemical attacks as the number of hexagons is greater than 20 and the pentagons occupy highly tensioned position resulting in the reduced stability of the molecule [12]. The inner as well as the outer surface of fullerenes are covered with homogeneous π electrons, which can strongly interact with organic contaminants via π–π stacking. The clusters of fullerene provide various adsorptive sites that are responsible for the adsorption of different contaminants like surface adsorption areas, grooves occurring between adjacent fullerenes and the interstitial voids between fullerenes in aggregate. Applications of fullerene as adsorbent has also been reported for the removal of organic, inorganic, organometallic compounds etc. [91–93].

2.3.1 Synthesis of Fullerenes

Fullerenes are conventionally synthesized by the arc-discharge vaporization of graphite, by CVD techniques, and by combustion processes (Fig. 6c). These techniques are not, however, extremely effective, thus further development and advancement in this area are required. The investigation of novel synthetic techniques is therefore essential to the development of the fullerene science. Kroto et al. synthesized an extremely stable cluster of C60 by using vaporized graphite via laser irradiation [89].

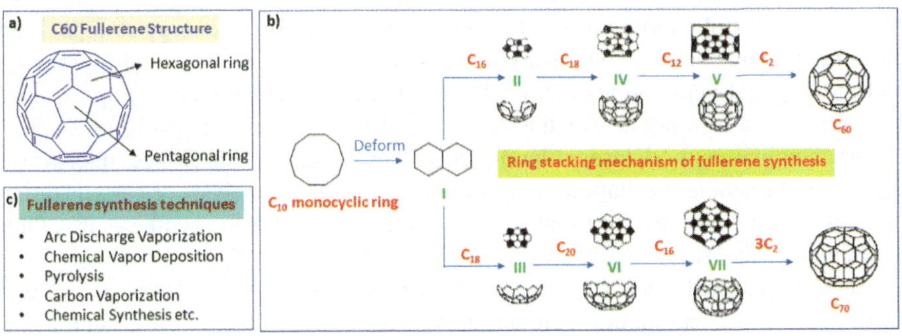

Fig. 6 **a** Structure of C60 fullerene **b** Ring stacking mechanism of fullerene synthesis (Reprinted from [94] © 1992 with permission from Elsevier) c) Fullerene synthetic techniques

A surge of interest in the chemistry of C60 was sparked by this experiment. Scientists have attempted to explain the process of fullerene creation since the discovery of fullerenes. As per a report, the carbon feedstock is initially vaporized to small components (like C-atoms or C-dimers), which lastly recombine under specified temperature and pressure conditions to form fullerenes after undergoing a series of reactions [90]. However, several theories have been put forth to explain the production process of fullerenes because this basic hypothesis is not an adequate one such as Goroff model, icospiral particle nucleation scheme, ring stacking model [90, 94]. The ring stacking model as shown in Fig. 6b, assumes that a closed cage of carbon is produced via sequential stacking with appropriate numbers and combinations of only even-numbered carbon rings without any loss of carbon atoms [94].

Fullerene's thermal and chemical characteristics, especially their potential to accept electrons, have attracted scientific and commercial interest for environment remediation. The BET specific surface area of fullerene depends on the method of synthesis and was evaluated experimentally ranging from approximately 1.1–176 $m^2\ g^{-1}$[12]. A fullerene's surface can be subjected to a variety of physical and chemical processes that alter the systems' overall qualities. For instance, the corrosion of carbon caused by heat treatment with CO_2 results in the creation of CO and the removal of the -OH and -CH_3 surface groups. In addition to chemical changes, heat treatment may cause tiny pores inside of the carbon particles to open, increasing the total surface area [12].

2.4 *Carbon Quantum Dots (CQDs)*

CQDs are one of the most important emerging 0D CNs. CQDs are usually just a few nanometers in size (generally smaller than 10 nm) and exhibit a plethora of attractive characteristics such as high luminescent, high surface area, high quantum yield, broad absorption spectra, long fluorescent life, low comparative toxicity, high photo stability, ease of surface modifications, chemical inertness, high emission tenability and biocompatibility [95]. With such useful properties, CQDs attracts great interest of researchers in applications various fields such as drug delivery, energy storage, light emitting diodes, photocatalytic activity, biosensing, chemical sensing, bio-imaging and bio-medical applications etc. [95]. Both top-down and bottom-up approaches have been widely reported for the synthesis of CQDs. Several protocols such as electrochemical method, laser ablation, plasma treatment, arc discharge, microwave digestion, ultrasonic oscillation, hydrothermal synthesis, pyrolysis or carbonization of small organic moieties or biomass have been reported for the synthesis of CQDs as shown in Fig. 7 [96, 97].

According to past studies, CQDs have been fabricated from various natural carbon feedstocks such as ammonium citrate, ginkgo leaf, citric acid, ascorbic acid, denatured milk, graphitic micro-particles, zein biopolymer, dried leaf, sodium hydroxide, gelatin, food wastes, grass, pomelo fruit, humic acid etc [98]. Also, CQDs can be exfoliated from activated carbon, coal, soot, carbon blak etc. by treating them with

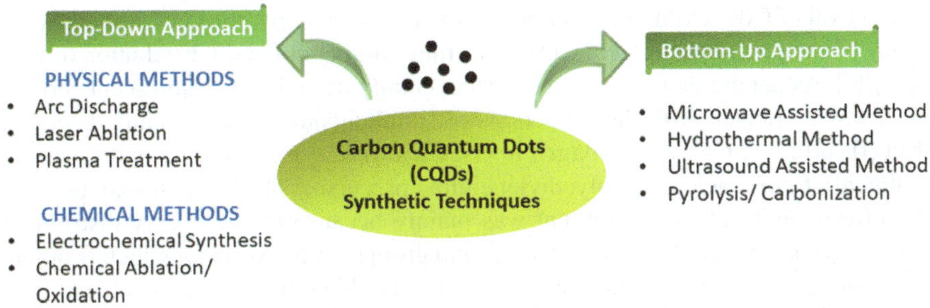

Fig. 7 Synthetic techniques for CQDs

strong oxidizing agents like H_2SO_4, HNO_3 and $KMnO_4$. This method has a potential of adoptability at a large-scale as these low-cost carbon sources are readily available. However, the major limitations associated with method are difficult removal of excess of oxidising agents and inorganic salts, evolution of highly toxic gases during the reaction [95]. Therefore, it is crucial that eco-friendly and simple CQD production procedures are devised in order for the successful large-scale development of these CDs to be implemented. When compared to previously described methods, the hydrothermal approach is the most popular because of its ease of use, mild reaction conditions, and high quantum yield. For example, a carbon and nitrogen-doped CQDs were prepared by hydrothermal method using ammonium citrate as precursor and combined with layered double hydroxide for the fast and efficient decontamination of Cd(II) [99]. Similarly, N, S co-doped CQDs were synthesized by the hydrothermal treatment of L-Histidine and L-Cysteine upto 180 °C for 10 h and used for the adsorption of methylene blue dye after loading the CQDs into hexagonal mesoporous silica [100]. CVD method has also been reported for the synthesis of CQDs using Fe-Co/CaO catalyst, where, defective CNTs and CQD-CNT composites were fabricated selectively at different catalyst reduction temperatures and used for the removal of methylene blue dye [101].

Sugarcane bagasse, an agricultural waste, was reported for the efficient and economic synthesis of CQDs, where, GO was chemically shredded to graphene quantum dots and CQDs were prepared by chemical oxidation followed by exfoliation [102]. Graphene quantum dots are small fragments with a diameter <100 nm along with a few-layers of graphene structure (one to ten). Very recently, graphene quantum dots have been reported to be synthesized via laser-induced polymerization of polyimide powder precursor in a single-step [103].

Recently, carbon dots (CDs) were reported as an efficient adsorbent for the removal of heavy metal from wastewater. Hydrothermal method was adopted for the fabrication of a photoluminescent CDs from petroleum coke waste as precursor. The as-prepared CDs was then well dispersed in a chitosan polymer matrix to form a fluorescent film with abundance of O, N and S functional groups which were responsible for the removal of Cd^{2+} from industrial wastewater. Interestingly, the Cd^{2+} removal efficiency by the photoluminescent CDs was 5 times higher under the irradiation of UV-light [96]. As per another report, an environmentally friendly, 3D fluorescent aerogel mesh was fabricated with dual functions of simultaneous adsorption and detection of Pb(II) utilizing CDs and collagen fibers. CDs were synthesized by hydrothermal treatment of citric acid and polyethyleneimine precursors at 180 °C for about 20 h. The effectiveness of this adsorbent was mainly because of the evenly distributed CDs on the aerogel and the numerous amino groups on the surface, which promote metal ion aggregation and chelation [104]. The application of microwaves has also been reported for the production of fluorescent CQDs using urea with bagasse, cellulose or carboxymethyl cellulose as precursors. The synthesized CQDs were utilized for the adsorption of Pb(II) from an aqueous solution [105]. As per some latest advances, CDs and rGO composites are also being prepared with the aid of ultrasound, where electron transfer occurs from CDs to GO which helps in its reduction to rGO. This composite of CDs and rGO was further used in the determination as well as removal of organochlorine pesticides [106]. These findings demonstrate the potential efficiency of CQDs synthesized from various methods as promising carbon-based nanomaterials for utilization in the field of wastewater remediation through adsorption, pollution detection and chemical sensing applications.

3 Conclusion

Ever-increasing interest in utilizing carbon-based nanomaterials for various applications in different fields has led to continued efforts in evaluation of their physicochemical properties, structural studies, surface modifications, composite formation and other significant advancements. This chapter provides an overview of various carbon-based nanomaterials which have been typically utilized as adsorbent for wastewater purification. The fabrication techniques for these CNs including CNTs, graphene and its derivatives, fullerenes and CQDs, involves both top-down and bottom-up approaches. With time, these techniques have been modified to develop an efficient, cost effective, facile, and environmentally benign synthetic strategy that can be implemented at a commercial scale. Overall, CNs, due to their exceptional qualities are excellent candidate to substitute the conventional techniques of water purification, if proper modifications are adopted.

References

1. Aruna, Bagotia, N., Sharma, A.K., Kumar, S.: A review on modified sugarcane bagasse biosorbent for removal of dyes. Chemosphere **268**, 129309 (2021)
2. Crini, G.: Non-conventional low-cost adsorbents for dye removal: a review. Bioresour. Technol. **97**, 1061 (2006)

3. Crini, G., Lichtfouse, E.: Advantages and disadvantages of techniques used for wastewater treatment. Environ. Chem. Lett. **17**, 145 (2019)
4. Pan, X., Gu, Z., Chen, W., Li, Q.: Preparation of biochar and biochar composites and their application in a Fenton-like process for wastewater decontamination: a review. Sci. Total. Environ. **754**, 142104 (2021)
5. Yadav, A., Bagotia, N., Sharma, A.K., Kumar, S.: Simultaneous adsorptive removal of conventional and emerging contaminants in multi-component systems for wastewater remediation: a critical review. Sci. Total. Environ. **799**, 149500 (2021)
6. Buzea, C., Pacheco, I.I., Robbie, K.: Nanomaterials and nanoparticles: sources and toxicity. Biointerphases. **2**, MR17 (2007)
7. Queiroz, R.N., Prediger, P., Vieira, M.G.A.: Adsorption of polycyclic aromatic hydrocarbons from wastewater using graphene-based nanomaterials synthesized by conventional chemistry and green synthesis: a critical review. J. Hazard. Mater. **422**, 126904 (2022)
8. Devi, N., Kumar, R., Chen, Y.-S., Singh, R.K.: Carbon-based nanomaterials: carbon nanotube, fullerene, and carbon dots. In: Nanomaterials. Springer Nature Singapore, Singapore (2023)
9. Yadav, A., Bagotia, N., Sharma, A.K., Kumar, S.: Advances in decontamination of wastewater using biomass-based composites: a critical review. Sci. Total. Environ. **784**, 147108 (2021)
10. Rasool, M.A., Sattar, R., Anum, A., Al-Hussain, S.A., Ahmad, S., Irfan, A., Zaki, M.E.A.: An insight into carbon nanomaterial-based photocatalytic water splitting for green hydrogen production. Catalysts **13**, 66 (2022)
11. Boota, M., Ahmad, I., Ahmad, J.: Production of graphene-derivatives using organic molecules for supercapacitors and beyond. Colloids Surfaces A Physicochem. Eng. Asp. **658**, 130693 (2023)
12. Machado, F.M., Fagan, S.B., da Silva, I.Z., de Andrade, M.J.: Carbon Nanoadsorbents. Presented at the (2015)
13. Yadav, A., Bagotia, N., Yadav, S., Sharma, A.K., Kumar, S.: Adsorptive studies on the removal of dyes from single and binary systems using saccharum munja plant-based novel functionalized CNT composites. Environ. Technol. Innov. **24**, 102015 (2021)
14. Chen, S., Shen, W., Wu, G., Chen, D., Jiang, M.: A new approach to the functionalization of single-walled carbon nanotubes with both alkyl and carboxyl groups. Chem. Phys. Lett. **402**, 312 (2005)
15. Ma, P.C., Siddiqui, N.A., Marom, G., Kim, J.K.: Dispersion and functionalization of carbon nanotubes for polymer-based nanocomposites: a review. Compos. Part A Appl. Sci. Manuf. **41**, 1345 (2010)
16. Dresselhaus, M.S., Dresselhaus, G., Saito, R.: Physics of carbon nanotubes. Carbon N. Y. **33**, 883 (1995)
17. Thostenson, E.T., Ren, Z., Chou, T.-W.: Advances in the science and technology of carbon nanotubes and their composites: a review. Compos. Sci. Technol. **61**, 1899 (2001)
18. Iijima, S.: Helical microtubules of graphitic carbon. Nature **354**, 56 (1991)
19. Sharma, S.: Current synthetic methodologies of carbon nanotubes: a review. Mini. Rev. Org. Chem. **20**, 55 (2022)
20. Sadek, E.M., Ahmed, S.M., Mansour, N.A.: Carbon nanotubes and other carbon nanomaterials: Prospects for functionalization. Elsevier Ltd. p. 107–147 (2023)
21. Chrzanowska, J., Hoffman, J., Małolepszy, A., Mazurkiewicz, M., Kowalewski, T.A., Szymanski, Z., Stobinski, L.: Synthesis of carbon nanotubes by the laser ablation method: Effect of laser wavelength. Phys. status solidi. **252**, 1860 (2015)
22. Purohit, R., Purohit, K., Rana, S., Rana, R.S., Patel, V.: Carbon nanotubes and their growth methods. Procedia Mater. Sci. **6**, 716 (2014)
23. Manafi, S., Nadali, H., Irani, H.R.: Low temperature synthesis of multi-walled carbon nanotubes via a sonochemical/hydrothermal method. Mater. Lett. **62**, 4175 (2008)
24. Makgabutlane, B., Nthunya, L.N., Maubane-Nkadimeng, M.S., Mhlanga, S.D.: Green synthesis of carbon nanotubes to address the water-energy-food nexus: a critical review. J. Environ. Chem. Eng. **9**, 104736 (2021)

25. Moses, J.C., Gangrade, A., Mandal, B.B.: Carbon nanotubes and their polymer nanocomposites. In: Nanomaterials and Polymer Nanocomposites. Elsevier (2019)
26. Kumar, M.: Carbon nanotube synthesis and growth mechanism. In: Carbon Nanotubes—Synthesis, Characterization, Applications. InTech (2011)
27. Jafarpour, S.M., Kini, M., Schulz, S.E., Hermann, S.: Effects of catalyst configurations and process conditions on the formation of catalyst nanoparticles and growth of single-walled carbon nanotubes. Microelectron. Eng. **167**, 95 (2017)
28. Hoyos-Palacio, L.M., García, A.G., Pérez-Robles, J.F., González, J., Martínez-Tejada, H. V: Catalytic effect of Fe, Ni, Co and Mo on the CNTs production. IOP Conf. Ser. Mater. Sci. Eng. **59**, 012005 (2014)
29. Wang, H., Zhu, S., Dai, Z., Li, X., Zhou, T.: Selective synthesis of large diameter single-walled carbon nanotubes on rice husk-derived catalysts. J. Environ. Chem. Eng. **11**, 109261 (2023)
30. Paul, S., Samdarshi, S.K.: A green precursor for carbon nanotube synthesis. New Carbon Mater. **26**, 85 (2011)
31. Suriani, A.B., Azira, A.A., Nik, S.F., Md Nor, R., Rusop, M.: Synthesis of vertically aligned carbon nanotubes using natural palm oil as carbon precursor. Mater. Lett. **63**, 2704 (2009)
32. Lotfy, V.F., Fathy, N.A., Basta, A.H.: Novel approach for synthesizing different shapes of carbon nanotubes from rice straw residue. J. Environ. Chem. Eng. **6**, 6263 (2018)
33. Tripathi, N., Pavelyev, V., Islam, S.S.: Synthesis of carbon nanotubes using green plant extract as catalyst: unconventional concept and its realization. Appl. Nanosci. **7**, 557 (2017)
34. Kumar, M., Ando, Y.: Carbon nanotubes from camphor: an environment-friendly nanotechnology. J. Phys. Conf. Ser. **61**, 643 (2007)
35. Kumar, M., Ando, Y.: Single-wall and multi-wall carbon nanotubes from camphor—a botanical hydrocarbon. Diam. Relat. Mater. **12**, 1845 (2003)
36. Ghosh, P., Afre, R.A., Soga, T., Jimbo, T.: A simple method of producing single-walled carbon nanotubes from a natural precursor: Eucalyptus oil. Mater. Lett. **61**, 3768 (2007)
37. Bernd, M.G.S., Bragança, S.R., Heck, N., Filho, L.C.P. da S.: Synthesis of carbon nanostructures by the pyrolysis of wood sawdust in a tubular reactor. J. Mater. Res. Technol. **6**, 171 (2017)
38. Kumar, R., Tiwari, R.S., Srivastava, O.N.: Scalable synthesis of aligned carbon nanotubes bundles using green natural precursor: neem oil. Nanoscale Res. Lett. **6**, 92 (2011)
39. Ghosh, P., Soga, T., Afre, R.A., Jimbo, T.: Simplified synthesis of single-walled carbon nanotubes from a botanical hydrocarbon: Turpentine oil. J. Alloys Compd. **462**, 289 (2008)
40. Awasthi, K., Kumar, R., Tiwari, R.S., Srivastava, O.N.: Large scale synthesis of bundles of aligned carbon nanotubes using a natural precursor: turpentine oil. J. Exp. Nanosci. **5**, 498 (2010)
41. Awasthi, K., Kumar, R., Raghubanshi, H., Awasthi, S., Pandey, R., Singh, D., Yadav, T.P., Srivastava, O.N.: Synthesis of nano-carbon (nanotubes, nanofibres, graphene) materials. Bull. Mater. Sci. **34**, 607 (2011)
42. Hamid, Z.A., Azim, A.A., Mouez, F.A., Rehim, S.S.A.: Challenges on synthesis of carbon nanotubes from environmentally friendly green oil using pyrolysis technique. J. Anal. Appl. Pyrolysis **126**, 218 (2017)
43. Tu, C., Luo, W., Peng, Y., Yu, P., Shi, C., Wu, Z., Shao, L., Zhan, P.: Preparation of lignin-based carbon nanotubes using micelles as soft template. Ind. Crops Prod. **191**, 116009 (2023)
44. Yadav, A., Bagotia, N., Yadav, S., Sharma, N., Sharma, A.K., Kumar, S.: Environmental application of saccharum munja biomass-derived hybrid composite for the simultaneous removal of cationic and anionic dyes and remediation of dye polluted water: a step towards pilot-scale studies. Colloids Surfaces A Physicochem. Eng. Asp. **650**, 129539 (2022)
45. Yadav, A., Bagotia, N., Yadav, S., Sharma, A.K., Kumar, S.: In-situ fabrication of surfactant modified CNT-based novel bio-composite and its performance evaluation for simultaneous removal of anionic dyes: optimization by Box-Behnken design. Sep. Purif. Technol. **284**, 120262 (2022)
46. Yadav, S., Yadav, A., Bagotia, N., Sharma, A.K., Kumar, S.: Novel composites of Pennisetum glaucum with CNT: preparation, characterization and application for the removal of safranine

O and methylene blue dyes from single and binary systems. Biomass Convers. Biorefinery (2022)
47. Yadav, S., Yadav, A., Bagotia, N., Sharma, N., Sharma, A.K., Kumar, S.: Simultaneous adsorption of three anionic dyes at neutral pH from their individual and multi-component systems on a CTAB modified Pennisetum glaucum based carbon nanotube green composite: adsorption mechanism and process optimization by Box-Behnken design. J. Mol. Liq. **358**, 119223 (2022)
48. Yadav, S., Yadav, A., Bagotia, N., Sharma, A.K.: Pennisetum glaucum lignocellulosic adsorbent modified with CNT for the removal of cationic dyes from individual and multi—component systems. Cellulose (2022)
49. Vashisth, N., Sharma, S.P., Kumar, S., Yadav, A.: Green synthesis of 3-(1-naphthyl), 4-methyl-3-(1-naphthyl) coumarins and 3-phenylcoumarins using dual-frequency ultrasonication. Green Process. Synth. **9**, 399 (2020)
50. Khurana, I., Saxena, A., Bharti, Khurana, J.M., Rai, P.K.: Removal of dyes using graphene-based composites: a review. Water. Air. Soil Pollut. **228**, (2017)
51. Bagotia, N., Choudhary, V., Sharma, D.K.: Synergistic effect of graphene/multiwalled carbon nanotube hybrid fillers on mechanical, electrical and EMI shielding properties of polycarbonate/ethylene methyl acrylate nanocomposites. Compos. Part B Eng. **159**, 378 (2019)
52. Novoselov, K.S., Geim, A.K., Morozov, S. V., Jiang, D., Zhang, Y., Dubonos, S. V., Grigorieva, I. V., Firsov, A.A.: Electric field effect in atomically thin carbon films. Science (80-.). **306**, 666 (2004)
53. Ali, I., Alharbi, O.M.L., Tkachev, A., Galunin, E., Burakov, A., Grachev, V.A.: Water treatment by new-generation graphene materials : hope for bright future. Environ. Sci. Pollut. Res. **25**, 7315 (2018)
54. Zhang, S., Li, B., Wang, X., Zhao, G., Hu, B., Lu, Z.: Recent developments of two-dimensional graphene-based composites in visible-light photocatalysis for eliminating persistent organic pollutants from wastewater. Chem. Eng. J. **390**, 124642 (2020)
55. Sturala, J., Luxa, J., Pumera, M., Sofer, Z.: Chemistry of graphene derivatives: synthesis, applications, and perspectives. Chem. A Eur. J. **24**, 5992 (2018)
56. De Silva, K.K.H., Huang, H.-H., Joshi, R.K., Yoshimura, M.: Chemical reduction of graphene oxide using green reductants. Carbon N. Y. **119**, 190 (2017)
57. Bhuyan, M.S.A., Uddin, M.N., Islam, M.M., Bipasha, F.A., Hossain, S.S.: Synthesis of graphene. Int. Nano Lett. **6**, 65 (2016)
58. Yi, M., Shen, Z.: A review on mechanical exfoliation for the scalable production of graphene. J. Mater. Chem. A. **3**, 11700 (2015)
59. Li, J., Yan, H., Dang, D., Wei, W., Meng, L.: Salt and water co-assisted exfoliation of graphite in organic solvent for efficient and large scale production of high-quality graphene. J. Colloid Interface Sci. **535**, 92 (2019)
60. Arao, Y., Mori, F., Kubouchi, M.: Efficient solvent systems for improving production of few-layer graphene in liquid phase exfoliation. Carbon N. Y. **118**, 18 (2017)
61. Narayan, R., Lim, J., Jeon, T., Li, D.J., Kim, S.O.: Perylene tetracarboxylate surfactant assisted liquid phase exfoliation of graphite into graphene nanosheets with facile re-dispersibility in aqueous/organic polar solvents. Carbon N. Y. **119**, 555 (2017)
62. Hu, Y., Su, M., Xie, X., Sun, C., Kou, J.: Few-layer graphene oxide with high yield via efficient surfactant-assisted exfoliation of mildly-oxidized graphite. Appl. Surf. Sci. **494**, 1100 (2019)
63. Kosynkin, D.V., Higginbotham, A.L., Sinitskii, A., Lomeda, J.R., Dimiev, A., Price, B.K., Tour, J.M.: Longitudinal unzipping of carbon nanotubes to form graphene nanoribbons. Nature **458**, 872 (2009)
64. Lim, J., Narayan Maiti, U., Kim, N.-Y., Narayan, R., Jun Lee, W., Sung Choi, D., Oh, Y., Min Lee, J., Yong Lee, G., Hun Kang, S., Kim, H., Kim, Y.-H., Ouk Kim, S.: Dopant-specific unzipping of carbon nanotubes for intact crystalline graphene nanostructures. Nat. Commun. **7**, 10364 (2016)

65. Manikandan, V., Lee, N.Y.: Reduced graphene oxide: Biofabrication and environmental applications. Chemosphere **311**, 136934 (2023)
66. Naushad, M.: A new generation material graphene: applications in water technology, pp. 1–471 (2018)
67. Brodie, B.C.: On the atomic weight of graphite. Philos. Trans. R. Soc. London. **149**, 249 (1859)
68. Staudenmaier, L.: Verfahren zur Darstellung der Graphitsäure. Berichte der Dtsch. Chem. Gesellschaft. **31**, 1481 (1898)
69. Hummers, W.S., Offeman, R.E.: Preparation of graphitic oxide. J. Am. Chem. Soc. **80**, 1339 (1958)
70. Mohan, V.B., Lau, K., Hui, D., Bhattacharyya, D.: Graphene-based materials and their composites: a review on production, applications and product limitations. Compos. Part B Eng. **142**, 200 (2018)
71. García-Argumánez, A., Llorente, I., Caballero-Calero, O., González, Z., Menéndez, R., Escudero, M.L., García-Alonso, M.C.: Electrochemical reduction of graphene oxide on biomedical grade CoCr alloy. Appl. Surf. Sci. **465**, 1028 (2019)
72. Sengupta, I., Chakraborty, S., Talukdar, M., Pal, S.K., Chakraborty, S.: Thermal reduction of graphene oxide: how temperature influences purity. J. Mater. Res. **33**, 4113 (2018)
73. Chua, C.K., Pumera, M.: Chemical reduction of graphene oxide: a synthetic chemistry viewpoint. Chem. Soc. Rev. **43**, 291 (2014)
74. Shin, H., Kim, K.K., Benayad, A., Yoon, S., Park, H.K., Jung, I., Jin, M.H., Jeong, H., Kim, J.M., Choi, J., Lee, Y.H.: Efficient reduction of graphite oxide by sodium borohydride and its effect on electrical conductance. Adv. Funct. Mater. **19**, 1987 (2009)
75. Becerril, H.A., Mao, J., Liu, Z., Stoltenberg, R.M., Bao, Z., Chen, Y.: Evaluation of solution-processed reduced graphene oxide films as transparent conductors. ACS Nano **2**, 463 (2008)
76. Wu, Z.-S., Ren, W., Gao, L., Liu, B., Jiang, C., Cheng, H.-M.: Synthesis of high-quality graphene with a pre-determined number of layers. Carbon N. Y. **47**, 493 (2009)
77. Wang, S., Chia, P.-J., Chua, L.-L., Zhao, L.-H., Png, R.-Q., Sivaramakrishnan, S., Zhou, M., Goh, R.G.-S., Friend, R.H., Wee, A.T.-S., Ho, P.K.-H.: Band-like transport in surface-functionalized highly solution-processable graphene nanosheets. Adv. Mater. **20**, 3440 (2008)
78. Noreen, S., Bhatti, H.N., Farrukh, Z., Ilays, S., Jamal, M.A.: Continuous fixed bed removal of Novacron Orange P-2R using sugarcane bagasse: prediction of breakthrough curves. Desalin. Water Treat. **57**, 12814 (2016)
79. Moon, I.K., Lee, J., Ruoff, R.S., Lee, H.: Reduced graphene oxide by chemical graphitization. Nat. Commun. **1**, 73 (2010)
80. Fan, Z., Wang, K., Wei, T., Yan, J., Song, L., Shao, B.: An environmentally friendly and efficient route for the reduction of graphene oxide by aluminum powder. Carbon N. Y. **48**, 1686 (2010)
81. Manousi, N., Deliyanni, E.A., Zachariadis, G.A.: Green extraction approaches utilizing graphene and graphene oxide derivatives in environmental analysis. In: Novel Materials for Environmental Remediation Applications. Elsevier (2023)
82. Zhu, C., Guo, S., Fang, Y., Dong, S.: Reducing sugar: new functional molecules for the green synthesis of graphene nanosheets. ACS Nano **4**, 2429 (2010)
83. Vusa, C.S.R., Berchmans, S., Alwarappan, S.: Facile and green synthesis of graphene. RSC Adv. **4**, 22470 (2014)
84. Tavakoli, F., Salavati-Niasari, M., Badiei, A., Mohandes, F.: Green synthesis and characterization of graphene nanosheets. Mater. Res. Bull. **63**, 51 (2015)
85. Haydari, I., Aziz, K., Kaya, S., Daştan, T., Ouazzani, N., Mandi, L., Aziz, F.: Green synthesis of reduced graphene oxide and their use on column adsorption of phenol from olive mill wastewater. Process. Saf. Environ. Prot. **170**, 1079 (2023)
86. Joshi, R., De Adhikari, A., Dey, A., Lahiri, I.: Green reduction of graphene oxide as a substitute of acidic reducing agents for supercapacitor applications. Mater. Sci. Eng. B **287**, 116128 (2023)

87. Ruti, I., Kumar, S.: Bamboo shoot extract as a novel and efficient reducing agent for graphene oxide and its supercapacitor application. J. Mater. Sci. Mater. Electron. **34**, 11 (2023)
88. El Essawy, N.A., Ali, S.M., Farag, H.A., Konsowa, A.H., Elnouby, M., Hamad, H.A.: Green synthesis of graphene from recycled PET bottle wastes for use in the adsorption of dyes in aqueous solution. Ecotoxicol. Environ. Saf. **145**, 57 (2017)
89. Kroto, H.W., Heath, J.R., O'Brien, S.C., Curl, R.F., Smalley, R.E.: C60: Buckminsterfullerene. Nature **318**, 162 (1985)
90. Mojica, M., Alonso, J.A., Méndez, F.: Synthesis of fullerenes. J. Phys. Org. Chem. **26**, 526 (2013)
91. Ballesteros, E., Gallego, M., Valcárcel, M.: Analytical potential of fullerene as adsorbent for organic and organometallic compounds from aqueous solutions. J. Chromatogr. A **869**, 101 (2000)
92. Gupta, V.K., Saleh, T.A.: Sorption of pollutants by porous carbon, carbon nanotubes and fullerene-an overview. Environ. Sci. Pollut. Res. **20**, 2828 (2013)
93. Cheng, X., Kan, A.T., Tomson, M.B.: Naphthalene adsorption and desorption from Aqueous C 60 fullerene. J. Chem. Eng. Data **49**, 675 (2004)
94. Wakabayashi, T., Achiba, Y.: A model for the C60 and C70 growth mechanism. Chem. Phys. Lett. **190**, 465 (1992)
95. Hu, S., Wei, Z., Chang, Q., Trinchi, A., Yang, J.: A facile and green method towards coal-based fluorescent carbon dots with photocatalytic activity. Appl. Surf. Sci. **378**, 402 (2016)
96. Jlassi, K., Eid, K., Sliem, M.H., Abdullah, A.M., Chehimi, M.M., Krupa, I.: Rational synthesis, characterization, and application of environmentally friendly (polymer–carbon dot) hybrid composite film for fast and efficient UV-assisted Cd2+ removal from water. Environ. Sci. Eur. **32**, 12 (2020)
97. Saha, A., Bhattacharjee, L., Bhattacharjee, R.R.: Synthesis of carbon quantum dots. In: Carbon Quantum Dots for Sustainable Energy and Optoelectronics. Elsevier (2023)
98. Rani, U.A., Ng, L.Y., Ng, C.Y., Mahmoudi, E.: A review of carbon quantum dots and their applications in wastewater treatment. Adv. Colloid Interface Sci. **278**, 102124 (2020)
99. Rahmanian, O., Dinari, M., Abdolmaleki, M.K.: Carbon quantum dots/layered double hydroxide hybrid for fast and efficient decontamination of Cd(II): the adsorption kinetics and isotherms. Appl. Surf. Sci. **428**, 272 (2018)
100. Teymoorian, T., Hashemi, N., Mousazadeh, M.H., Entezarian, Z.: N, S doped carbon quantum dots inside mesoporous silica for effective adsorption of methylene blue dye. SN Appl. Sci. **3**, 305 (2021)
101. Chen, Q., Wang, H., Tang, X., Ba, Z., Zhao, X., Wang, Y., Deng, H.: One-step synthesis of carbon quantum dot-carbon nanotube composites on waste eggshell-derived catalysts for enhanced adsorption of methylene blue. J. Environ. Chem. Eng. **9**, 106222 (2021)
102. Baweja, H., Jeet, K.: Economical and green synthesis of graphene and carbon quantum dots from agricultural waste. Mater. Res. Express. **6**, 0850g8 (2019)
103. Liu, J., Qin, Y., Li, X., Wang, R., Guo, H., Tang, J., Liu, J., Liu, L.: One-step synthesis of graphene quantum dots by laser-induced polymer. Mater. Sci. Semicond. Process. **153**, 107146 (2023)
104. Scaria, J., Karim, A. V., Divyapriya, G., Nidheesh, P.V., Suresh Kumar, M.: Carbon-supported semiconductor nanoparticles as effective photocatalysts for water and wastewater treatment. In: Nano-Materials as Photocatalysts for Degradation of Environmental Pollutants. Elsevier (2020)
105. Tohamy, H.S., El-Sakhawy, M., Kamel, S.: Eco-friendly synthesis of carbon quantum dots as an effective adsorbent. J. Fluoresc (2022)
106. Xu, S., Liu, H., Chen, C., Feng, S., Fan, J.: Ultrasound-assisted one-step reduction and self-assembly of carbon dots-reduced graphene oxide: Mechanism investigation and solid phase microextraction of ultra-trace organochlorine pesticides. Chem. Eng. J. **451**, 138569 (2023)

Carbon Nanomaterials and Their Composites as Adsorbents

Khlood A. Alrefaey, A. B. Abdel-Aziz, Lobna A. Said, Irene Samy Fahim, and Ahmed G. Radwan

Abstract Carbon nanomaterials with various nanostructures (carbon nanotubes, graphene, graphene oxide, fullerene, nano diamonds, carbon quantum dots, carbon nanofibers, graphitic carbon nitrides, and nano porous carbons) are the decade's most distinguishing and popular materials. They have distinctive physicochemical qualities such as chemical stability, mechanical strength, hardness, thermal and electrical conductivities, and so on. Furthermore, they are easily surface functionalized and tweaked, modifying them for high-end specific applications. Carbon nanostructures' properties and surface characteristics are determined by the synthesis method used to create them. Nanoscience and nanotechnology have the potential to create materials with unexpected functions and qualities, which are transforming all industries. Carbon nanoparticles such as fullerene, carbon nanotubes, and graphene stand out among the various kinds of nanomaterials. These nanoparticles offer a wide range of practical applications, particularly in adsorption processes. Carbon nanoparticles exhibit unique structures that could be used in the construction of extremely sensitive, selective, and effective adsorbent devices for the removal of inorganic, organic, and biological pollutants from water solutions, as well as nano sensors and drug delivery systems. In this chapter, we demonstrated the number of studies published in recent years that used carbon nanomaterials as adsorbents. Furthermore, this chapter discusses essential features of adsorption and different nanocarbon carbon composite material, such as the contrast between physical and chemical absorption. Furthermore, diverse carbon nanomaterial synthesis such as

K. A. Alrefaey · I. S. Fahim (✉)
SESC Center, School of Engineering and Applied Sciences, Nile University, Giza, Egypt
e-mail: isamy@nu.edu.eg

A. B. Abdel-Aziz
October High Institute for Engineering & Technology, Giza, Egypt

L. A. Said · A. G. Radwan
Nanoelectronics Integrated Systems Center (NISC), Nile University, Giza, Egypt

A. G. Radwan
Engineering Mathematics and Physics Dept, Faculty of Engineering, Cairo University, Giza 12613, Egypt

J. Tharini and S. Thomas (eds.), *Carbon Nanomaterials and their Composites as Adsorbents*, Carbon Nanostructures, https://doi.org/10.1007/978-3-031-48719-4_3

AC–FeO –Cu and Bimetallic FeO –Cu/algae activated carbon composites AC–Fe0 –Cu methodologies, functionalization, and characteristics are provided and logically addressed.

Keywords Carbon nanomaterials · Adsorption · Activated carbon · Commercial · Nano adsorbents

1 Introduction

Nanotechnology is defined as the design and manufacture of materials, electronics systems with control at nanoscale dimensions. As a result, size and control are at the heart of nanotechnology. The term "nanotechnologies" is chosen by some because of the wide range of applications, yet they all share the common attribute of control at the nanoscale scale [1].

Nanobiotechnology and bio nanotechnology—they are essentially synonyms—refer to materials and processes at the nanometer scale that are based on biological, biomimetic or biologically-inspired molecules, and nanotechnological devices used to monitor or control biological processes, e.g., in medicine (Laurent et al. 2010). An example of the former is the optically switched optical switch incorporating the biomolecule bacteriorhodopsin3 and an example of the latter is the biochip—an array of known DNA fragments used to capture unknown DNA from a sample [2].

The potential impact of nanoparticles on enhancing our quality of life as well as their potential involvement in environmental preservation, particularly with regard to water treatment, have generated a great deal of interest in nanotechnology. Recent advances in nanotechnology have produced a wide range of materials that are now being considered as potential nano adsorbents for use in biological and environmental applications. Carbon nanoparticles (CN), which include fullerene, carbon nanotubes (CNT), and members of the graphene family, stand out among the several groups of nanomaterials [3].

Recent research suggests that the application of CN may be able to address many of the current issues with water quality. Using CN as nano adsorbents, several research teams are presently concentrating on the treatment and purification of wastewater. The majority of the projects created by these teams aim to compare the effectiveness of CN to that of conventional adsorbents, particularly activated carbon. Conducting a survey of the papers that have been written on the issue is one way to properly understand the significance that has been or is accorded to CN for this purpose. The survey's results, shownCNT have been the most thoroughly studied for adsorption applications among the three classes of possible carbon nano adsorbents that will be analyzed in this book. According to numerous studies, CNT has a good potential for adsorbing many of the organic and inorganic contaminants found in aqueous phases [4]. Although the graphene family's members (such as graphene oxide and reduced graphene oxide) have not yet undergone as much research as CNT, several authors have shown that these materials are capable of effectively

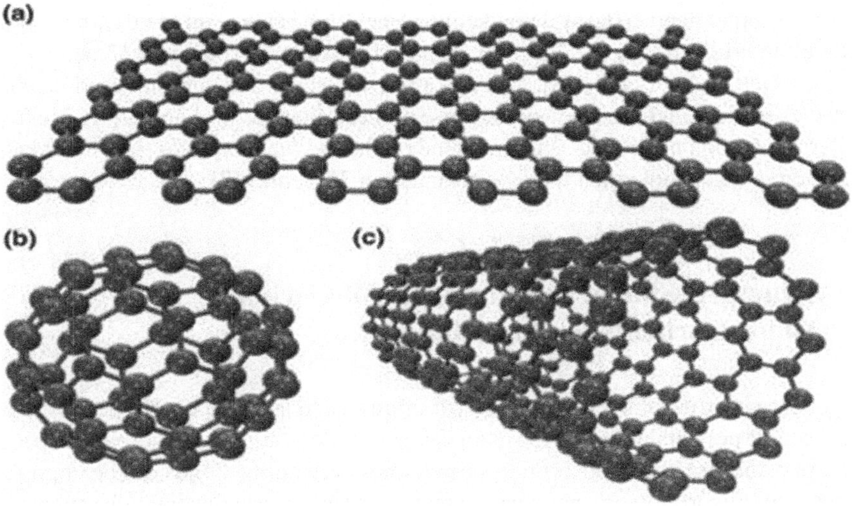

Fig. 1 CN structures with potential for adsorption applications: a graphene, b fullerene andc [3]

removing a range of contaminants, such as antibiotics, heavy metals, and synthetic dyes, from aqueous solutions. Although less intensively than for the other two carbon allotropes, fullerenes' adsorptive ability for organic pollutant chemicals in water has also been studied [5].

There has been a lot of study on new methods of diagnosing and treating diseases employing CN for medication delivery as well as for biosensor monitoring in terms of biological applications. Finding the perfect materials to guarantee the best properties of nanostructured sensors and drug delivery systems is currently the subject of much research. Given the growing number of articles in this field, CN offer enormous promise for such applications (Fig. 1).

2 Carbon Nanomaterial

Nanomaterials, especially carbon nanostructures, are promising systems for a number of uses, especially molecule adsorption, because of the way they are shaped, how they feel, and how they behave electrically. Allotropes of carbon are different because of the number of bonds between the carbon atoms or the way the layers are arranged in a crystal lattice. Carbon nanomaterials with different atomic bonds and arrangements are made by sp2 hybridization at different levels. For example, a 2D carbon allotropic form like graphene [4] is a hypothetical infinite aromatic surface with sp2 hybridization and a thickness of one carbon atom. Single-walled (SWCNT) [6], and multiwalled (MWCNT) [7] carbon nanotubes (both 1D systems) can be simplified as the result of one or more graphene sheets rolled up in a concentric shape.

On the other hand, a 0D system like fullerene, C60, is made up of 60 carbon atoms that are hybridized in a sp2-like way. It has 20 hexagonal rings and 12 pentagonal rings, which make it curve [8]. The adsorbate molecule and carbon-based nanoadsorbents interact in different ways because of these changes in shape [9]. Aside from fullerene, carbon nanotubes, and graphene on their own, these materials should also be taken into account when they are arranged in 3D (bulk) [8]

3 Structure, Shape, and Properties of Carbon Nanoadsorbents for Absorbing Gases

Textural properties of the carbon nanostructures, such as surface area, pore volume, and average pore diameter, are the main factors that determine the adsorption capacity [3]. Also, the structures' adsorption surface sites or dangling bonds can be important parts of the adsorption process because they are usually very reactive. The surface of carbonaceous nanomaterials can also be thought of as amphoteric, which means that they can be protonated (charged positively) or deprotonated (charged negatively). Also, oxygen can be used to add other functional groups, which can make new sites for chemical adsorption. In this case, the goal of this section is to give information about how graphene, carbon nanotubes, fullerene, and other related materials absorb things [8].

3.1 Fullerene' Adsorption Properties

Due to the curvature effect, the hybridization of fullerenes is similar to that of sp2 molecules, even though they are made of carbon. The C60 molecule is made up of 12 pentagons and 20 hexagons of carbon atoms. It has a diameter of about. The number of carbon atoms on a fullerene can be 20, 60, 70, 82, 100, 180, or even 960 [10]. This is always an even number. As the number of hexagons goes above 20 (as in the case of C60), the molecule becomes less stable as the pentagons move into more and more tense positions, making them more vulnerable to chemical attack. Laser ablation [9], carbon vaporization, or high-temperature heating of a graphite rod followed by an arc discharge between graphite electrodes are all ways to make fullerenes. Other ways to make it include synthesis in combustion [10], hybrid plasma, and thermal plasma, among others [11]. Say that the thermal and chemical properties of fullerene, including the fact that it could be an electron acceptor, make it of interest to business and science for use in environmental applications. By the Brunauer, Emmett, Teller–BET method, the specific surface area of fullerene found in experiments ranges from about 1.1 to 176 m^2 g1 [12]. The different ways of synthesizing can explain this difference. As shown in Fig. 2, adsorption can happen at three different places on fullerene aggregates: the surface (A), the groove (B) that forms when two fullerenes touch, and

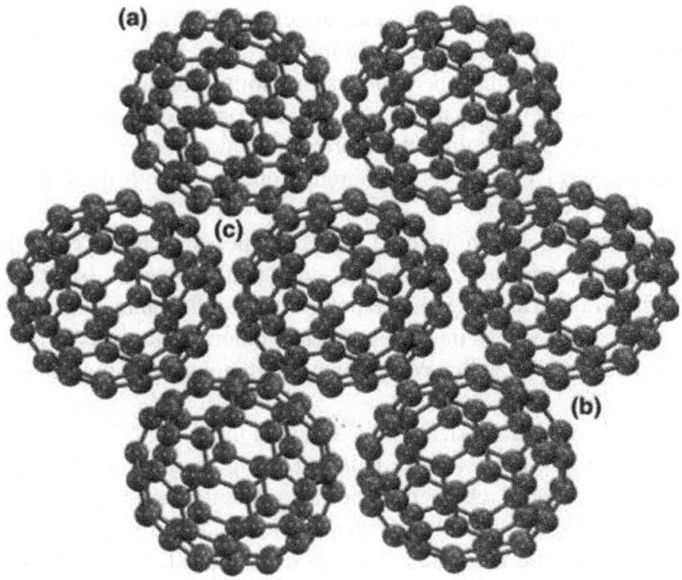

Fig. 2 Schematic model for the possible adsorption sites of fullerene aggregates: surface area [17]

the spaces (C) between the fullerenes in aggregates. Researchers have looked into how fullerenes absorb organic pollutants like organic and organometallic compounds [13], polycyclic aromatic hydrocarbons [14], and naphthalene and 1,2-dichlorobenzene in aqueous solutions. Also, the adsorption of organochlorine compounds in water solutions was studied and compared to the adsorption of activated carbon with and without fullerenes [15, 16]. The way that fullerene absorbs things was found to be better than that of activated carbon. It has been decided that most physical adsorption on fullerenes happens through dispersive interaction forces [14].

3.2 Carbon Nanotubes Adsorption Properties

Similar to fullerenes, carbon nanotubes (CNT) are carbon allotropes with an aromatic surface when the carbon atoms are rolled up in a tubular structure (1D system). This carbon allotrope is a unique nanostructure with amazing electronic and mechanical properties that depend directly on its chirality and diameter [14]. These unique properties and shapes make them very appealing for many practical applications, such as the development of devices for energy storage [7] and adsorption, with high sensitivity, selectivity, and efficiency [18].

The main ways to make CNT can be broken down into high-temperature methods, like arc discharge and laser ablation synthesis, and low-temperature methods, like chemical vapor deposition. When thinking about how CNTs attract things, it is important to think about both their outer and inner surfaces. In particular, rolling up the graphene sheet to make the tube changes the nonplanar sp2 configuration of the carbon orbitals, which changes the charge surface compared to the original graphene sheet. Because of this curvature, the thermodynamic properties of the adsorption process with the CNT can be very different from those with the flat carbon sheet [19].

Several studies [18, 19] show that CNTs have pores. In MWCNTs, the porosity can be divided into inner hollow cavities with small diameters (ranging from 3 to 6 nm in a narrow range) and aggregated pores with a wide range of sizes (20–40 nm), which are made when CNTs interact with each other. Several MWCNTs that stick to each other because of van der Waals forces get tangled up. This makes a group of pores [20]

Many methods of characterization have shown that SWCNTs are better at being microporous and that MWCNTs are better at being mesoporous [19]. So, the first one (SWCNT) usually has a higher surface area than the other ones. Figure 3 shows that adsorption on SWCNT bundles can happen in four different places: I the grooves formed at the contact between adjacent tubes; (ii) outside of the bundles; (iii) the interstitial channels between the tubes in bundles; and (iv) inside the nanotubes (pores) with open ends. Small molecules, like gases, might be able to stick to the spaces between the SWCNTs when they are bundled together [7]. In contrast to SWCNTs, MWCNTs don't come in bundles. In this last case, defects should also be thought of as places where adsorption can happen. Large biological contaminants like bacteria and viruses stick to the aggregated pores more than to other types of pores [21]. CNT's structure lets it interact strongly with organic molecules through noncovalent forces like hydrogen bonding,—stacking, electrostatic forces, van der Waals forces, and hydrophobic interactions. Also, the structure of CNTs makes it possible to add one or more chemical functional groups, which could make the system more selective and stable Functionalization or purification processes can be used to attach the chemical functional groups to the CNT surface [22]. Chemical methods of purification can change the specific CNT surface area, the volume of its micropores and mesopores, and the average diameter of its pores in a big way. Chemical purification, which is also called "oxidative purification," can be done with acid solutions (like HCl, HNO_3, and H_2SO_4), basic compounds (like KOH and NH_3), and gaseous compounds (like O_2, CO_2, and O_3), among other things. The CNT's texture may also change during the thermal purification process.

3.3 Nano Carbon's Electronic and Optical Characteristics

The ability of carbon nanoadsorbents to generate -interactions and other charge-transfer driven mechanisms is highly controlled by these potentials. In turn, environmental factors (such as pH and ionic strength), oxygen content or doping of the carbon nanostructure, as well as the adsorbate specie type, have an impact on

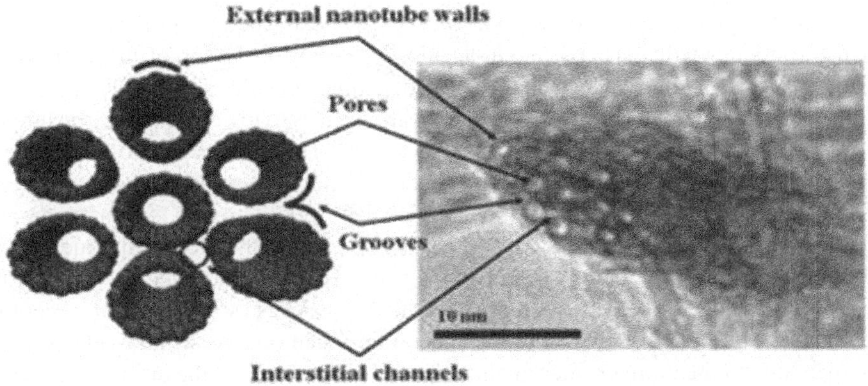

Fig. 3 Schematic model for the possible adsorption sites of SWCNT bundle (left) and TEM [17]

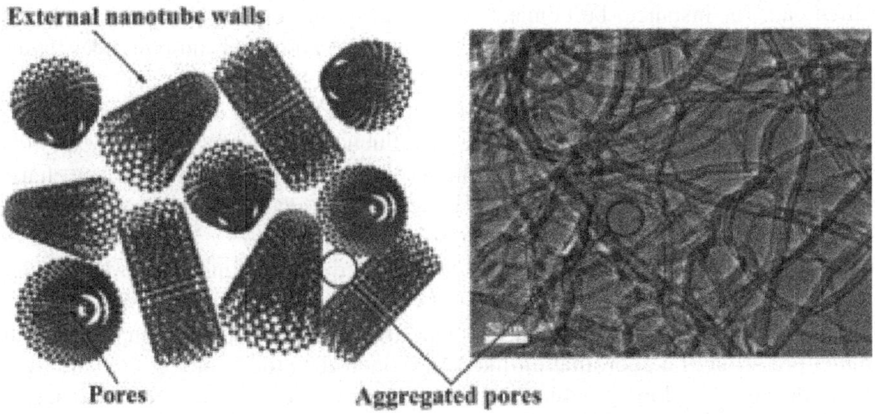

Fig. 4 Chematic model for the pores aggregated of MWCNT (left) and TEM image of aggregated MWCNTs (right) [17]

"–" interactions and other charge transfer-induced mechanisms (charge, presence of aromatic rings, etc.) (Fig. 4).

3.4 Effect of Electronic Properties on Adsorption Capability

It is well known that fullerene, CNT, and graphene interact with various molecules through charge-transfer. Raman spectroscopy is frequently used to acquire data on carbon nanomaterial properties, such as molecular charge transfer in graphene or CNTs. Raman maps that infer information on the adsorption of species onto carbon nanoadsorbents or the potential for it can be obtained by scanning the energy of the excitation light. There would be a shift on G or 2D band frequencies between before and after contact with the particular molecule, depending on the sort of charge transfer between the CNTs or graphene and the provided molecule (electron donor or acceptor induced mechanism [23]. Raman G band position shifts to lower frequency positions when electron donor species are adsorbed, but shifts to high frequency positions when electron acceptor species are adsorbed. Charge transfer causes the band gap area near Dirac point to tune, resulting in mid-gap molecular levels. Through "–" interactions, organic compounds with aromatic rings can also change the electrical structure of graphene. Metal–semiconductor transitions may result from the selective interaction of electron donor molecules with semiconducting SWCNTs and electron acceptor molecules with metallic ones. For instance, it has been observed that—interactions (depending on the number of aromatic rings, size and shape of the aromatic system, and molecular substitution units) are the reason why the sorption affinity of phenolic compounds on MWCNTs increases as the number of aromatic rings increases [24]. Adsorption is influenced by the type of substance and the strength of the interaction, which can, for instance, be connected to polarity and charge distribution. Strong chemical connections can be formed between CNTs and other radionuclides thanks to their free electrons.

The electronic states and consequent interactions between a specific molecule or ion and the carbon nanospecies can also be influenced by environmental factors like pH and ionic strength. When the pH is high (pH > pKa), the negative surface charge promotes electrostatic interaction that is helpful for adsorbing cations, whereas a drop in pH causes the surface charge to be neutralized [3].

CNT adsorption properties can also be impacted by metal/nonmetal doping. For instance, B- and N-doping reduce the energy required for hydrogen to bind to CNTs. B-doped materials have an electron-deficient structure, and H adsorption would result in the formation of a coordination-like B-H bond. The atomic adsorption energy of H is reduced by N-doping, which forms an electron-rich structure. N-doped and B-doped CNTs were found to promote Pt adsorption, but via distinct processes. While the B-doped case's improvement was linked to a substantial hybridization between the Pt d orbital and boron p orbital, the N-doped case's rise in Pt adsorption was due to activation of nitrogen adjacent carbon atoms [17].

3.5 *Effect of Adsorption on the Optoelectronic Properties*

It is possible to achieve varied band gaps by converting metals into semiconductors or semiconductors into other semiconductors by altering the structure of SWCNTs caused by atom adsorption [14]. As might be expected, the partial charge of the C atom and its gap energy can be greatly influenced by the tube-molecule distance [25]. For example, significant modifications in the density of states (DOS) close to the Fermi level may produce candidates for sensor applications. High binding energies can result in considerable alterations of electronic states around Fermi energy for both the zigzag and armchair of the SWCNT after interaction with a particular specie, even though low binding energies (0.5 eV) suggest that the form of interaction is most likely physisorption. For instance, Ti adsorption changes semiconducting Zigzag SWCNTs into metals. Generally speaking, the transition metal atoms' d orbitals are the source of their significantly higher binding energies, which modify the number of filled d states [26].

Even intense adsorption (chemisorption) of particular molecules, however, is occasionally unable to produce noticeable changes in the density of states close to the Fermi level. Exciton quenching in semiconducting SWCNT has already been observed using photoluminescence imaging [26].

Adsorption of a particular molecule can also have an impact on the field emission current from CNT tips, based on variations in the Fermi level and DOS close to the Fermi level [4]. One can anticipate that the Fermi level will shift toward the conduction band upon the adsorption of a specific molecule.

4 Adsorption Mechanism

Adsorption is a physicochemical treatment method in which dissolved molecules in impure water are chemically and physically linked to the adsorbent surface. Adsorption technology was referred to as a powerful technique for pollutants removal from filthy wastewater due to its higher pollutant's removal performance than other conventional approaches. Adsorption is recognized as a process that produces high-quality products at a reasonable price. The four processes of dye adsorption include dyes diffusion/convection, dyes diffusion across a diffusion boundary layer, dyes diffusion from the surface into the interior of the adsorbent material, and dyes diffusion from the surface into the interior of the adsorbent material [27]. But the procedure comes to a close with the fourth and final stage. It denotes the adsorption of dye molecules on a substance's surface brought on by interactions between the materials. The second phase could be impacted by the dye concentration and agitation period. Due to its impact on dye adsorption on the surface of the substrate, step three was also referred to as the rate-determining step. Porous diffusion, in which the adsorbate diffuses through the liquid and fills the pores, and adsorbed in the process are two more phenomena that could occur in step three. Surface diffusion is the second phenomenon, where the adsorbate is first adsorbed and then diffused from one site to another. The type of dye molecules—which can be either anionic or cationic—determines step 4 of the process. Several adsorbents are efficient at removing colours from wastewater that has been contaminated by textiles [4].

5 Activated Carbon

Activated carbon is a porous carbonaceous substance with various uses in desalination, water and wastewater treatment, and air purification due to its unique properties [28]. It is a versatile adsorbent material with a high porosity degree and a large surface area, with up to 90% of its carbon surface area. The outer surface structure of activated carbon, which contains many functional groups such as carbonyl, carboxyl, phenol, quinone, and lactone, was responsible for its perfect adsorption qualities. In addition to the appearance of hydrogen, oxygen, Sulphur, and nitrogen, those functional groups were reported responsible for the contaminant's adsorption process into the activated carbon adsorbent material [29]. These functional groups were created by precision, activation procedures, and thermal purification [30]. Agricultural leftovers such as biomass, olive corn, corn stalks, rice rolls, bagasse, fruit stones, hard shells, fruit pulp, bones, and coffee beans can make activated carbon [31]. The raw materials used to make activated carbon should be inexpensive, plentiful, and safe [32]. Furthermore, the mineral concentration of this material and its biodegradability during initial storage should be kept to a minimum. Recently, cellulosic materials have become one of the most widely used materials in activated carbon production. The activated carbon has a high specific surface area with a highly porous structure of about 800 m^2/g, indicating a greater adsorption capacity [23]. However, due to the high cost of commercial activated carbon, considerable amounts of naturally existing wastes 2 and plants have received more attention in recent years for pollution control applications. Algae is naturally renewable biochar used to produce activated carbon. Organic cellulose, alginate, lignin, starch, and other inorganic elements which are found in marine algae were investigated for their potential to be converted into activated carbon. Because of their low cost and simple supply, marine algae gained popularity as raw materials for making activated carbon in recent years. A variety of marine algae, including Ulva Lactuca [27], Systoceira strict, Turbinaria turbinata [33], Enteromorpha prolifera, Pterocladia capillacea, and Euphorbia rigida, were used to produce activated carbon [28].

6 Commercial Activated Carbon

Due to its versatility in removing a variety of contaminants from contaminated media, commercial activated carbon, also known as charcoal, is a form of industrial adsorbent that is frequently used. Additionally, the thermal stability, enormous surface area of 500 to 2000 m^2/g, high porosity, resistance to bases and acids, low cost, and customizable pore structure of activated carbon made them stand out [34].

7 Biochar

To make activated carbon, materials with a high carbon content can be used, including wood, coal, peat, tobacco, nutshells, and lignite [35, 36]. A further carbonaceous source for the formation of activated carbon is agriculture wastes, such as sugarcane bagasse, olive stones, cotton residues, soybean hulls, corn straws, peach stones, pinecones, rice hulls, rice straw, banana peels, apricot stones, corn cobs, bamboo, pith, etc. [37, 38]. However, earlier research claimed that 40% of activated carbon was removed from charcoal and that it could get rid of taste and odor. To produce activated carbon, two continuous stages of carbonization and activation were typically used. The first stage was carried out at a high temperature during the pyrolysis process, and the second stage was used to improve the pore structure of the activated carbon [39, 40]. Other research showed that the steps included a pretreatment procedure that involved crushing, grinding, and screening to provide the ideal size for an effective succeeding operation [41]. Figure 5 depicts the physical and chemical activation procedures for activated carbons. Due to the lower cost, higher surface area, and improved porosity of chemical activation, it is preferable to physical activation [42]. Chemical activation, however, necessitates additional expenditures for the chemical activation agents and additional steps in the chemical agent washing procedures [43]. The efficacy of pollutant adsorption is increased by microwave-assisted chemical activation by increasing surface area. Activated carbon can also be impregnated with chemicals like metal oxides and hydroxides to increase the effectiveness of removal.

Various characterization methods, including the scanning electron microscopy (SEM) and transmission electron microscope, can be used to examine the prepared materials' surface and physical features (TEM).

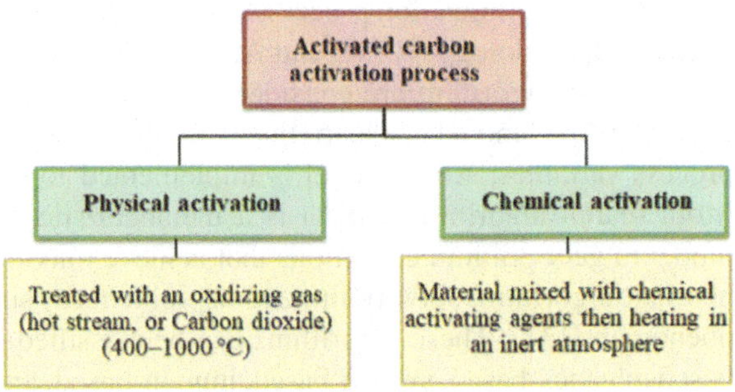

Fig. 5 Schematic production process of activated carbons by physical or chemical activation [44]

8 Bimetallic F E0 −Cu/algae Activated Carbon Composites AC–F E0 −Cu

There is another type of adsorbent bimetallic nanoparticle composite with activated carbon used for adsorption such as **AC–F e0 −Cu** which used for treating industrial wastewater contaminated with anionic methyl orange (M.O) dye. The bimetallic nanoparticles were effectively used as adsorbents. When two metals are combined, they can keep their own characteristics while also gaining new ones as a result of the joining process, creating a bimetallic molecule [45]. The properties of bimetallic nanoparticles were affected by their size, structure, and appearance. The combination of a small fraction of transition metals with nZVI functions as a reducing agent, increasing the surface reaction to remove organic molecules and other impurities [46]. Various materials, including activated carbon, were used as adsorbents to remove contaminants such as dyes from liquid effluents. The activated carbon has a high specific surface area with a highly porous structure of about 800 m^2/g, indicating a greater adsorption capacity [47]. However, due to the high cost of commercial activated carbon, considerable amounts of naturally existing wastes and plants have received more attention in recent years for pollution control applications. Algae is naturally renewable biochar used to produce activated carbon [48]. Organic cellulose, alginate, lignin, starch, and other inorganic elements which found in marine algae were investigated for their potential to be converted into activated carbon. Because of their low cost and simple supply, marine algae gained popularity as raw materials for making activated carbon in recent years. A variety of marine algae, including Ulva Lactuca [49], Systoceira strict, Turbinaria turbinata [50], Enteromorpha prolifera [51] Pterocladia [52], and Euphorbia rigida [53], were used to produce activated carbon. Most of the previous research used one factor at a time (OFAT) experiments for the optimization process, although the interacted models such as Response Surface Methodology (RSM) and Artificial Neural Network (ANN) combined with optimization algorithm are considered more effective for the optimization process [54]. The ANN model can learn linear, nonlinear and complex relationships between process variables. ANN model is implemented to predict removal efficiency. The Moth search algorithm (MSA) is a metaheuristic algorithm applied to the ANN model to get optimum conditions that achieve maximum removal efficiency. Metaheuristic algorithms solve optimization problems by simulating biological or physical phenomena [31]. These algorithms can be classified into five main classes, nature-based, physics-based, swarm-based, human-based, and animal-based. Metaheuristic algorithms explore optimal solutions in a reasonable time and use sufficient computational resources. Many applications use metaheuristic algorithms such as detecting the edges of the brain tumor from a patient's MR scan image of the brain by Genetic Algorithm (GA). Furthermore, metaheuristic algorithms were used to determine the parameters and orders of three fractional-order chaotic systems [28]. Methyle Orange (M.O) removal by polyaniline-based nano-adsorbent was investigated in [55]. RSM was implemented to predict removal efficiency. Furthermore, ANN was integrated with differential evolution optimization (DEO) for the prediction of removal efficiency. ANN-DEO had better accuracy in prediction than RSM. However, the maximum removal efficiency was not evaluated. Furthermore, the polyaniline nano-adsorbent was used to adsorb M.O in [56]. ANN was implemented to predict the removal efficiency. OFAT was implemented to get the maximum removal efficiency. The M.O removal by lignin-derived zeolite templated carbon materials was optimized by RSM with Box–Behnken design. RSM was used to find the most influential factor on the removal efficiency. This paper aims to develop low-cost adsorbents of bimetallic $Fe0$ −Cu/algae activated carbon composites AC–$Fe0$ −Cu for treating industrial wastewater contaminated with anionic methyl orange (M.O) dye. The results of the time and the concentration experiments were used to test the isotherm and the kinetic models [57].

8.1 Preparation of Bimetallic ($FE0$ −Cu)/algae Activated Carbon Composites.

The collected raw algae have been washed, dried, and ground into fine particles. The dried raw material is then utilized to make physical and chemical activated carbon. First, physical activated carbon is prepared by placing the actual weight of the algae in the muffle at 400 °C for 1 h and then increasing the temperature to 600 °C for 2 h. The temperature is then reduced to 300 °C for 15 min [58]. On the other hand, chemical activated carbon is prepared by adding chemical activators (250 ml of 85% H_3PO_4 or 150 ml of 4 M H_2SO_4) to 50 g of algae and mixing at ambient temperature to create a homogeneous mixture.

The mixture is then placed in an oven at 110◦C for two hours, then carbonized for three hours at 600 °C, rinsed with deionized water till pH = 7 is attained, and dried in an oven at 110 °C [59, 60]. The synthesis of bimetallic $Fe0$ −Cu, Additionally, bimetallic ($Fe0$ −Cu)/algae-activated carbon composites (AC–$Fe0$ −Cu) are generated in-situ in a single step. The 0.1 M $FeCl3.6H_2O$ solution, 0.002 M $CuCl_2.2H_2O$ solution, and 1 g/L of the prepared activated carbon (AC) are constantly stirred at 250 rpm in the conical flask. A 0.5 M sodium borohydride $NaBH_4$ solution is added drop by drop from the burette. The solution is agitated for an additional 20 min after adding all of the $NaBH_4$ solution to ensure that the iron ion reduction reaction in the solution has been completed. The generated precipitate is then separated via filtration method. The separated precipitate is washed many times with deionized water and ethanol. Finally, the product is dried at 70◦C and stored at room temperature under an ethanol layer [61, 62].

9 Fava Bean Activated Carbon-Supported Bimetallic Nano Zero-Valent Iron Copper (AC–F E0 –Cu):

Researchers used different transition elements in recent years to see the most efficient for removing colours and heavy metal ions from industrial effluent. Bimetallic nano zero-valent iron-copper (F e0 –Cu) attracted much attention due to the high efficiency seen in their performance to remove colours [62]. The initial concentration of dyes was one of the most important aspects for selecting the adsorption technique used to treat industrial wastewater. The adsorption efficiency decreased with increasing the dye content. Pores, charges, hydrophilic or hydrophobic character, size, and distribution all affected adsorption effectiveness. It also depended on the surface area and whether functional groups were present or not [63]. The magnetic property inherent in particular adsorbents provided outstanding and remarkable efficiency. This found in various treatments, such as using nano ilmenite $FeTiO_3$ to remove cationic dyes like methylene blue with high effectiveness of up to 71.9 mg/g [64]. The contact time of the adsorbent materials with the industrial wastewater was critical in removing as many contaminants as feasible. By extending the contact time, the adsorption capacity of the adsorbent increased [65, 66]. Fava bean was widely consumed in several regions, including the Mediterranean region of Europe and Africa, as well as Latin America, China, and India. Fava bean peels were thrown out in large amounts. Different activation techniques were used to produce activated carbon with different properties. In prior research, fava bean activated carbon powder was employed to remove heavy metal ions with high removal efficiencies [67]. Response Surface Methodology (RSM) and Artificial Neural Network (ANN) combined with optimization algorithm are considered more effective for the optimization process [54]. The ANN model can learn linear, nonlinear and complex relationships between process variables [68]. ANN model is implemented to predict removal efficiency. The Moth search algorithm (MSA) is a metaheuristic algorithm applied to the ANN model to get optimum conditions that achieve maximum removal efficiency. Methyle Orange (M.O) removal by polyaniline-based nano-adsorbent was investigated in [9]. ANN was implemented to predict the removal efficiency. OFAT was implemnted to get the maximum removal efficiency. The M.O removal by lignin-derived zeolite templated carbon materials was optimized by RSM with Box–Behnken design [57]. RSM was used to find the most influential factor on the removal efficiency.

10 Activated Carbon Application

Due to its large surface area and high pore number, activated carbon was used to remove metals from contaminated wastewater. Polychlorinated biphenyls (PCBs), humanistic compounds, pesticides, detergents, and organic pollutants can all be removed from contaminated streams using activated carbon [69]. Activated carbon's capacity to absorb color and other contaminants from textile and dye wastes was also mentioned and commercially available in the application. For instance, utilizing filtrasorb circledR activated carbon, acidic, dispersion, and basic contaminants from textile-polluted wastewater were effectively removed, but direct dyes were not [70]. Activated carbon has been shown in numerous experiments to be capable of removing both cationic and anionic dyes, including reactive black and methylene blue. Numerous compounds, including sodium dodecyl sulfate, sodium dodecylbenzene sulfonate, and sodium dioctyl sulfosuccinate, interact with activated carbon to improve its ability to bind to heavy metals [71].

11 Activated Carbon Activation Methods

The dry raw precursor was first carbonized at high temperatures, followed by activation, to create activated carbon. To extract the hydrocarbons from materials like wood, waste, and coal during the carbonization process, they should be exposed to a red spot in the distillation apparatus that is less than 700 C in temperature. The substance, also known as carbonized material, char, or biochar, was carbonized via a pyrolytic process [72]. After the activation procedure, different activation techniques were used to develop the porosity and build the structure that enables the activation of the fine solid cavities in the activated carbon that was created.

11.1 Physical Activation

Commercially, physical activation is applied in a two-step process that begins with carbonization (pyrolysis) in a neutral atmosphere and ends with activation in atmospheric oxidizing gases like steam, carbon dioxide, and nitrogen or air mixtures at temperatures ranging from 400 to 900 C [73]. However, some drawbacks were noted for the physical activation of activated carbon, including a lengthy activation time, low prepared activated carbon adsorption capacity, and high energy consumption. This method can produce porous structure activated carbon with high physical power; however, it is an expensive approach for producing activated carbon [58, 74]

11.2 Chemical Activation

In order to chemically activate materials with a high cellulose content, such as wood, fruit pits, and sawdust, wet oxidation is frequently used. Starting with organic precursors, chemicals, and high temperatures, the chemical activation process is prepared [74]. The raw components were heavily dried and soaked with oxidizing agents before the chemical activation occurred. The remaining mixture was then heated for a set period of time after the suspension had been dried. Depending on the activation

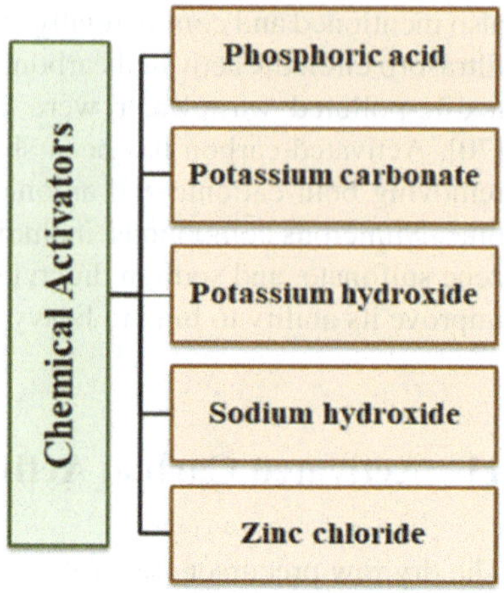

Fig. 6 Different applied activators for the activation of activated carbon [44]

components and the ultimate properties of the produced product, activation could take place at a temperature between 300 and 800 circC. The produced activated carbon was then obtained by repeatedly washing the mixture. Another objective of the last rinse is the recovery of active chemicals [75]. Dehydrating agents that affect pyrolytic are chemical activators.

11.3 Activated Carbon Activators

When obtaining activated carbon, some chemical substances were used as activation agents. The precursor undergoes a variety of reactions depending on the activating agent, which alters the adsorption behavior. Alkaline or acidic groups are the primary chemical types used as potential activator agents, as depicted in Fig. 6.

11.3.1 Activation by Phosphoric Acid

It was common practice to use phosphoric acid, which has the chemical formula H_3PO_4, in the activation of various lignocellulosic materials [76]. It is the chemical that is used most frequently in the activation of activated carbon because it can turn raw materials into high-porous activated carbon. Compared to potassium, zinc chloride, and hydroxide, H_3PO_4 has fewer toxicological and environmental pollutants. Additionally, H_3PO_4 may produce vast quantities of alkaline or acid-soluble phosphates with substances like nickel, iron, boron, and others that can be incorporated into carbon precursors. It also has a low activation temperature and is not volatile [77]. When H_3PO_4 was used as an activator, the pores that were formed on the activated carbon surface area were shown to be tunnel-shaped in carbon electron micrographs and images from scanning electron microscopes. Additionally, it typically featured a honeycomb structure, which was completely developed as the chambers' corners could be seen [78]

11.3.2 Activation by Zinc Chloride

Among other activating agents, zinc chloride was frequently used to create activated carbon, particularly from cellulosic and lignocellulosic precursors [28]. For the samples that have been impregnated with this substance during activation, zinc chloride serves as a dampening agent. The passage of volatile chemicals through the saturated pores of zinc chloride remains unaffected, and after the activation procedure, the volatile substances are released from the surface of the activated carbon. Zinc chloride's mass ratio is increased, which makes volatile compounds more easily released and promotes the absorption of nitrogen by activated carbon [79]. The activation of zinc chloride causes swelling, an electrolytic reaction, in the cellulose molecular structure. Additionally, expansion increases the number of intra- and inter-coated cavities and breaks down cellulose molecules, increasing the surface area of the activated carbon [80]. As lignocellulosic materials are activated, carbon, hydrogen atoms, oxygen, carbon dioxide, methane, and aldehydes are generated, along with diatomaceous distillates [81].

11.3.3 Activation by Potassium Carbonate

Potassium carbonate, which is a well-known activating ingredient in the synthesis of activated carbon, has the chemical formula K_2CO_3 [82]. Unlike potassium and sodium hydroxide, which have negative effects, using potassium carbonate as an activation agent for nutritional supplements is safe [83]. Furthermore, potassium carbonate was shown to be a better activating agent than potassium hydroxide since it increased the yield, surface area, and pore volume of activated carbon. Additionally, potassium carbonate-produced activated carbon has reduced ash and sulfur content and a higher capability for adsorbing big molecules like methylene blue [84]

11.3.4 Activation by Sodium Hydroxide

A large number of microspores are produced on the surface of the activated carbon as a result of chemical activation using alkaline chemicals like potassium hydroxide and sodium hydroxide. Numerous investigations have demonstrated the effectiveness of sodium hydroxide as an activating agent in the production of activated carbon. As a result of probable interactions between the active components and the organic precursor surface, the surface of the activated carbon generates micropores. This is due to the emission of CO, CO_2, and H_2 gases as a result of the alkali metal's interaction with the carbon structure. These gases are produced when Na_2CO_3 and hydroxyl reduction, respectively, decompose at high temperatures. Activated carbon with a rough surface and different pore diameters is what remains after the sodium hydroxide and other ingredients have evaporated during the preparation process. This process focuses on creating activated carbon's porous structure [85].

11.3.5 Activation by Potassium Hydroxide

Recently, many potassium compounds, including K_2CO_3 and KOH, were widely used in the synthesis of activated carbon [86]. Because of its enormous surface area and ability to form activated carbon, potassium hydroxide has been used the most widely of all activators. Additionally, under the same circumstances, the fine pore size distribution, low environmental contamination, less corrosiveness, and lower cost [87]. When lignocellulosic materials were in their uncarbonized state, phosphoric acid and zinc chloride were used as chemical activators. On the other hand, activating coal precursor materials was done using metal compounds like potassium hydroxide [88]. Numerous investigations have demonstrated the efficiency of KOH-activated carbon in the adsorption of phenols, heavy metals, dyes, and insecticides, among other organic compounds [89]. The evaporation of potassium hydroxide from areas where this activator had previously been present caused the cavities that were observed on the surface of activated carbon after the activation process [90]. In contrast to other activators like $ZnCl_2$ and H_3PO_4, KOH-activated carbon has a higher pore volume and surface area but is noted for having a lower yield (10–40%) [91].

11.3.6 Factors Affecting the Adsorption Process

The adsorption process can be changed by time, pH, the concentration of the pollutant at the start, and the amount of adsorbent material used. To get a high adsorption capacity for removing pollutants from polluted wastewater, these parameters should be optimized. In the process of treating wastewater, pH is one of the things that affects how well activated carbon absorbs things [92]. Say that a change in pH changes the degree of ionization of the adsorptive molecules and the surface properties of the adsorptive material. The pH of the dye solution might not be very important to the adsorption process if the dye binds to the surface through van der Waals' force, hydrophobic-hydrophobic interaction, or a hydrogen bond. Also, the best pH for dyes to stick to activated carbon depends on how it was activated and what kind of carbon precursors were used [93]. It was said that the concentration of dyes as an adsorbent material is a factor that affects how well dyes absorb. For example, as the number of effluent dyes goes up, the dye's ability to absorb water goes up until the process is saturated. At this point, there is no more adsorption because all of the binding sites on the dyes are full. On the other hand, other research has shown that the dye's ability to stick to a surface decreases as its concentration rises. This is because the adsorption sites on the surface of the adsorbent are filled up quickly [94]. Adsorbent dose: The amount of adsorbent is an important parameter because it affects the process of adsorption. In general, increasing the amount of adsorbent material makes dye removal work better. This is because the adsorbent surface has a lot of places where dye can stick to it. The effect of adsorbent dosage is an important factor that shows how well a material can absorb dyes. If adsorption happens with a small amount, it shows how important the material is from an economic point of view [95]. To get the best dye removal, the different factors that affect the adsorption process were optimized. Many optimized models were used to do this. For the optimization process, most studies used OFAT models, which stands for "one factor at a time." But optimization with interactive models like RSM and ANN is a better way to do things.

12 Carbon Nanomaterials as Adsorbents for Environmental and Biological Applications.

CN have been studied a lot for adsorption applications because they are chemically stable, have different structures, are light, and can be made in large quantities. Their textural properties, such as average pore diameter, total pore volume, and infinitely high surface-to-volume ratio, have led to a lot of research into the use of CN as potential adsorbents for water purification and biological applications [46, 60, 72]. Because CN is a chemical, its surfaces are easy to change with chemical and physical treatments. This lets its properties be improved in a specific way [39]. Also, the fact that biofunctionalizations are possible suggests that there are a lot of ways they could be used in biological systems. Because of these things, CN is an interesting subject for developing highly sensitive, selective, and efficient adsorbent devices for removing inorganic, organic, and biological contaminants from water solutions, as well as drug delivery systems and nano sensors [40].

13 Conclusion

Remarks Carbon nanoparticles, as illustrated in this chapter, can be effective in the filtration and treatment of water. These nanomaterials, which include fullerene, carbon nanotubes, and graphene, have particularly intriguing structure, morphology, and adsorption properties, making them potentially useful in industry as nano adsorbents with high affinity, capacity, and selectivity. Sargassum dent folium, a low-cost natural alga, was effectively used to produce another composite nanomaterial with activated carbon, such as AC–F e0 −Cu supported on algal-activated carbon (AC–F e0 −Cu), for treating anionic methyl orange-contaminated industrial wastewater. Composites of unprocessed algae and $H_3P\ O_4$ chemical activated carbon demonstrated the most effective decolorization. Furthermore, activated carbon from fava beans and bimetallic F e0 −Cu were used to treat industrial wastewater for the anionic methyl orange (M.O) dye, with a high removal efficiency expected. The removal efficiency is predicted using RSM and ANN models for different values of the components. By optimizing these models, the highest removal efficiency can be determined. ANN is improved by the Moth Search Algorithm (MSA). The highest removal efficiency and associated factor values are projected using these optimization strategies.

References

1. Poole, C.P., Owens, F.J.: Introduction to Nanotechnology. Wiley (2003)
2. Tiwari, J.N., Tiwari, R.N., Kim, K.S.: Zero-dimensional, one-dimensional, two-dimensional and three-dimensional nanostructured materials for advanced electrochemical energy devices. Prog. Mater. Sci. **57**(4), 724–803 (2012)
3. Ren, X., et al.: Carbon nanotubes as adsorbents in environmental pollution management: a review. Chem. Eng. J. **170**(2–3), 395–410 (2011)
4. Kim, C., et al.: The effect of gas adsorption on the field emission mechanism of carbon nanotubes. J. Am. Chem. Soc. **124**(33), 9906–9911 (2002)
5. Bethune, D.S., et al.: Cobalt-catalysed growth of carbon nanotubes with single-atomic-layer walls. Nature **363**(6430), 605–607 (1993)
6. Iijima, S.: Helical microtubules of graphitic carbon. Nature **354**(6348), 56–58 (1991)
7. Iijima, S., Ichihashi, T.: Single-shell carbon nanotubes of 1-nm diameter. Nature **363**(6430), 603–605 (1993)
8. Kroto, H.W., et al.: C60: Buckminsterfullerene. Nature **318**(6042), 162–163 (1985)
9. Chowdhury, S., Balasubramanian, R.: Recent advances in the use of graphene-family nanoadsorbents for removal of toxic pollutants from wastewater. Adv. Colloid Interface Sci. **204**, 35–56 (2014)
10. Inagaki, M., Kang, F.: Materials Science and Engineering of Carbon: Fundamentals. Butterworth-Heinemann (2014)
11. Yoshida, T., et al.: Characterization of a hybrid plasma and its application to a chemical synthesis. J. Appl. Phys. **54**(2), 640–646 (1983)
12. Wang, C., et al.: Synthesis of fullerenes from carbon powder by using high power induction thermal plasma. Thin Solid Films **390**(1–2), 31–36 (2001)
13. Ballesteros, E., Gallego, M., Valcarcel, M.: Analytical potential of fullerene as adsorbent for organic and organometallic compounds from aqueous solutions. J. Chromatogr. A **869**(1–2), 101–110 (2000)
14. Yang, X., Ni, J.: Metal-semiconductor and semiconductor-semiconductor transitions in carbon nanotubes induced by intercalating alkali atoms. Phys. Rev. B **71**(16), 165438 (2005)
15. Cheng, X., Kan, A.T., Tomson, M.B.: Uptake and sequestration of naphthalene and 1, 2-dichlorobenzene by C60. J. Nanopart. Res. **7**, 555–567 (2005)
16. Berezkin, V.I., et al.: A comparative study of the sorption capacity of activated charcoal, soot, and fullerenes for organochlorine compounds. Tech. Phys. Lett. **28**(11), 967–969 (2002)
17. Bergmann, C.P., Machado, F.M. (eds.): Carbon Nanomaterials as Adsorbents for Environmental and Biological Applications. Springer International Publishing (2015)
18. Yang, Q.H., et al.: Adsorption and capillarity of nitrogen in aggregated multi-walled carbon nanotubes. Chem. Phys. Lett. **345**(1–2), 18–24 (2001)
19. Serp, P., Corrias, M., Kalck, P.: Carbon nanotubes and nanofibers in catalysis. Appl Catal A: Gen. **253**(2), 337–358 (2003)
20. Upadhyayula, V.V., et al.: Application of carbon nanotube technology for removal of contaminants in drinking water: a review. Sci. Total. Environ. **408**(1), 1–13 (2009)
21. Machado, F.M., et al.: Adsorption of Reactive Blue 4 dye from water solutions by carbon nanotubes: experiment and theory. Phys. Chem. Chem. Phys. **14**(31), 11139–11153 (2012)

22. Sze, M.F.F., Lee, V.K.C., McKay, G.: Simplified fixed bed column model for adsorption of organic pollutants using tapered activated carbon columns. Desalination **218**(1–3), 323–333 (2008)
23. Dash, B.: Competitive adsorption of dyes (Congo red, Methylene blue, Malachite green) on activated carbon. [Doctoral dissertation]. Rourkela: National Institute of Technology (2010)
24. Chin, C.J.M., Shih, M.W., Tsai, H.J.: Adsorption of nonpolar benzene derivatives on single-walled carbon nanotubes. Appl. Surf. Sci. **256**(20), 6035–6039 (2010)
25. Rafati, A.A., Hashemianzadeh, S.M., Nojini, Z.B.: Electronic properties of adsorption nitrogen monoxide on inside and outside of the armchair single wall carbon nanotubes: a density functional theory calculations. J. Phys. Chem. C **112**(10), 3597–3604 (2008)
26. Crochet, J.J., et al.: Photoluminescence imaging of electronic-impurity-induced exciton quenching in single-walled carbon nanotubes. Nat. Nanotechnol. **7**(2), 126–132 (2012)
27. Tan, K.B., et al.: Adsorption of dyes by nanomaterials: recent developments and adsorption mechanisms. Sep. Purif. Technol. **150**, 229–242 (2015)
28. Yousefi, M., et al.: Modification of pumice with HCl and NaOH enhancing its fluoride adsorption capacity: kinetic and isotherm studies. Hum. Ecol. Risk Assess. **25**(6), 1508–1520 (2019)
29. Morin-Crini, N., et al.: Hemp-based adsorbents for sequestration of metals: a review. Environ. Chem. Lett. **17**, 393–408 (2019)
30. AboBakr, A., et al.: Experimental comparison of integer/fractional-order electrical models of plant. AEU-Int J Electron Commun. **80**, 1–9 (2017)
31. Abdollahzadeh, B., Gharehchopogh, F.S., Mirjalili, S.: Artificial gorilla troops optimizer: a new nature-inspired metaheuristic algorithm for global optimization problems. Int. J. Intell. Syst. **36**(10), 5887–5958 (2021)
32. Jolly, G., et al.: Improved Cu and Zn sorption on oxidized wheat lignocellulose. Environ. Chem. Lett. **4**, 219–223 (2006)
33. Özacar, M., Şengil, İA.: Adsorption of reactive dyes on calcined alunite from aqueous solutions. J. Hazard. Mater. **98**(1–3), 211–224 (2003)
34. Mukherjee, S., et al.: Contemporary environmental issues of landfill leachate: assessment and remedies. Crit. Rev. Environ. Sci. Technol. **45**(5), 472–590 (2015)
35. Torretta, V., et al.: Novel and conventional technologies for landfill leachates treatment: a review. Sustainability. **9**(1), 9 (2017)
36. Li, W., Zhang, J.H.: Biosorption of hexavalent chromium by immobilized activated sludge in a continuous flow reactor. Process Biochem. **43**(3), 341–347 (2008)
37. Chan, L.S., et al.: Error analysis of adsorption isotherm models for acid dyes onto bamboo derived activated carbon. Chin. J. Chem. Eng. **20**(3), 535–542 (2012)
38. Skodras, G., et al.: Enhanced mercury adsorption in activated carbons from biomass materials and waste tires. Fuel Process. Technol. **88**(8), 749–758 (2007)
39. Hidayu, A.R., Muda, N.J.P.E.: Preparation and characterization of impregnated activated carbon from palm kernel shell and coconut shell for CO_2 capture. Procedia Eng. **148**, 106–113 (2016)
40. Zubair, M., et al.: Recent progress in layered double hydroxides (LDH)-containing hybrids as adsorbents for water remediation. Appl. Clay Sci. **143**, 279–292 (2017)
41. Alslaibi, T.M., et al.: A review: production of activated carbon from agricultural byproducts via conventional and microwave heating. J. Chem. Technol. Biotechnol. **88**(7), 1183–1190 (2013)
42. Allwar, A.: Characteristics of micro-and mesoporous structure and surface chemistry of activated carbons produced by oil palm shell. In: International Conference on Chemical, Ecology and Environmental Sciences Proceedings (2012)
43. Shehzad, A., et al.: An overview of heavily polluted landfill leachate treatment using food waste as an alternative and renewable source of activated carbon. Process Safety Environ Prot. **98**, 309–318 (2015)
44. Husien, Sh. et al.: Review of activated carbon adsorbent material for textile dyes removal: preparation, and modelling. Curr. Res. Green Sustain. Chem, 100325 (2022)
45. Bokare, A.D., et al.: Iron-nickel bimetallic nanoparticles for reductive degradation of azo dye Orange G in aqueous solution. Appl. Catal. B Environ. **79**(3), 270–278 (2008)

46. Scaria, J., Nidheesh, P.V., Kumar, M.S.: Synthesis and applications of various bimetallic nanomaterials in water and wastewater treatment. J. Environ. Manage. **259**, 110011 (2020)
47. Jung, S.H., Kim, J.S.: Production of biochars by intermediate pyrolysis and activated carbons from oak by three activation methods using CO_2. J. Anal. Appl. Pyrol. **107**, 116–122 (2014)
48. Hock, P.E., Ahmad Zaini, M.A.A.: Activated carbons by zinc chloride activation for dye removal–a commentary. Acta Chim. Slov. **11**(2), 99–106 (2018)
49. El Nemr, A., Abdelwahab, O., Khaled, A., El Sikaily, A.: Biosorption of Direct Yellow 12 from aqueous solution using green alga Ulva lactuca. Chem. Ecol. **22**(4), 253–266 (2006)
50. Altenor, S., et al.: Textural characteristics, physiochemical properties and adsorption efficiencies of Caribbean alga Turbinaria turbinata and its derived carbonaceous materials for water treatment application. Biochem. Eng. J. **67**, 35–44 (2012)
51. Wang, M., et al.: Preparation of Enteromorpha prolifera-based cetyl trimethyl ammonium bromide-doped activated carbon and its application for nickel (II) removal. Ecotoxicol. Environ. Saf. **104**, 254–262 (2014)
52. El-Sikaily, A., El Nemr, A., Khaled, A.: Copper sorption onto dried red alga Pterocladia capillacea and its activated carbon. Chem. Eng. J. **168**(2), 707–714 (2011)
53. Kılıç, M., Apaydın-Varol, E., Pütün, A.E.: Preparation and surface characterization of activated carbons from Euphorbia rigida by chemical activation with $ZnCl_2$, K_2CO_3, NaOH and H_3PO_4. Appl. Surf. Sci. **261**, 247–254 (2012)
54. Nahalia, L., et al.: Assessment of Brilliant Green and Eriochrome Black T dyes adsorption onto fava bean peels: kinetics, isotherms and regeneration study. Desalin. Water Treat. **245**, 255–269 (2022)
55. Karri, R.R., et al.: Optimization and modeling of methyl orange adsorption onto polyaniline nano-adsorbent through response surface methodology and differential evolution embedded neural network. J. Environ. Manage. **223**, 517–529 (2018)
56. Tanzifi, M., et al.: Artificial neural network optimization for methyl orange adsorption onto polyaniline nano-adsorbent: kinetic, isotherm, and thermodynamic studies. J. Mol. Liq. **244**, 189–200 (2017)
57. Saini, K., et al.: Preparation and characterization of lignin-derived hard templated carbon (s): statistical optimization and methyl orange adsorption isotherm studies. Bioresour. Technol. **342**, 125924 (2021)
58. Pallarés, J., González-Cencerrado, A., Arauzo, I.: Production and characterization of activated carbon from barley straw by physical activation with carbon dioxide and steam. Biomass Bioenergy **115**, 64–73 (2018)
59. Kushwaha, J.P., Srivastava, V.C., Mall, I.D.: Treatment of dairy wastewater by commercial activated carbon and bagasse fly ash: parametric, kinetic and equilibrium modelling, disposal studies. Bioresour. Technol. **101**(10), 3474–3483 (2010)
60. Charoensook, K., et al.: Preparation of porous nitrogen-doped activated carbon derived from rice straw for high-performance supercapacitor application. J. Taiwan Inst. Chem. Eng. **120**, 246–256 (2021)
61. Eljamal, O., et al.: Investigating the design parameters for a permeable reactive barrier consisting of nanoscale zero-valent iron and bimetallic iron/copper for phosphate removal. J. Mol. Liq. **299**, 112144 (2020)
62. Sharma, G., et al.: Novel development of nanoparticles to bimetallic nanoparticles and their composites: a review. J. King Saud Univ. Sci. **31**(2), 257–269 (2019)
63. Mehdizadeh, S., et al.: Removal of heavy metals from aqueous solution using platinum nanopartcles/Zeolite-4A. J. Environ. Health Sci. Eng. **12**, 1–7 (2014)
64. Chen, Y.H.: Synthesis, characterization and dye adsorption of ilmenite nanoparticles. J. Non-Cryst. Solids **357**(1), 136–139 (2011)
65. Hallajiqomi, M., Eisazadeh, H.: Adsorption of manganese ion using polyaniline and its nanocomposite: Kinetics and isotherm studies. J. Ind. Eng. Chem. **55**, 191–197 (2017)
66. Roghani, M., et al.: Adsorption and oxidation study on arsenite removal from aqueous solutions by polyaniline/polyvinyl alcohol composite. J. Water Process. Eng. **14**, 101–107 (2016)

67. Etorki, A.M. et al.: Removal of some heavy metals from wastewater by using of fava beans. Am. J. Anal. Chem. (2014)
68. Mitra, S., Mukherjee, T., Kaparaju, P.: Prediction of methyl orange removal by iron decorated activated carbon using an artificial neural network. Environ. Technol. **42**(21), 3288–3303 (2021)
69. Al-Degs, Y.S., et al.: Sorption of Zn (II), Pb (II), and Co (II) using natural sorbents: equilibrium and kinetic studies. Water Res. **40**(14), 2645–2658 (2006)
70. McKay, G., Porter, J.F., Prasad, G.R.: The removal of dye colours from aqueous solutions by adsorption on low-cost materials. Water Air Soil Pollut. **114**, 423–438 (1999)
71. Hameed, B.H., Din, A.T.M., Ahmad, A.L.: Adsorption of methylene blue onto bamboo-based activated carbon: kinetics and equilibrium studies. J. Hazard. Mater. **141**(3), 819–825 (2007)
72. Byamba-Ochir, N., et al.: Highly porous activated carbons prepared from carbon rich Mongolian anthracite by direct NaOH activation. Appl. Surf. Sci. **379**, 331–337 (2016)
73. Peláez-Cid, A.A., et al.: Elimination of textile dyes using activated carbons prepared from vegetable residues and their characterization. J. Environ. Manage. **181**, 269–278 (2016)
74. Yahya, M.A., Al-Qodah, Z., Ngah, C.W.Z.: Agricultural bio-waste materials as potential sustainable precursors used for activated carbon production: a review. Renew. Sustain. Energy Rev. **46**, 218–235 (2015)
75. Samsuri, A.W., Sadegh-Zadeh, F., Seh-Bardan, B.J.: Characterization of biochars produced from oil palm and rice husks and their adsorption capacities for heavy metals. Int. J. Environ. Sci. Technol. **11**, 967–976 (2014)
76. Batzias, F.A., Sidiras, D.K.: Dye adsorption by prehydrolysed beech sawdust in batch and fixed-bed systems. Bioresour. Technol. **98**(6), 1208–1217 (2007)
77. Chen, Y., et al.: Removal of coke powders in coking diesel distillate using recyclable chitosan-grafted Fe3O4 magnetic nanoparticles. Fuel **238**, 345–353 (2019)
78. De Rossi, A., et al.: Chromium (VI) biosorption by Saccharomyces cerevisiae subjected to chemical and thermal treatments. Environ. Sci. Pollut. Res. **25**, 19179–19186 (2018)
79. Arami-Niya, A., Wan Daud, W.M.A., Mjalli, F.S.: Using granular activated carbon prepared from oil palm shell by ZnCl2 and physical activation for methane adsorption. J. Anal. Appl. Pyrolysis **89**(2), 197–203 (2010)
80. Reddy, K.S., Al Shoaibi, A., Srinivasakannan, C.: A comparison of microstructure and adsorption characteristics of activated carbons by CO_2 and H_3PO_4 activation from date palm pits. New Carbon Mater. **27**(5), 344–351 (2012)
81. Anisuzzaman, S.M., et al.: Removal of chlorinated phenol from aqueous media by guava seed (Psidium guajava) tailored activated carbon. Water Resour Ind. **16**, 29–36 (2016)
82. Abbas, A.F., Ahmed, M.J.: Mesoporous activated carbon from date stones (Phoenix dactylifera L.) by one-step microwave assisted K_2CO_3 pyrolysis. J. Water Process Eng. **9**, 201–207 (2016)
83. Budinova, T., et al.: Use of an activated carbon from antibiotic waste for the removal of Hg (II) from aqueous solution. J. Environ. Manage. **88**(1), 165–172 (2008)
84. Karagöz, S., et al.: Activated carbons from waste biomass by sulfuric acid activation and their use on methylene blue adsorption. Bioresour. Technol. **99**(14), 6214–6222 (2008)
85. Martins, A.C., et al.: Removal of tetracycline by NaOH-activated carbon produced from macadamia nut shells: kinetic and equilibrium studies. Chem. Eng. J. **260**, 291–299 (2015)
86. Hui, T.S., Ahmad Zaini, M.A.A.: Potassium hydroxide activation of activated carbon: a commentary. Carbon Lett. **16**(4), 275–280 (2015)
87. Zuo, L., et al.: Enhanced removal of sulfonamide antibiotics by KOH-activated anthracite coal: batch and fixed-bed studies. Environ. Pollut. **211**, 425–434 (2016)
88. Yakout, S.M., Sharaf, E.-D.: Characterization of activated carbon prepared by phosphoric acid activation of olive stones. Arab. J. Chem. **9**, S1155–S1162 (2016)
89. Tounsadi, H. et al.: Experimental design for the optimization of preparation conditions of highly efficient activated carbon from Glebionis coronaria L. and heavy metals removal ability. Process Saf. Environ. Prot. **102**, 710–723 (2016)
90. Uddin, M.K.: A review on the adsorption of heavy metals by clay minerals, with special focus on the past decade. Chem. Eng. J. **308**, 438–462 (2017)

91. Ahmed, M.J., Theydan, S.K.: Optimization of microwave preparation conditions for activated carbon from Albizia lebbeck seed pods for methylene blue dye adsorption. J. Anal. Appl. Pyrolysis **105**, 199–208 (2014)
92. Nandi, B.K., Goswami, A., Purkait, M.K.: Removal of cationic dyes from aqueous solutions by kaolin: kinetic and equilibrium studies. Appl. Clay Sci. **42**(3–4), 583–590 (2009)
93. Hassan, M.M., Carr, C.M.: Biomass-derived porous carbonaceous materials and their composites as adsorbents for cationic and anionic dyes: a review. Chemosphere **265**, 129087 (2021)
94. Travlou, N.A., et al.: Graphite oxide/chitosan composite for reactive dye removal. Chem. Eng. J. **217**, 256–265 (2013)
95. Salleh, M.A.M., et al.: Cationic and anionic dye adsorption by agricultural solid wastes: a comprehensive review. Desalination **280**(1–3), 1–13 (2011)

Biological Properties of Carbon Based Materials for Biomedical Applications

A. Joel Amalan, M. Kanagalakshmi, K. Hemkumar, and Anitha Pius

Abstract The study of carbon-based materials (CBMs) have grasped the attention of scientists from various field of science all around the world, like bio-sensing, cancer therapy, drug delivery, etc. This is due to the unique physical, chemical, and biological properties which includes thermal, mechanical, electrical, optical, antibacterial and antimicrobial activity. Various types of carbon-based materials like carbon fiber (CF), carbon nanotubes (CNT), graphene, carbon black (CB), activated carbon (AC), carbon dots/graphene quantum dots (CDs/GQDs), graphene oxide (GO), etc. have been investigated for various bio-medical applications which shows unique biological properties.

Keywords Graphene oxide · Activated carbon · Low cytotoxicity · Biocompatibility · Antibacterial activity · Biodegradability

1 Introduction

Carbon-based materials (CBMs) is the fashioning materials for the recent research in the field of materials science and nanotechnology. Carbon is one of the most basic nanomaterials, which attracts wide attention due to its outstanding properties in terms of electrical conductivity; fast charge transfer, high stability and ease of modification. The morphology of the material can be easily altered for high surface area and large pore sizes, which resulted in the amorphous carbon material. Experiments have been done not only to alter the morphological structure of carbon but also to modify the elements. The following are a few advantages of carbon nanomaterials:

- Their supramolecular pi-pi stacking allows them to absorb a large amount of drugs.

A. J. Amalan · M. Kanagalakshmi · K. Hemkumar · A. Pius (✉)
Department of Chemistry, The Gandhigram Rural Institute-Deemed to be University,,
Gandhigram, Dindigul, Tamil Nadu 624 302, India
e-mail: dranithapius@gmail.com

J. Tharini and S. Thomas (eds.), *Carbon Nanomaterials and their Composites as Adsorbents*, Carbon Nanostructures, https://doi.org/10.1007/978-3-031-48719-4_4

- CBMs can be used as theranostics materials due to their unique optical properties and ease of amalgamation with luminous chemicals.
- CBMs possess high heat conversion capacity in the near-infrared region, making them appropriate for photo-thermal therapy.
- Tunable surface chemistry can be used for the controlled release of therapeutics (Fig. 1).

Various investigations on the alteration of carbon materials have been studied. Extensive studies in the past decades reveal that the carbon-based materials could be exploited to mimic physiological micro environments to promote cell adhesion, proliferation, differentiation and even more to control their eventual fate. They possess many superior properties like chemical stability, biocompatibility, low toxicity and flexibility for surface functionalization. Because of their excellent biocompatibility, solubility and selectivity, carbon-based materials have shown great potential as the bio-sensing and bio-imaging materials. Carbon-based materials play the promising candidates for the bio-delivery carriers with unique physicochemical properties, such as large surface area, high purity, good bio-functional ability, good solubility, high drug loading capacity, the capability of cell membrane penetration, etc. There are many challenges in developing effective treatment, sensing, and

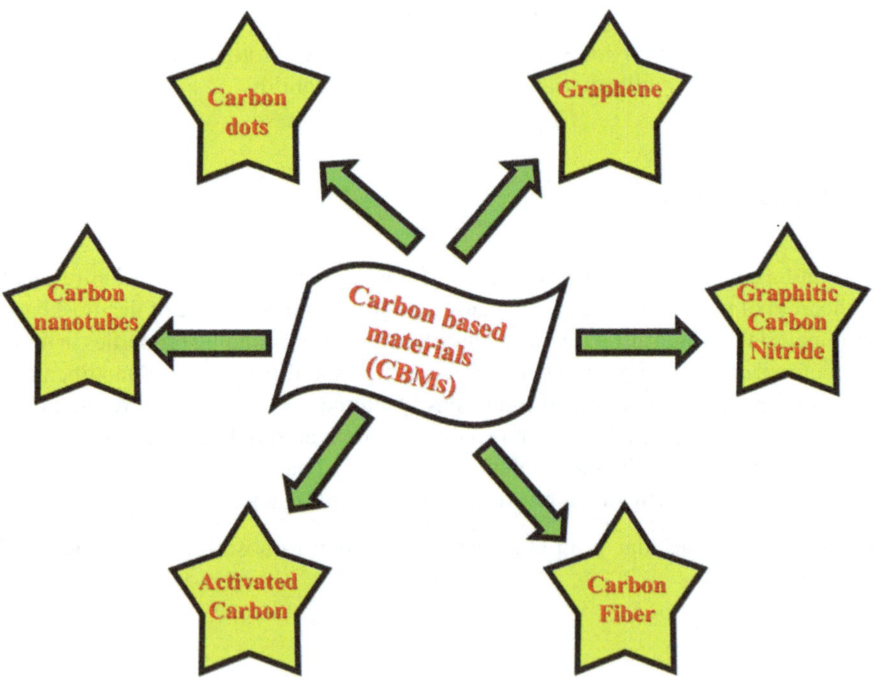

Fig. 1 Classification of various carbon based materials (CBMs)

imaging technologies in the biomedical applications. Here, we briefly review relevant studies of carbon-based materials to explore their potential biological properties and applications, including in vitro, in vivo, cancer therapy, etc.

2 Carbon Dots

In 2004, Scrivens made the first carbon dots (CD) discovery using arc discharge technique [1]. They are typically of zero-dimensional (0D) nanoparticles with the particle size ranging between 0 and 10 nm. Different synthetic methods namely electrochemical synthesis, combustion method, hydrothermal method, oxidation method, microwave method, ultrasonication, arc discharge, laser ablation and plasma method have been adopted [2]. They have many surface functional groups, such as carboxyl, amino, hydroxyl, and other groups, which makes it easier to alter their optical properties. The absorption band may be controlled through various surface passivation or modification techniques. The optical absorption of CDs is typically strongest in the UV region, with a tail that extends into the visible range. The majority of one-step manufactured CDs have absorbance between 280 and 360 nm [3].

2.1 Biological Properties and Applications of Carbon Dots:

CDs have outstanding properties such as good biocompatibility, penetrability, low toxicity, weak interactions with proteins, resistance to photobleaching, fast clearance, low cost, and simple synthesis. CDs have shown tremendous applications in vitro and in vivo imaging due to their robust emission fluorescence and low cytotoxicity. CDs can easily pass through biological membranes and accumulate in the cytoplasm or nucleus of cells which serve as the fluorescent probe. The cytotoxicity of luminous CDs has so far been thoroughly studied in vitro [4]. There is the increasing number of studies to investigate the possible uses of CDs in vivo [5].

Recently, both in vitro and in vivo studies on the cytotoxicity of PEG1500N-passivated CDs and PEG1500N (PEGylated 1500 N) were conducted. Human colorectal adenocarcinoma HT-29 cell and breast cancer MCF-7 cell line were both treated with PEG1500N-passivated CDs and PEG1500N [6]. It was demonstrated that PEG1500N had more effects on cell viability and proliferation than CDs. Additionally, using intravenous injections in mice, in- *vivo* bio-distribution and cytotoxicity of CDs were studied. CDs shows good photoluminescence (PL) bio-imaging at the lower concentration (Fig. 2).

Along with having exceptional biocompatibility, CDs have good electron-donor and electron-acceptor capabilities. CDs could be utilized for the intracellular ion detection, protein and enzyme identification, vitamin detection and nucleic acid detection, etc. The ion detection methods are identical, even though CDs made from diverse raw materials were used to sense variety of ions, including Cu^{2+}, Hg^{2+},

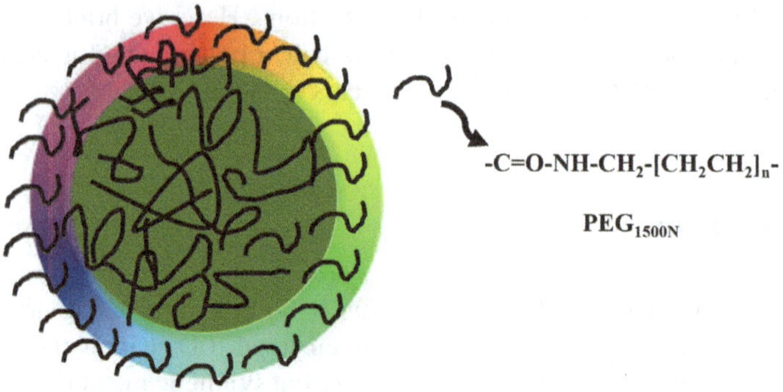

Fig. 2 Structure of carbon dots with surface passivation agents—PEG1500N

Ag^{2+}, Cr^{3+}, Fe^{3+}, K^+, Cl^-, and H^+. The surface functional groups on CDs shows distinct affinities to various target ions, which causes the PL intensity to be quenched by an electron or energy transfer mechanism and results in the good selectivity to other ions. To detect Hg^{2+}, a brand-new optical sensor made of graphene oxide and oligodeoxyribonucleotide (ODN)-CDs was developed [7].

Graphene oxide could effectively suppress the fluorescence of ODN-CDs using fluorescence resonance energy transfer. The luminescence was restored with the addition of Hg^{2+} by releasing ODN-CDs. An ultrasensitive CDs-based sensor have been developed to sense phosphate in the micromolar concentration [8]. CDs have received much attention because of their good photo-stability and reduced cytotoxicity. CDs serve as the excellent candidates in the bio-applications due to their visible excitation, emission wavelengths and high brightness at the individual dot level.

3 Graphene

Graphene is nothing but the thick sheet of carbon atoms organized in a hexagonal arrangement and bound together by sp^2 hybridization. It shows high thermal conductivity, high current, chemical inertness, optical transmittance and extremely high hydrophobicity, [9]. It is the thinnest material created so far and it consists of multiple graphite layers. The crystal structure of graphene shows densely packed honeycomb of carbon atoms [10]. GO is produced from the oxidation of graphene [11] which has superior surface functionalities, amphiphilicity, fluorescence quenching ability and surface-enhanced Raman's scattering properties [12]. Graphene is the promising platform for tissue engineering, molecular drug delivery, cancer treatment, biosensing and bioimaging due to its biocompatibility and speedy functionalization.

3.1 Biological Properties and Applications of Graphene

It can be employed for biosensing, bioimaging, hypothermia capabilities, biomolecules recognition, bioassays, molecular medicine and small-molecule drug delivery [13] due to its wide surface area, oxygen-containing activity, improved conductivity and good biocompatibility. Scientists have discovered that nanoparticles based on graphene exhibit outstanding antibacterial capabilities. The enormous surface area of graphene, particularly monolayer graphene and GO could leads to the surface phenomena, such as physical adsorption or catalytic chemical reaction.

Zhang and colleagues prepared GO/polyaniline nanocomposite for the biosensing of DNA at the concentration range between 1×10^{-6} and 1×10^{-14} [14]. Bone tissue does not possess sufficient regeneration ability because of the nonvascular and cell-free properties. Hence graphene based materials and biopolymers together are often adopted in bone tissue engineering to accelerate the growth and differentiation of bone cells (Fig. 3).

Biomaterials used in the bone tissue engineering must be biodegradable in most cases, like collagen, gelatin (natural degradable biopolymers), polylactic acid, polyglycolic acid (PGA) (synthetic degradable biopolymers). Li et al. designed Poly (L-lactic acid) (PLLA) and graphene composite [15] for the biomedical applications. When GO is modified with PLA to form GO-g-PLLA, the tensile strength and elongation at the break of the prepared GO-g-PLLA nano-composite were improved. GO forms hydrogen bonds with PLLA and it is equally distributed in the polylactic acid and hence resulting in the polylactic acid/graphene oxide nanocomposites with significantly increased elastic modulus and tensile strength. Graphene may emerge as a distinct nanoparticle for application in biomedical research through effective collaboration with other disciplines of science.

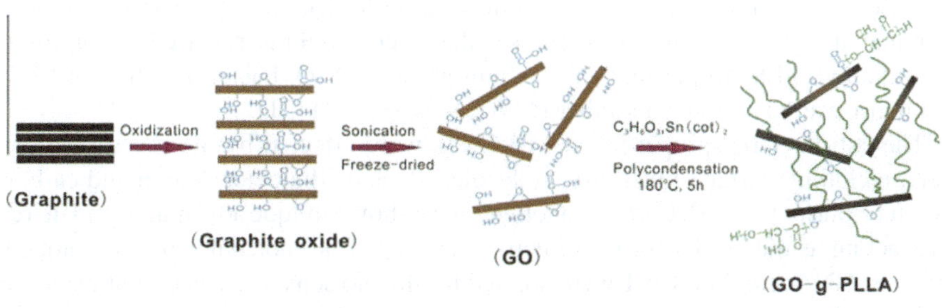

Fig. 3 Scheme illustration for the preparation and functionalization of GO-g-PLLA

4 Carbon Nanotubes (CNT)

CNT has a cylindrical and long tubular structure which consists of rolled-up sheets of graphene. It comprises of single-walled CNTs (SWCNTs) which are made of a single layer of carbon atoms and multi-walled CNTs (MWCNTs) which are formed by several nanotubes interlinked concentrically [16] Several biomedical applications of CNTs includes drug vectors, gene delivery to the cells or organs, tissue regeneration, and biosensor diagnostics and analysis.

4.1 Biological Properties and Applications of Carbon Nanotubes

CNTs shows excellent fast electron transfer kinetics, ultra-light weight, chemical inertness, high tensile strength, good antibacterial and antifungal properties. MWCNTs' antibacterial capabilities were influenced by the surface functional molecules, carbon nanotube density, diameter, length, and purity [17]. CNT derivatives shows superior action against gram-positive bacteria and comparably lesser activity against gram-negative bacteria [18, 19]. CNTs have numerous favorable features that make them perfect candidate for the optical detection. It gives wide applications in the biomedical applications due to the low photo absorption and low auto-fluorescent background [20].

CNTs have attracted immense attention as a highly competent vehicle for transporting various drug molecules into the living cells. Generally, drug molecules are attached to CNT sidewalls via covalent or non-covalent bonding between the drug molecules and functionalized CNT. Functionalization of carbon nanotubes might increases the antimicrobial activity while helping to decrease the toxicity towards mammalian cells. The current work uses the electric field to release the ibuprofen from the hybrid hydrogel made of sodium alginate (SA), bacterial cellulose (BC) and multi-walled carbon nanotubes (MWCNTs) (Fig. 4) [21].

The anti-bacterial properties of MWCNTs were affected by the several parameters, including surface functional molecules, density, diameter, length and carbon nanotube purity [22, 23]. Carbon nanotubes have shown unique advantages and therefore, acting as the carrier for the effective delivery of biomolecules like antibiotics, proteins, DNA, etc. MWCNT were applied for the bio sensing devices, antibacterial actives and cancer diagnosis.

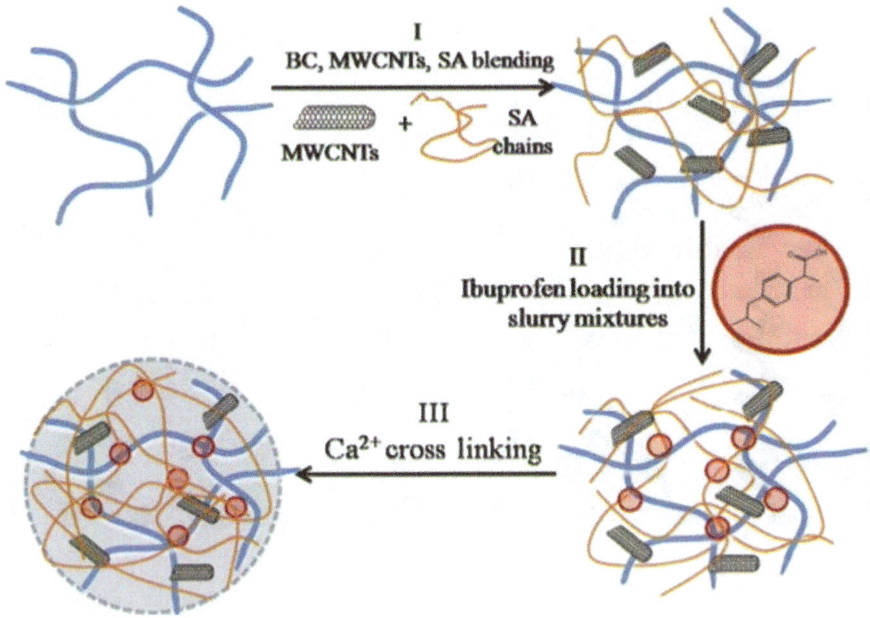

Fig. 4 Schematic of the preparation of the drug loaded BC/SA/MWCNTs hybrid hydrogels

5 Graphitic Carbon Nitride

Graphitic carbon nitride (g-C_3N_4)-based materials are burgeoning fluorescent polymeric materials mainly due to the presence of C and N atoms. They find applications in the diagnosis and therapy, bio-sensing and antibacterial application [24]. Graphitic carbon nitride (g-C_3N_4) shows outstanding biocompatibility and unique fluorescent feature which find applications in the biosensor applications. To accomplish target-specific biomedical applications, we need to prepare g-C_3N_4-based materials with superior water-solubility, biodegradability, thin nanosheets and enhanced light absorption. Many attempts have been made to prepare synthesis novel g-C_3N_4-based hybrids.

5.1 Biological Properties and Applications of Graphitic Carbon Nitride

G-C_3N_4 nanostructures possess high fluorescent quantum and quenching yields, prominent photo stability, outstanding biocompatibility, low cost and nontoxicity, which have been used in in vitro and in vivo fluorescent biosensors [25]. Hydrogen-bonded g-C_3N_4supramolecular quantum dots were synthesized by the engineering

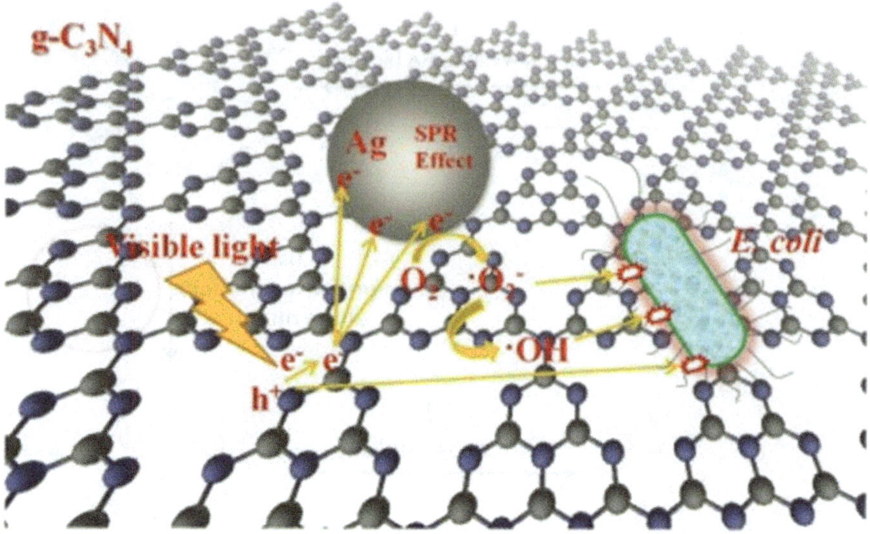

Fig. 5 The mechanism of *E.coli* inactivation in the presence of Ag/g-C_3N_4 under visible light

of supramolecular chemistry which shows low cytotoxicity, biocompatibility and outstanding biodegradability [26]. G-C_3N_4 are more stable than conventional photosensitizers, which facilitates further chemical engineering modification for a stronger therapeutic impact and multi-functionality [27].

G-C_3N_4-based photocatalysts have attracted attention in the antibacterial applications due to their high stability, bulk availability, low cost, and eminent optical performance. Pure g-C_3N_4 without any additives shows outstanding antibacterial properties under visible light illumination due to its unique photo catalytic property [28]. By modifying g-C_3N_4 with Ag NPs, the photocatalytic performance of the g-C_3N_4 could be significantly enhanced. The Ag/g-C_3N_4 heterostructures were reported to have excellent disinfection effects for Escherichia coli due to the surface plasmon resonance effect, good adsorption of visible light and decreased recombination of free charges [29] (Fig. 5).

6 Activated Carbon

Activated carbon, also known as activated charcoal, is a refined form of carbon which shows high capacity for adsorbing different types of organic and inorganic chemicals, filtering out impurities, purifying antibiotics, removing color from sugar, purifying gases and more. As the result, there are hundreds of different variants of commercially produced carbons are produced. Activated carbon is produced in the three primary categories namely extruded activated carbon, granular activated carbon, and powdered activated carbon.

6.1 Biological Properties and Applications of Activated Carbon

Nitrifying bacteria or heterotrophic bacteria are responsible for the metabolizing biodegradable organic materials which are aided by the activated carbon (AC) [30]. Numerous elements favor the adhesion, growth, and metabolism of these bacteria. Some of them are mentioned below:

- Extracellular polymers secreted by the bacteria.
- Grain surface unevenness coupled with a physical–chemical type adsorption capability (similar to that of organic molecules); these properties allow the bacteria to remain attached to the AC even in the presence of a high sheer force.
- Bacteria will be protected against chlorine when chlorine is found in the water.
- Organic molecules are used as the bacteria nutrients which are trapped in the AC pores.

7 Carbon Fiber

Carbon fibers are fibers of about 5–10 μm in diameter and are composed mostly of carbon atoms. Carbon fibers have several advantages including high stiffness, high tensile strength, low weight, high chemical resistance, high temperature tolerance and low thermal expansion. These properties have made carbon fiber very popular in aerospace, civil engineering, military and motorsports and other competition sports. However, they are relatively expensive when compared with similar fibers, such as glass fibers or plastic fibers.

7.1 Biological Properties and Applications of Carbon Fiber

Since the late 1970s, carbon fibers are utilized extensively as the biomaterials due to the light weight, high strength and flexibility. There have been several studies on carbon fiber composites and the majority of them includes the incorporation of polymers and other materials. Carbon fibers are typically used in the conjunction with other materials to enchanes their properties. Cell therapy, cytokine treatment and gene therapy are a few examples of techniques used in the regenerative medicine.

Numerous studies on these techniques, either alone or in combination, are moving forward quickly [31]. To obtain outstanding regenerative performance, a strong scaffold material is required that can keep cells or cytokines at the spot and aid in the synthesis of new tissue [32]. Carbon fibres composites have shown good biocompatibility in both in vivo implantation study and an in vitro cell adhesion study [33] Carbon fibres have gained lot of attention in the field of tissue regeneration. Carbon fibres were successfully used by Cameron et al. to repair the damaged abdominal walls in the rats [34].

8 Conclusion

This chapter has dealt with the most recent developments in carbon nanoparticles' therapeutic and imaging in the biomedicine. Carbon nanomaterials are the novel class of materials with intriguing chemical, optical and mechanical as well as biological characteristics. These carbon based materials shows excellent antibacterial applications. These nanomaterials, which are made from the same chemical element but have different allotropic forms of carbon, displays diverse properties and behaviors depending on the size and how the carbon atoms are bound to form the bigger structures on the nanoscale. It is now possible to create water-soluble and biocompatible carbon nanomaterials with minimal in vitro and in vivo toxicity and to use these nanostructures for a variety of biomedical imaging and therapeutic applications.

References

1. Xu, X., Ray, R., Gu, Y., Ploehn, H.J., Gearheart, L., Raker, K., et al.: Electrophoretic analysis and purification of fluorescent single-walled carbon nanotube fragments. J. Am. Chem. Soc. **126**(40), 12736–12737 (2004)
2. Khairol Anuar, N.K., Tan, H.L., Lim, Y.P., So'aib, M.S., Abu Bakar, N.F.: A review on multifunctional carbon-dots synthesized from biomass waste: design/fabrication, characterization and applications. Front. Energy Res. **9**, 626549 (2021)
3. Multi-applications of carbon dots and polydopamine-coated carbon dots for Fe3+ detection, bioimaging, dopamine assay and photothermal therapy, Discover Nano **18**, 30 (2023)
4. Cao, L., Wang, X., Meziani, M.J., Lu, F., Wang, H., Luo, P.G., et al.: Carbon dots for multiphoton bioimaging. J. Am. Chem. Soc. **129**(37), 11318–11319 (2007)
5. Li, Q., Ohulchanskyy, T.Y., Liu, R., Koynov, K., Wu, D., Best, A., et al.: Photoluminescent carbon dots as biocompatible nanoprobes for targeting cancer cells in vitro. J. Phys. Chem. C **114**(28), 12062–12068 (2010)
6. Wang, Y., Anilkumar, P., Cao, L., Liu, J.H., Luo, P.G., Tackett, K.N., et al.: Carbon dots of different composition and surface functionalization: cytotoxicity issues relevant to fluorescence cell imaging. Exp. Biol. Med. **236**(11), 1231–1238 (2011)
7. Li, N., Liang, X., Wang, L., Li, Z., Li, P., Zhu, Y., et al.: Biodistribution study of carbogenic dots in cells and in vivo for optical imaging. J. Nanoparticle Res. **14**(10), 1177 (2012)
8. Cui, X., Zhu, L., Wu, J., Hou, Y., Wang, P., Wang, Z., et al.: A fluorescent biosensor based on carbon dots-labeled oligodeoxyribonucleotide and graphene oxide for mercury (II) detection. Biosens. Bioelectron. **63**, 506–512 (2015)

9. Xu, J., Zhou, Y., Cheng, G., Dong, M., Liu, S., Huang, C.: Carbon dots as a luminescence sensor for ultrasensitive detection of phosphate and their bioimaging properties: carbon dots as a luminescence sensors. Luminescence **30**(4), 411–415 (2015)
10. Choi, W., Lahiri, I., Seelaboyina, R., Kang, Y.S.: Synthesis of graphene and its applications: a review. Crit. Rev. Solid State Mater. Sci. **35**(1), 52–71 (2010)
11. Graphene oxide, highly reduced graphene oxide, and graphene - versatile building blocks for carbon-based materials. Small **6**(6), 711–723 (2010)
12. Sanchez, V.C., Jachak, A., Hurt, R.H., Kane, A.B.: Biological interactions of graphene-family nanomaterials: an interdisciplinary review. Chem. Res. Toxicol. **25**(1), 15–34 (2012)
13. Chung, C., Kim, Y.K., Shin, D., Ryoo, S.R., Hong, B.H., Min, D.H.: Biomedical applications of graphene and graphene oxide. Acc. Chem. Res. **46**(10), 2211–2224 (2013)
14. Cheng, D., Yang, L., Li, X., Zhou, J., Chen, Q., Yan, S., et al.: An electrochemical DNA sensing platform using carboxyl functionalized graphene as the electrode modified material. J. Electrochem. Soc. **164**(6), H345–H351 (2017)
15. Li, W., Xu, Z., Chen, L., Shan, M., Tian, X., Yang, C., et al.: A facile method to produce graphene oxide-g-poly (L-lactic acid) as an promising reinforcement for PLLA nanocomposites. Chem. Eng. J. **237**, 291–299 (2014)
16. Aqel, A., El-Nour, K.M.M.A., Ammar, R.A.A., Al-Warthan, A.: Carbon nanotubes, science and technology part (I) structure, synthesis and characterisation. Arab. J. Chem. **5**(1), 1–23 (2012)
17. Smart, S.K., Cassady, A.I., Lu, G.Q., Martin, D.J.: The biocompatibility of carbon nanotubes. Carbon **44**(6), 1034–1047 (2006)
18. Patel, K.D., Singh, R.K., Kim, H.W.: Carbon-based nanomaterials as an emerging platform for theranostics. Mater. Horiz. **6**(3), 434–469 (2019)
19. Thermo conductive carbon nanotube-framed membrane for skin heat signal responsive transdermal drug delivery. Polym. Chem. **8**, 3154–3163
20. Welsher, K., Sherlock, S.P., Dai, H.: Deep-tissue anatomical imaging of mice using carbon nanotube fluorophores in the second near-infrared window. Proc. Natl. Acad. Sci. **108**(22), 8943–8948 (2011)
21. Shi, X., Zheng, Y., Wang, C., Yue, L., Qiao, K., Wang, G., et al.: Dual stimulus responsive drug release under the interaction of pH value and pulsatile electric field for a bacterial cellulose/sodium alginate/multi-walled carbon nanotube hybrid hydrogel. RSC Adv. **5**(52), 41820–41829 (2015)
22. Kang, S., Pinault, M., Pfefferle, L.D., Elimelech, M.: Single-walled carbon nanotubes exhibit strong antimicrobial activity. Langmuir. **23**(17), 8670–8673 (2007)
23. Hirschfeld, J., Akinoglu, E.M., Wirtz, D.C., Hoerauf, A., Bekeredjian-Ding, I., Jepsen, S., et al.: Long-term release of antibiotics by carbon nanotube-coated titanium alloy surfaces diminish biofilm formation by Staphylococcus epidermidis. Nanomedicine Nanotechnol Biol Med. **13**(4), 1587–1593 (2017)
24. Ismael, M.: A review on graphitic carbon nitride (g-C3N4) based nanocomposites: synthesis, categories, and their application in photocatalysis. J. Alloys Compd. **846**, 156446 (2020)
25. Wu, J., Yang, S., Li, J., Yang, Y., Wang, G., Bu, X., et al.: Electron injection of phosphorus doped g-C_3N_4 quantum dots: controllable photoluminescence emission wavelength in the whole visible light range with high quantum yield. Adv. Opt. Mater. **4**(12), 2095–2101 (2016)
26. Zhao, F., Li, Z., Wang, L., Hu, C., Zhang, Z., Li, C., et al.: Supramolecular quantum dots as biodegradable nano-probes for upconversion-enabled bioimaging. Chem. Commun. **51**(67), 13201–13204 (2015)
27. Fluorescent graphene-like carbon nitrides - synthesis, properties and applications. J. Mater. Chem. C **4**, 8146–8160 (2016)
28. Thurston, J.H., Hunter, N.M., Cornell, K.A.: Preparation and characterization of photoactive antimicrobial graphitic carbon nitride (g-C_3N_4) films. RSC Adv. **6**(48), 42240–42248 (2016)
29. Ma, S., Zhan, S., Jia, Y., Shi, Q., Zhou, Q.: Enhanced disinfection application of Ag-modified g-C3N4 composite under visible light. Appl. Catal. B Environ. **186**, 77–87 (2016)

30. Soonmin, H., Kabbashi, N.A.: Review on activated carbon: synthesis, properties and applications. Int. J. Eng. Trends Technol. **69**(9), 124–139 (2021)
31. Takahashi, K., Tanabe, K., Ohnuki, M., Narita, M., Ichisaka, T., Tomoda, K., et al.: Induction of pluripotent stem cells from adult human fibroblasts by defined factors. Cell. **131**(5), 861–872 (2007)
32. Takahashi, K., Okita, K., Nakagawa, M., Yamanaka, S.: Induction of pluripotent stem cells from fibroblast cultures. Nat. Protoc. **2**(12), 3081–3089 (2007)
33. Bačáková, L., Stary, V.: Polishing and coating carbon fiber-reinforced carbon composites with a carbon–titanium layer enhances adhesion and growth of osteoblast-like MG63 cells and vascular smooth muscle cells in vitro. J. Biomed. Mater. Res. **54**, 567–578 (2001)
34. Cameron and Taylor - Carbon-fibre versus Marlex mesh in the repair of experimental abdominal wall defects in rats. Br. J. Surg. **72**(8), 648–650 (1985)

Synthesis Strategies of Various Carbon Materials

Seema Panicker, Javad B. M. Parambath, and Ahmed A. Mohamed

Abstract Carbon materials have been known since early human activities for heating purposes and basic daily life. With the development and the industrial revolution, the need for carbon materials for energy and scientific developments became inevitable. The diverse forms of carbon materials include cheap allotropy such as graphite and expensive such as diamond. The synthesis strategies are divided into natural and laboratory. The synthesis of carbon nanotubes in different forms has added many applications to the field. Moreover, graphene and its doped forms are considered a huge expansion into the chemistry of carbon materials. Here in this chapter, we discuss the fundamentals of the synthesis strategies of carbon materials.

Keywords Carbon materials · Carbon nanotubes · Graphene · Carbon quantum dots · Synthesis strategies

1 Introduction

Carbon (Latin *carbo,* meaning "coal") is one of the most abundant elements on the earth's surface. It has a total of six electrons with four valence electrons which makes carbon unique as it enables the ability for catenation i.e., it can form C–C bonds to any extent as well as it also forms covalent bonds with other atoms that give rise to hundreds of different stable organic compounds and a wide range of carbon materials. Carbon has been there through civilizations on earth, and the most common carbon materials that have been used by humans since the pre-historic era are diamond, graphite, and charcoal. These materials consist of only carbon, but they have a broad spectrum of properties that were utilized to create a variety of materials. With the advancement of science and invent of modern technologies, carbon is now

S. Panicker
Fatima College of Health Sciences, Ajman 3798, UAE

J. B. M. Parambath · A. A. Mohamed (✉)
Department of Chemistry, College of Sciences, Sharjah 27272, UAE
e-mail: ah.mohamed@sharjah.ac.ae

J. Tharini and S. Thomas (eds.), *Carbon Nanomaterials and their Composites as Adsorbents*, Carbon Nanostructures, https://doi.org/10.1007/978-3-031-48719-4_5

used to synthesize a wide variety of materials like fullerenes, graphitic onions, carbon nanotubes, fibers, carbon foils, etc. These carbon materials are built using different physical and chemical processes in a well-controlled system. Advancements in the field of synthesis and applications of carbon materials particularly carbon nanomaterials have been an active area of research in recent times. Carbon nanomaterials exhibit applications ranging from the formation of nanocomposites, sensors, electronics, and catalysis to energy storage and production devices, as adsorption and purification materials, in wastewater treatment, and many more. Carbon materials can be used in multidisciplinary fieldsonly because of their unique micromorphological structures, crystallinity, and multifaceted properties, which are entirely dependent on the methods and strategies used in their synthesis [1].

Carbon is abundantly available on earth's crust and is being utilized by mankind in numerous ways. Out of the different forms of carbon, graphite is one of the most thermodynamically stable ones and it has shown a remarkably high conductivity, making it suitable for use in batteries, solar panels, electrodes, etc. Graphite is made up of layers of sheets that consist of graphene. It is strong and is recently being used to synthesize nanomaterials after conversion to graphene oxide. Graphene nanoparticles are used in several applications [2]. Another popular carbon material used nowadays is carbon nanotubes (CNTs). The chemical and physical properties of graphene, CNTs, and fullerenes are related to each other, and they exhibit tremendous scope in research and applications. The catalytic and adsorption properties of carbon materials are well-recorded, Figs. 1 and 2 [3, 4]. In recent times, different synthesis strategies have been developed for making heterostructures of carbon-based nanomaterials, Fig. 1 [4]. Their synthesis under strictly regulated conditions has been successfully used in biomedical applications, drug delivery systems, nanoelectronics, optoelectronics as well as sensors and storage purposes [5]. Here, we discuss some of the crucial synthesis strategies of carbon materials and explore their dynamic development into multifaceted structures.

2 Carbon Nanotubes (CNTs)

CNTs are multipurpose and one of the most important types of carbon materials, that have been developed in recent times. They have a cylindrical and tubular structure made up of sheets of graphene. There are single-walled CNTs (SWCNTs) and multi-walled CNTs (MWCNTs) consisting of interlinked nanotubes. As MWCNTs are multi-layered with carbon atoms, they have very high mechanical strength and possess higher tensile strength than common metals like iron and steel [6]. CNTs are most popularly used in making vacuum microelectronics, and electron field emission systems, used for storing energy and in electrochemical functions. There are various methods of synthesizing CNTs like the plasma-based methods which include the arc discharge and laser ablation methods and thermal preparation methods that use chemical vapor deposition (CVD) and hydrothermal methods.

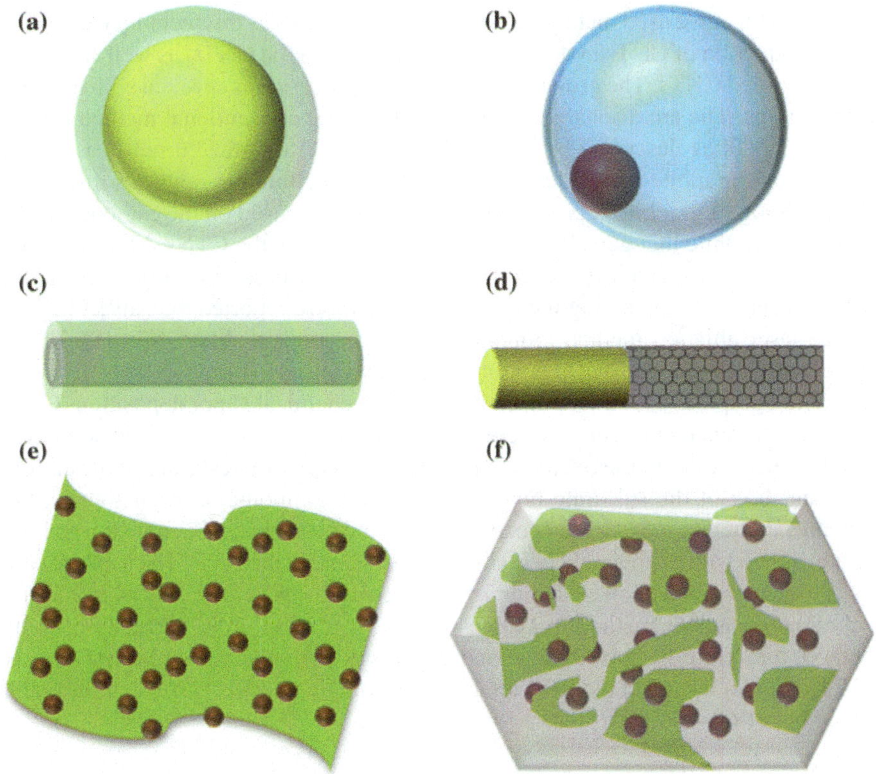

Fig. 1 Examples of heterostructures: zero-dimensional **a** core–shell and **b** core-cage nanoparticles, one-dimensional **c** core–shell nanowires and **d** multi-segment nanowires nanotubes, **e**, **f** multi-dimensional sheet or network nanocomposites [4]

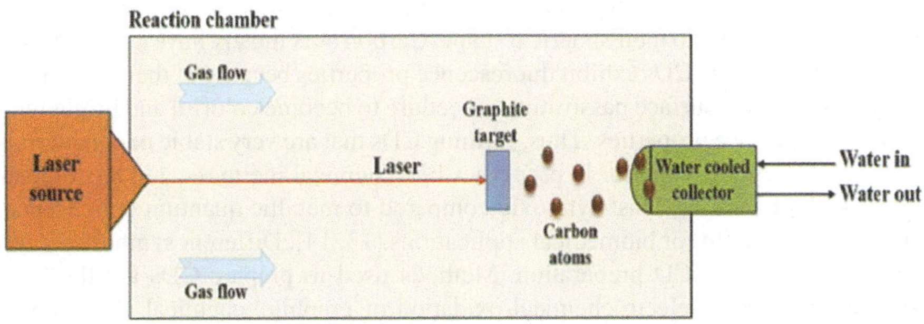

Fig. 2 Laser ablation method for the synthesis of SWCNTs [3]

CVD is the most used method in the synthesis of CNTs as this technique gives rise to the production of CNTs in different forms like powdered CNTs, as well as straight chain or coiled CNTs. The growth of CNTs in this process was checked with different temperatures. The arc discharge method was the most conventional method used to produce CNTs as described by Iijima [7]. In this method, the current was passed between two graphite electrodes in the presence of helium gas, and this resulted in the formation of graphite vapors. The vapors then condensed on the cathode and the well. The graphite that gets deposited on the walls forms the CTNs. Metals like Co and Ni can be used as anodes. Several other methods like passing a carbon-containing gas over a catalyst containing metal nanoparticles were reported by Calvert [8].

The laser ablation method is mainly used to synthesize MWCNTs [9]. In this process of synthesis, a pulsed or continuous laser was used to form vapors of a mixture containing graphite with traces of Co or Ni. In the chemical vapor deposition (CVD) method, different metal catalysts were used, and they are deposited on a substrate. The metal catalysts initiate the nucleation process, subsequently resulting in the etching and annealing of the substrate by compounds like ammonia. Carbon sources like methane, carbon monoxide, and acetylene were then used and were transformed into carbon atoms with the help of energy from plasma or a heated coil. The substrate and the carbon were then allowed to interact, giving rise to CNTs [10]. The hydrothermal technique is another method to synthesize CNTs, primarily the MWCNTs. In this method, a mixture of polyethylene and water was prepared and a catalyst e.g., Ni was added and heated at elevated temperatures ranging between 700 and 800 °C at a pressure of 60–100 MPa [11].

3 Carbon Quantum Dots and Carbon Sphères

The carbon quantum dots (CDs) and carbon spheres are grouped as zero-dimensional carbon materials due to their spherical shape. Carbon dots mostly have a size ranging from 2 to 10 nm. The CDs exhibit fluorescence properties because of their small size, and they undergo a surface passivating procedure to become colorful and bright with photoluminescence properties. Thus, forming CDs that are very stable nanomaterials [12]. CDs are highly superior in properties like chemical inertness, and biocompatibility, and they are the least cytotoxic compared to metallic quantum dots. Hence, CDs are widely used for biomedical applications [13, 14]. Different synthesis strategies are applied for CD preparation. Methods used to prepare CDs are the laser ablation of graphite, electrochemical oxidation of graphite, chemical oxidation of a suitable precursor, proton-beam irradiation of nanodiamonds, microwave-assisted method, and thermal oxidation of suitable precursors. Carbon spheres with desired size and surface modifications were synthesized for a wide range of applications such as drug delivery, electrodes, catalysts, gas storage purposes, etc. Precise and highly controlled synthesis strategies were used to obtain carbon spheres with large monodispersity [1]. The synthesis approaches include hydrothermal reduction, emulsion, self-assembly, and templating methods. It is challenging to prepare uniform carbon spheres with sizes within the range of 200 nm, as smaller sizes are required for their bio-medical applications. Carbonization of polymer analogs was conducted to prepare carbon-based spheres. These polymers should be thermally stable and able to form carbon residues after elevated temperature pyrolysis. Hence, many various chemical synthesis methods were reported for the preparation of polymers and carbon spheres. The carbon nanospheres were synthesized by the carbonization of a polymer of resorcinol and formaldehyde in the presence of the basic amino acid L-lysine, which acted as a catalyst in this process. The size of the resulting carbon nanospheres was tuned in the desired range by adjusting the concentration of catalyst and resorcinol [1].

4 Buckminsterfullerene

Buckminsterfullerene is the most popular and common fullerene, with 60 carbon atoms that are arranged in a cage-like structure. Recently C60 based nanoparticles of various shapes like nanorods, nanotubes, and nanosheets are being extensively used in the field of nanoscience.

Fullerenes can be synthesized by the arc discharge method and the process was conducted under elevated-temperature furnaces. The carbon clusters were attached with a buffer and the reaction resulted in the condensation of carbon. Fullerenes thus synthesized were separated using chromatographic techniques [15]. Wei [16] reported another method for the synthesis of fullerenes by detonation method. In this process, trinitrotoluene (TNT) and graphite mixture were placed in a mold and converted into a column shape by applying pressure. HMX was used as a charge booster and the detonator was connected to a firing box and the TNT/graphite mixture, HMX, and a detonator were reacted in a vacuum vessel. The vessel was then put in cooling water to facilitate cooling. The products obtained were treated with acetone.

5 Graphene Oxide

Graphene is one of the most researched materials and is considered the "material of the future" [17]. Graphene is unique in the sense that it contains only carbon atoms in its structure, where each carbon atom is attached to three other carbon atoms. Because of its unique properties, graphene has large applications as gas and biomolecule sensors, transparent conductive films (TCFs), field effect transistors (FETs), and graphene batteries, Fig. 3 [18]. Graphene oxide is a carbon layered structure with functional groups such as = O, –OH, –O–, –COOH where oxygen is common in all of them. These groups are attached at both ends of the graphene oxide layers as well as on the edges of the plane [19]. Graphene oxides were synthesized by the process of oxidation of graphite into graphite oxide and then subjected to exfoliation to form graphene oxide. Because of its hydrophilic nature, graphene

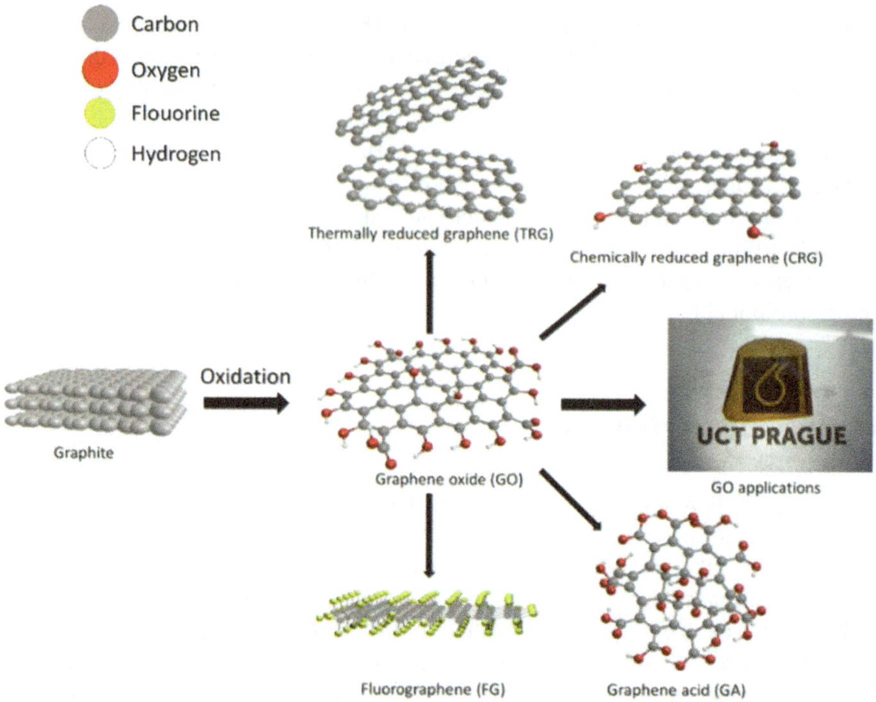

Fig. 3 Graphene oxide as the precursor of different derivatives [18]

oxides are easily prepared as water or organic-based suspensions. Graphene oxide can be used as the precursor due to the presence of oxygen in it, for the synthesis of graphene derivatives such as fluorographene, bromographene, graphene, and many others, Fig. 3 [18].

The first attempt to synthesize graphene oxide was performed as early as 1859 by Brodie [20] who studied the reactivity of flake graphite. Nowadays, several routes of synthesis methods are used to prepare graphite oxide/graphene oxide. The use of an oxidizing agent in an acidic medium is one of the most common methods of synthesis. Electrochemical and microbial methods are also used very commonly. In one of the studies reported by Tour, Fig. 4 [21], phosphoric acid was mixed with sulfuric acid in a 1:9 ratio, and into this, a solution of potassium permanganate and graphite in the ratio of 6:1 was added in freezing conditions. After that, the mixture was heated at 50 °C and stirred for 12 h. The mixture was then poured into ice and finally 30% H_2O_2 was added to remove the excess potassium permanganate. Phosphoric acid works as a dispersive and etching agent, as well as a stabilizer of the oxidation process, which makes the synthesis of graphene oxide safe. This route produced a higher yield of graphene oxide with a higher level of oxidation and a more regular structure [21]. The eco-friendly electrochemical method is another popular method

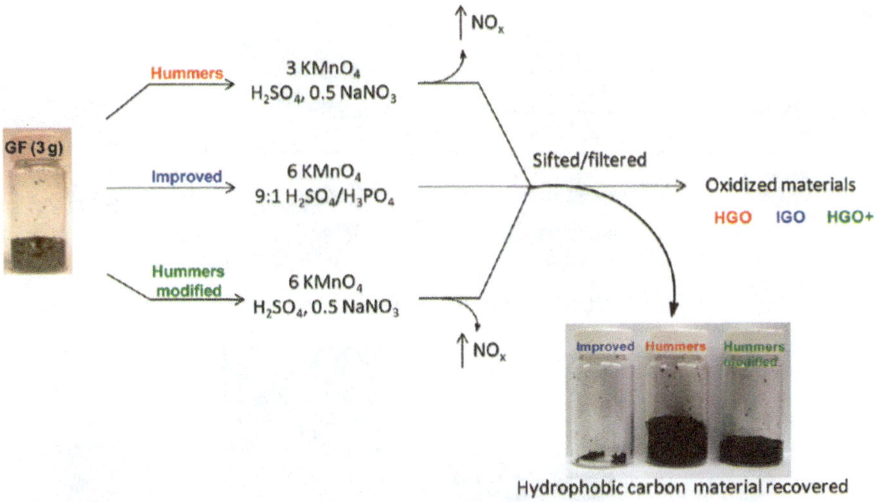

Fig. 4 Representation of the procedures followed starting with graphite flakes (GF). Under-oxidized hydrophobic carbon material recovered during the purification of IGO, HGO, and HGO+. The increased efficiency of the IGO method is indicated by the very small amount of under-oxidized material produced [21]

of synthesis and is used for large-scale production of graphene oxide [22]. The use of biological systems also produced eco-friendly graphene oxide, Fig. 5 [23].

6 Nano-Diamond

Nanoscale diamonds were discovered long before the plethora of nanomaterials began. Nanodiamonds (NDs) moreover, were naturally present in stars and meteorites before their first artificial synthesis [24, 25]. However, extensive applications and synthetic strategies have only been developed recently. Momentarily, NDs with diverse structures, small sizes, and relatively low toxicity occupy a special niche in the carbon nanomaterials class. First, synthetic NDs were formed by detonating carbon-rich explosives such as trinitrotoluene (TNT) and hexogen (RDX) in a closed chamber in an inert condition named detonation nanodiamond (DND) [26]. Considerable progress was achieved since then in NDs synthesis, discoveries such as laser ablation [27], high-effect grinding ball milling [28], chemical vapor deposition (CVD) [29], micro plasma-aided gas phase synthesis [30], chlorination of carbides [31], heavy ion irradiation on graphite [32], carbon onions [33], and ultrasonic cavitation [34]. Either top-down or bottom-up approaches are used to synthesize NDs as illustrated in Fig. 6. However, thus far we are unable to precisely control the NDs composition and still, the major synthetic strategy utilized at the production level is

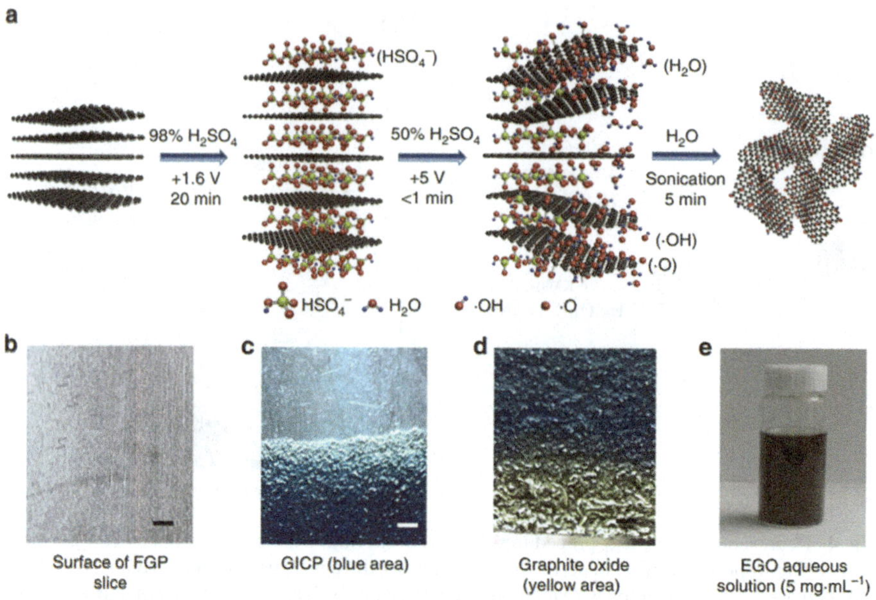

Fig. 5 Synthesis of EGO by water electrolytic oxidation. **a** Schematic illustration of the synthesis process of EGO by water electrolytic oxidation. **b–e** Photos of the raw material and the products obtained at each step. **b** FGP. c, GICP (blue area) obtained after EC intercalation of FGP in 98 wt.% H_2SO_4 at 1.6 V for 20 min. **d** Graphite oxide (yellow area) obtained by water electrolytic oxidation of the GICP in 50 wt.% H_2SO_4 at 5 V for 30 s. **e** Well-dispersed EGO aqueous solution (5 mg·mL^{-1}) obtained by sonication of the graphite oxide in water for 5 min. Scale bars in **b–d**: 1 mm [23]

the detonation method. Contextually, here we discuss recent advances in the synthesis of nanodiamonds.

Academic and industrial interest in NDs synthesis led to the exploration of a wide range of applications, including additives in lubricants, optoelectronics, drug delivery, bio labels or biomarkers in the biomedical community, etc., owing to their outstanding mechano-chemical resistance, biocompatibility, and optronic properties. Shenderova and Gruen provided a detailed discussion of the NDs' synthetic strategies [35]. At present, high-pressure high-temperature (HPHT) synthesized by the "top-down" method and detonation strategy a "bottom-up" method are the two major synthetic procedures for NDs [36].

DND NDs are formed after slow cooling of supersaturated carbon vapors generated by controlled inert condition explosion, resulting in 3–5 nm diamond particles [37]. Controlled air oxidation steps were reported to tune the particle sizes [38], and layer-by-layer C-atom removal via oxidation resulted in reduced size (5.2–4.8 nm) of NDs from DNDs [39]. Recently, NDs less than 2 nm were reported by one-step-controlled oxidation in the air [40]. On the other hand, bottom-up methods for NDs synthesis provide a finer tunability of size, purity, and better surface chemistry. HPHT NDs were obtained by grinding micro diamonds, which used high pressure

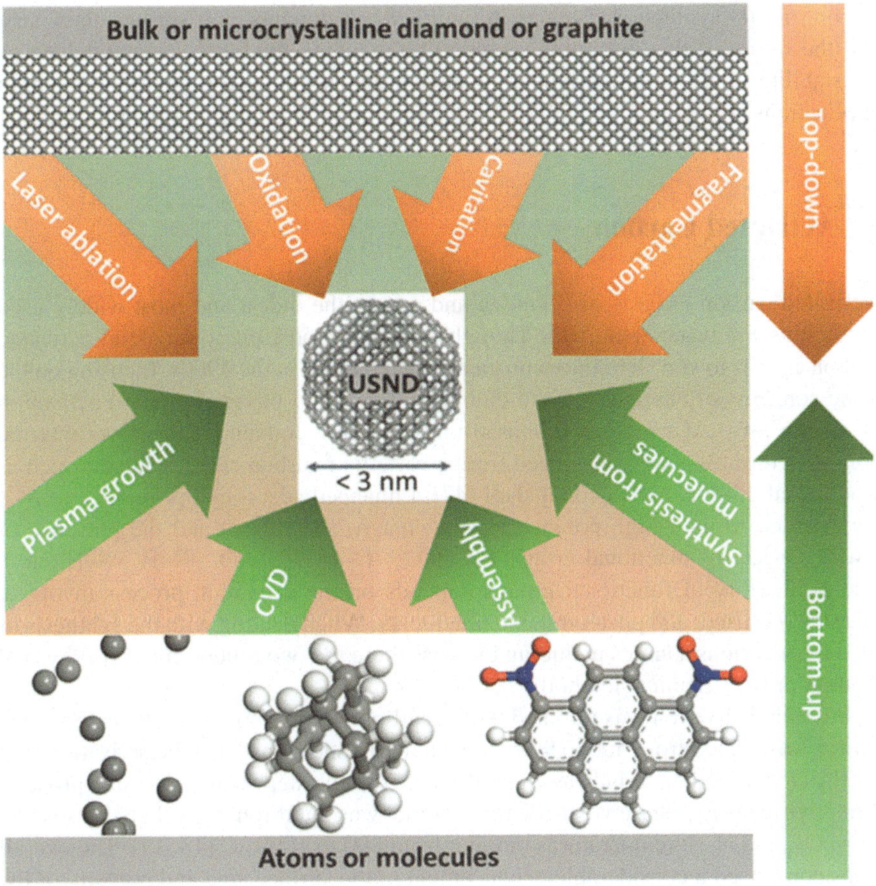

Fig. 6 Schematic illustration of NDs synthesis using top-down and bottom-up approaches [36]

and temperature in a hydraulic press in the presence of a metal catalyst and carbon precursor solution. These modern NDs are mostly far beyond 10 μm in size. Particle sizes reduced below 10 nm in the HPHT method are possible but aren't commercially scalable until now [41, 42]. Generally, synthesis of NDs should be possible in relatively low pressure and temperature, including variation from the theoretical graphite-diamond equilibrium line. However, the right choice of precursors and reaction conditions can thermodynamically and kinetically favor NDs formation at milder conditions, including a selection of a proper catalyst. For instance, a recent report showed the low-temperature, hydrothermal method to form NDs by using graphene-oxide or nitrated polycyclic aromatic hydrocarbons as a precursor [43]. A crystalline ND with an average size of 2–3 nm range was achieved even at a moderate 423 K temperature. A few other examples include the CVD technique used in diamond film synthesis. In general, adamantane and diamondoid are used as nucleation seeds for NDs film growth in CVD [44, 45]. Overall, material research has the potential to control ND synthetic strategies in tuning sizes, dopants, and surface chemistry, and the future of this discipline seems to be major in the basic science debate on the stability of sp^2 vs sp^3 carbon at the nanoscale, as well as for the development of applications.

7 Activated Carbon

Activated carbon (AC) is well-known and one of the oldest and most widely used adsorbents for water treatment. Though, the first report on commercial activated carbon enrolls to von Ostreijko who gained two patents in the 1900s, focusing on the basic concepts of chemical (metal chlorides, CO_2) and physical (steam) activation of carbon [46]. ACs are carbon-based materials that contain a full-grown internal pore structure. ACs were produced from a variety of carbon-rich precursors such as wood, coal, lignite, and coconut shell [47]. Enhanced adsorption properties of ACs are credited to their high porosity (meso-, macro and micro) and the presence of a wide variety of functional groups ($-COO^-$, –O–H, –C = O, –N–H, –S–H, etc.). The availability of functional groups depends on the activation process involved, precursor(s) used, and post-synthetic treatments. When dealing with the synthesis of ACs from various sources including biomass there are two routes, either (i) Physical method or (ii) Chemical method.

Physical activation was initiated with a carbonization step where the carbon-rich source was heated from 500 to 900 °C in an inert condition. Then the activation step for the carbonized material was either done in a single step or by a two-step process. After pyrolysis to release volatiles, the material was put through oxidative gases like CO_2, O_2, and N_2 at temperatures ranging from 800 to 1000 °C [48, 49]. The overall steps comprised a considerable enhancement in the surface area and porosity of the carbon material. However, an increase in temperature above 1200 °C reduced the ACs yield drastically along with the crumbling of pore structure [50]. Nonetheless, the properties of ACs depend on several synthetic parameters such as the nature of the activating agent, activation time, temperature, etc. [51]. Mostly, physical activation opens isolated/new pores and generates interconnected/widened pores [52].

Chemical methods of activation involve chemical agents such as H_3PO_4, $ZnCl_2$, KOH, etc. for treatment after physical activation. The major advantage of chemical activation includes well-grown pores, better yield, and low activation temperatures, despite its demerit of high cost, toxicity, and tedious synthetic steps. This helps more control over functional groups and surface functionalization, resulting improve performance on various applications [53, 54]. Potassium hydroxide was used in the activation as one of the most common methods. Which involves complex solid-state reactions and solid–liquid interface reactions including the reduction of K^+ to metallic K [51]. Li et al. reported the effect of KOH on surface area and pore volumes, both increasing up to 1:4 coal: KOH ratio and thereafter decreasing. Similar reports suggested the KOH activation in generating controlled pore sizes [55] and K^+ intercalation [56]. Alkali metal carbonates such as K_2CO_3 used activation significantly increased surface area/pore volume ratio with increased activation temperature [57, 58]. Metal chlorides such as $FeCl_3$ [59], $MgCl_2$, and $ZnCl_2$ [60, 61], and Bronsted acids like H_2SO_4, HNO_3, and H_3PO_4were also used to activate the carbon [62, 63]. An important objective of varying synthesis methodologies is to control pore shape and size, along with functional group modifications.

8 Conclusion

To summarize, some of the most exemplary formsof carbon-based nanomaterials like nanotubes, fullerenes, graphene oxide, nanodiamonds, and activated carbon-have undergone rapid development and new methods of synthesis strategies are being applied. The synthesis of carbon materials with well-defined morphology and nanostructure and adjustable surface area and pore size can be synthesized in a controlled manner.

References

1. Lu, A., Hao, G., Sun, Q., Zhang, X., Li, W.: Chemical synthesis of carbon materials with intriguing nanostructure and morphology. Chem. Synth. Appl. Graphene Carbon Mater. 115–157 (2017)
2. Smith, A.T., LaChance, A.M., Zeng, S., Liu, B., Sun, L.: Synthesis, properties, and applications of graphene oxide/reduced graphene oxide and their nanocomposites. Nano Mater. Sci. **1**, 31–47 (2019)
3. Rao, N., Singh, R., Bashambu, L.: Carbon-based nanomaterials: synthesis and prospective applications. Mater. Today Proc. **44**, 608–614 (2021)
4. Li, Y., Wu, J., Chopra, N.: Nano-carbon-based hybrids and heterostructures: progress in growth and application for lithium-ion batteries. J. Mater. Sci. **50**, 7843–7865 (2015)
5. Willner, I., Willner, B.: Biomolecule-based nanomaterials and nanostructures. Nano Lett. **10**, 3805–3815 (2010)
6. Kaur, R., Vatta, P., Kaur, M.: Carbon nanotubes: a review article. Int. J. Res. Appl. Sci. Eng. Technol. **6**, 5075–5077 (2018)
7. Iijima, S.: Helical microtubules of graphitic carbon. Nature **354**, 56–58 (1991)
8. Calvert, P.: Strength in disunity. Nature **357**, 365–366 (1992)
9. Sinnott, S.B., Andrews, R.: Carbon nanotubes: synthesis, properties, and applications. Crit. Rev. solid state Mater. Sci. **26**, 145–249 (2001)
10. Arunkumar, T., Karthikeyan, R., Ram Subramani, R., Viswanathan, K., Anish, M.: Synthesis and characterisation of multi-walled carbon nanotubes (MWCNTs). Int. J. Ambient Energy **41**, 452–456 (2020)
11. Gogotsi, Y., Libera, J.A., Yoshimura, M.: Hydrothermal synthesis of multiwall carbon nanotubes. J. Mater. Res. **15**, 2591–2594 (2000)
12. Anilkumar, P., et al.: Toward quantitatively fluorescent carbon-based "quantum" dots. Nanoscale **3**, 2023–2027 (2011)
13. Zong, J., Zhu, Y., Yang, X., Shen, J., Li, C.: Synthesis of photoluminescent carbogenic dots using mesoporous silica spheres as nanoreactors. Chem. Commun. **47**, 764–766 (2011)
14. Yang, S.-T., et al.: Carbon dots for optical imaging in vivo. J. Am. Chem. Soc. **131**, 11308–11309 (2009)

15. Yadav, J.: Fullerene: properties, synthesis and application. Res. Rev. J. Phys **6**, 1–6 (2018)
16. Wei, X., Han, Y., Long, X.: Preparation of C 60 by detonation technique. Mod. Phys. Lett. B **26**, 1250182 (2012)
17. Rümmeli, M.H., et al.: Graphene: piecing it together. Adv. Mater. **23**, 4471–4490 (2011)
18. Jiříčková, A., Jankovský, O., Sofer, Z., Sedmidubský, D.: Synthesis and applications of graphene oxide. Materials (Basel). **15**, 920 (2022)
19. Shin, D.S., et al.: Distribution of oxygen functional groups of graphene oxide obtained from low-temperature atomic layer deposition of titanium oxide. RSC Adv. **7**, 13979–13984 (2017)
20. Brodie, B.C.: XIII. On the atomic weight of graphite. Philos. Trans. R. Soc. London 249–259 (1859)
21. Marcano, D.C., et al.: Improved synthesis of graphene oxide. ACS Nano **4**, 4806–4814 (2010)
22. Ranjan, P., et al.: A low-cost non-explosive synthesis of graphene oxide for scalable applications. Sci. Rep. **8**, 12007 (2018)
23. Pei, S., Wei, Q., Huang, K., Cheng, H.-M., Ren, W.: Green synthesis of graphene oxide by seconds timescale water electrolytic oxidation. Nat. Commun. **9**, 145 (2018)
24. Goto, M., et al.: Spatially resolved 3 μm spectroscopy of Elias 1: origin of diamonds in protoplanetary disks. Astrophys. J. **693**, 610 (2009)
25. Daulton, T.L.: Extraterrestrial nanodiamonds in the cosmos. In: Ultrananocrystalline diamond. Elsevier, pp. 23–78 (2006)
26. Volkov, K.V., Danilenko, V.V., Elin, V.I.: Diamond synthesis from detonation carbon. Fiz. Goren. Vzryva **26**, 123–125 (1990)
27. Yang, G.-W., Wang, J.-B., Liu, Q.-X.: Preparation of nano-crystalline diamonds using pulsed laser induced reactive quenching. J. Phys. Condens. Matter **10**, 7923 (1998)
28. Boudou, J.-P., et al.: High yield fabrication of fluorescent nanodiamonds. Nanotechnology **20**, 235602 (2009)
29. Frenklach, M., et al.: Induced nucleation of diamond powder. Appl. Phys. Lett. **59**, 546–548 (1991)
30. Kumar, A., et al.: Formation of nanodiamonds at near-ambient conditions via microplasma dissociation of ethanol vapour. Nat. Commun. **4**, 2618 (2013)
31. Welz, S., Gogotsi, Y., McNallan, M.J.: Nucleation, growth, and graphitization of diamond nanocrystals during chlorination of carbides. J. Appl. Phys. **93**, 4207–4214 (2003)
32. Daulton, T.L., Kirk, M.A., Lewis, R.S., Rehn, L.E.: Production of nanodiamonds by high-energy ion irradiation of graphite at room temperature. Nucl. Instruments Methods Phys. Res. Sect. B Beam Interact. with Mater. Atoms **175**, 12–20 (2001)
33. Banhart, F., Ajayan, P.M.: Carbon onions as nanoscopic pressure cells for diamond formation. Nature **382**, 433–435 (1996)
34. Galimov, E.M., et al.: Experimental corroboration of the synthesis of diamond in the cavitation process. In: Doklady Physics. Springer, vol. 49, pp. 150–153 (2004)
35. Shenderova, O.A., Gruen, D.M.: Ultrananocrystalline diamond: synthesis, properties and applications. William Andrew (2012)
36. Chang, S.L.Y., Reineck, P., Krueger, A., Mochalin, V.N.: Ultrasmall Nanodiamonds: perspectives and questions. ACS Nano **16**, 8513–8524 (2022)
37. Galimov, E.M., Kudin, A.M., Skorobogatskii, V.N., Plotnichenko, V.G., Bondarev, O.L., Zarubin, B.G., Strazdovskii, V.V., Aronin, A.S., Fisenko, A.V., Bykov, I.V., Barinov, A.Y.: Experimental corroboration of the synthesis of diamond in the cavitation process. In: Doklady Physics **49**(3), 150–153 (2004)
38. Gaebel, T., et al.: Size-reduction of nanodiamonds via air oxidation. Diam. Relat. Mater. **21**, 28–32 (2012)
39. Etzold, B.J.M., et al.: Layer-by-layer oxidation for decreasing the size of detonation nanodiamond. Chem. Mater. **26**, 3479–3484 (2014)
40. Stehlik, S., et al.: High-yield fabrication and properties of 1.4 nm nanodiamonds with narrow size distribution. Sci. Rep. **6**, 38419 (2016)
41. Stehlik, S., et al.: Size and purity control of HPHT nanodiamonds down to 1 nm. J. Phys. Chem. C **119**, 27708–27720 (2015)

42. Boudou, J.-P., et al.: Fluorescent nanodiamonds derived from HPHT with a size of less than 10 nm. Diam. Relat. Mater. **37**, 80–86 (2013)
43. Shen, Y., et al.: Sub-4 nm nanodiamonds from graphene-oxide and nitrated polycyclic aromatic hydrocarbons at 423 K. ACS Nano **15**, 17392–17400 (2021)
44. Tzeng, Y.-K., et al.: Vertical-substrate MPCVD epitaxial nanodiamond growth. Nano Lett. **17**, 1489–1495 (2017)
45. Tiwari, R.N., Tiwari, J.N., Chang, L., Yoshimura, M.: Enhanced nucleation and growth of diamond film on Si by CVD using a chemical precursor. J. Phys. Chem. C **115**, 16063–16073 (2011)
46. Sontheimer, H., Crittenden, J.C., Summers, R.S.: Activated carbon for water treatment. American Water Works Association (1988)
47. Cookson, J.T., Cheremishinoff, P.N., Eclerbusch, F.: Carbon adsorption handbook. Publ, Michigan Ann Arbor Sci (1978)
48. Klei, H.E., Sahagian, J., Sundstrom, D.W.: Kinetics of the activated carbon-steam reaction. Ind. Eng. Chem. Process. Des. Dev. **14**, 470–473 (1975)
49. Pallarés, J., González-Cencerrado, A., Arauzo, I.: Production and characterization of activated carbon from barley straw by physical activation with carbon dioxide and steam. Biomass Bioenerg. **115**, 64–73 (2018)
50. Olorundare, O.F., Msagati, T.A.M., Krause, R.W.M., Okonkwo, J.O., Mamba, B.B.: Activated carbon from lignocellulosic waste residues: effect of activating agent on porosity characteristics and use as adsorbents for organic species. Water Air Soil Pollut. **225**, 1–14 (2014)
51. Sarathchandran, C., Devika, M.R., Prakash, S., Sujatha, S., Ilangovan, S.A.: Activated carbon: synthesis, properties, and applications. In: Handbook of Carbon-Based Nanomaterials. Elsevier, pp. 783–827 (2021)
52. Guo, S., et al.: Effects of CO_2 activation on porous structures of coconut shell-based activated carbons. Appl. Surf. Sci. **255**, 8443–8449 (2009)
53. Snoeyink, V.L., Weber, W.J.: The surface chemistry of active carbon; a discussion of structure and surface functional groups. Environ. Sci. Technol. **1**, 228–234 (1967)
54. Üçer, A., Uyanik, A., Aygün, ŞF.: Adsorption of Cu (II), Cd (II), Zn (II), Mn (II) and Fe (III) ions by tannic acid immobilised activated carbon. Sep. Purif. Technol. **47**, 113–118 (2006)
55. Cai, X., Tan, G., Deng, Z., Liu, J., Gui, D.: Preparation of hierarchical porous carbon aerogels by microwave assisted sol-gel process for supercapacitors. Polymers (Basel). **11**, 429 (2019)
56. Kim, B.-J., Lee, Y.-S., Park, S.-J.: A study on pore-opening behaviors of graphite nanofibers by a chemical activation process. J. Colloid Interface Sci. **306**, 454–458 (2007)
57. Hayashi, J., Yamamoto, N., Horikawa, T., Muroyama, K., Gomes, V.G.: Preparation and characterization of high-specific-surface-area activated carbons from K2CO3-treated waste polyurethane. J. Colloid Interface Sci. **281**, 437–443 (2005)
58. Li, W., Zhang, L., Peng, J., Li, N., Zhu, X.: Preparation of high surface area activated carbons from tobacco stems with K_2CO_3 activation using microwave radiation. Ind. Crops Prod. **27**, 341–347 (2008)
59. Xu, Z., et al.: Insights into the pyrolysis behavior and adsorption properties of activated carbon from waste cotton textiles by FeCl3-activation. Colloids Surfaces A Physicochem. Eng. Asp. **582**, 123934 (2019)
60. Guo, F., et al.: Synthesis of biomass carbon electrode materials by bimetallic activation for the application in supercapacitors. J. Electroanal. Chem. **844**, 105–115 (2019)
61. Arroyo-Gómez, J.J., Villarroel-Rocha, D., de Freitas-Araújo, K.C., Martínez-Huitle, C.A., Sapag, K.: Applicability of activated carbon obtained from peach stone as an electrochemical sensor for detecting caffeine. J. Electroanal. Chem. **822**, 171–176 (2018)
62. Jia, Y.F., Thomas, K.M.: Adsorption of cadmium ions on oxygen surface sites in activated carbon. Langmuir **16**, 1114–1122 (2000)
63. Jagtoyen, M., Derbyshire, F.: Activated carbons from yellow poplar and white oak by H3PO4 activation. Carbon N. Y. **36**, 1085–1097 (1998)

Surface Modification and Engineering of Nanoscale Absorbent and Their Composite

D. Aravind, P. Diwahar, M. Bharathi, K. Prakalathan, M. S. Prasanth, and Sridhar Jayavel

Abstract Surface modification and engineering is the science that deals with modification involving the surface area of any material especially solids. The act of surface modification is done by bringing changes in physical, chemical or biological characteristics that are different from the original ones. The main objective of this technique is to improve performance of materials that interacts with the environment as the interaction can degrade the surface over a period of time. The significant improvement required would be to resist damages that are caused by wear, corrosion, fatigue, creep, etc. A rising demand for an alternative material for applications in engineering sectors for the want of minimizing structural weight has resulted in a prospective growth of composites. The demand is due to requirement of reduction in structural weight that would create a tremendous positive impact on energy efficiency. The major challenge in engineering sector is to achieve a competitive structure as light as possible. The materials that are micro nano-sized, incorporates many special properties that are widely used in various fields. An adsorbent is a substance used to adsorb particles from liquid or gas resulting in the betterment of the materials, which can harm the environment. The structure of activated carbon adsorbents with its pore size and surface properties aids to provide high efficacy and safety.

D. Aravind
University Science Instrumentation Centre, Madurai Kamaraj University, Madurai, Tamil Nadu, India

P. Diwahar · M. S. Prasanth
Central Institute of Plastics Engineering and Technology, Chennai, India

M. Bharathi
Department of Mechanical Engineering, KIT, Coimbatore, India

K. Prakalathan
Department of Plastic Technology, CIPET, Chennai, India

S. Jayavel (✉)
Department of Biotechnology (DDE), Madurai Kamaraj University, Madurai, India
e-mail: jsridharbiotech@mkuniversity.ac.in

J. Tharini and S. Thomas (eds.), *Carbon Nanomaterials and their Composites as Adsorbents*, Carbon Nanostructures, https://doi.org/10.1007/978-3-031-48719-4_6

1 Introduction

One definition of a nanometer describes it as "a mystical spot on the dimension scale," describing it as "the point where the tiniest man-made gadgets meets the molecules and atoms of the physical world." Researchers and technologists have paid a lot of attention to the nanomaterials sector over the last decade, and with good reason: nanometer solid materials have become increasingly significant because of their unique qualities [1]. The advancement of nanoscale science and technology faces four significant obstacles: the synthesis of materials with controlled structure and accuracy at the molecular, or atomic level; the property characterization of structurally well-known components; the manufacturing of nanodevices; and the manipulation and integration of systems. Nanomaterials may vary greatly in their qualities due to the vast range of sizes and structures they can assume. These attributes are highly dependent on the nanomaterial's overall size, as well as its size distribution, form, and chemical make-up [2]. Nanomaterials are distinguished from traditional materials by several distinctive qualities, both chemical and physical. One of the most essential aspects of nanoparticles is the fact that the vast majority of the atoms within them have a high degree of chemical reactivity as well as the capacity to retain a large number of metal ions on the particles' surfaces [3]. Because saturation levels of the atoms present on the surface are low, they are susceptible to interaction with ions of different elements through static electricity [4]. As a result, nanoparticles have a high capacity for adsorption, therefore rendering them applicable to a diverse array of uses and purposes [5]. As a result, nanometer-sized materials may adsorb metal ions at high rates. Nanomaterials have recently been exploited as absorbing substances owing to their superior inherent qualities like high reactivity and tiny size of the particles in comparison to traditional substances such as normal sized titanium dioxide, alumina, and others [7, 8]. Since nanomaterials have large surface areas, they may strongly adsorb a wide range of chemicals, including trace metals and polar organic compounds [2, 5]. Nanomaterials may be produced using a variety of processes, including chemical vapour deposition and the sol-gel method [9–11].

Carbon stands out from the rest of the elements in the periodic table; since it seems to communicate with everyone on every level. Carbon is a very unique element. An atom of carbon may form a two electron bond, a four electron bond, or a six electron bond due to its electronic configuration of $1s^2\ 2s^2\ 2p^2$. Since carbon can hybridise its 2s and 2p orbitals in three separate ways, it has a high degree of versatility when it comes to forming connections with other atoms. This adaptability allows carbon to create bonds with atoms that have different properties. These ways are denoted by the following notations: sp^3 (for two electron bonding, tetrahedral), sp^2 (for four bonding, trigonal planar), and sp (for 6 electron bonding, linear). As of more recent times, fullerenes and carbon nanotubes have been added to the classic members of the carbon family tree, which include stable form of carbon (graphite), solid carbon (diamond), and carbon in amorphous state. Yet, in actuality, as a result of the exceptional bonding adaptability of carbon, the authentic carbon nanoforms exhibit a significantly more varied range of shapes and sizes than those depicted here [12].

2 Nanomaterials

Nanomaterials are distinguished from traditional materials by a broad range of chemical as well as physical characteristics. The majority of atoms have significant levels of chemical activity and adsorption capacity, which is a crucial factor. The surface of the nanomaterials has an excessive number of metal ions [3, 13]. Because the surface atoms are unsaturated, they can be combined by static charge with ions from other elements. As a direct consequence of this, nanoparticles have the ability to strongly adsorb a diverse spectrum of molecules, including polar chemical compounds and trace metals.

3 Classification of Nanoadsorbent

Nanomaterials are primarily categorized into categories in accordance with their inherent skin characteristic and the role they serve in adsorption applications and subsequent external functionalization (Fig. 1).

Metallic nanoparticles such as gold NPs, metallic oxide NPs, such as aluminium trioxide or titanium dioxide, nanostructured mixed oxides, such as nanostructured binary iron-titanium mixed oxide particles, and magnetic NPs, such as iron di and trioxide, are all examples of nanoparticles. According to their sorbent qualities, carbon nanoparticles, carbon nanosheets, and carbon nanotubes make up the category of carbonaceous nanomaterials (CNMs), which is another significant one (CNSs). Similarly, silicon nanomaterials (SiNMs) comprise silicon nanoparticles (SiNPs), silicon nanosheets, and silicon nanotubes (SiNTs) (SiNSs). Nanofibers

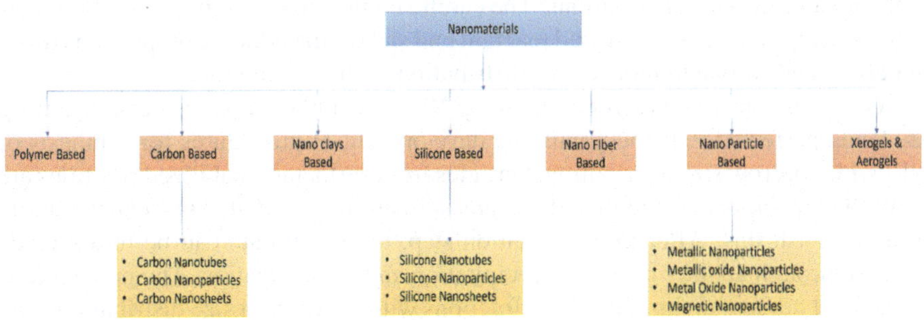

Fig. 1 Classification of nanoadsorbent

(NFs), nanoclays, polymer-based nanomaterials (PNMs), xerogels, and aerogels are additional nanomaterials for adsorption processes [4].

The physical, material, and chemical properties of NPs are all directly influenced by their intrinsic components, perceived sizes, and extrinsic skin structures [14–18]. Hence, the architecture, preparation, analysis, and uses of nanostructures have a significant impact on the advancement of the domain of nanomaterials. Many analytical techniques make significant uses of sample separation and preconcentration helped by nanoparticles [19]. Since their introduction to the field, nanoparticles (NPs) have been the subject of much research in the area of separation science, leading to several advancements in chromatographic and electrophoretic techniques for a select number of separation constituents or preconcentration medium. NPs can be added to the buffer and employed as a pseudo stationary phase in a form that are either continuously filled or partially filled, or they may serve as an interior surface covering in either a static or a dynamic state [20]. A nanoparticle is a Simplest-zero-dimensional (0 D) nano-object that has all of its distinct linear dimensions with the same magnitude [21]. NPs are typically spherical in form. Nano crystallites are NPs with an atom (or ion) configuration that is clearly organized.

4 Synthesis Techniques of Nanomaterials

Nanostructured materials are used for a variety of purposes, including IT, magnetocaloric refrigeration, medical field, automotive, optical electronic devices, and many others [22, 23]. As a result, the synthesis of such materials is becoming more important in the field of material science. Existing methods rely on high pressure, solgel, salt solvent mediated high temperature, thermal vaporization, coprecipitation using surfactants, surface capping agents, or organometallic precursor mediated growth processes, and the types of oxides that can be synthesized are rather limited [24, 25].

Physical properties of NPs are known to be highly dependent on their dimensions. Unfortunately, the majority of the currently known synthesis procedures provide NPs with rather broad size distributions. The careful control of reaction settings (Temperature, stirring speed, duration, and reactant and stabilizing additive proportions) does not always allow one to narrow this distribution to the desired range.

As a result, techniques for separating NPs into rather monodisperse fractions are being perfected in tandem with the development of procedures for synthesis of NPs with a narrow size distribution. Particles from surfactant-stabilized solutions are controlled precipitated, and then the coarsest fraction is separated by centrifugation, to accomplish this. After the decantation step, the precipitate can be re-dissolved and put through another round of precipitation and centrifugation. The procedure is carried out numerous times until NP fractions with the desired size distributions and dispersion degrees are obtained [26].

Key methods for the formation of NPs can be distinguished based on the type of precursor used and the characteristics of their processing, as follows.

(i) transformations of NPs with changes in composition (ii) dispersion preparation from macroscopic materials and (iii) chemical synthesis, i.e., targeted change in the substance composition with termination (in some way) of the nascent phase growth at the nanosized stage [4, 27].

4.1 Synthesis of Carbon Nanotubes

Carbon nanotube synthesis processes are divided into three categories: laser ablation, catalytic arc discharge, and Chemical Vapour Discharge [28]. The latter is perhaps the most intriguing for analytical purposes, as a result of the fact that it makes it possible to produce vertically aligned carbon nanotubes on a surface. For the purpose of scaling up production, several preparation methods have been tried and tested, all of which include carrying out the synthesis at temperatures that are relatively low while maintaining high yields and purity [29–31]. In most cases, the formation of carbon nanotubes (CNTs) is achieved by the catalytic dissociation of organic precursors or graphite. This process also results in the formation of nontubular carbons (NTCs). Because the majority of the time the NTCs are created after the catalyst is depleted, most CNT preparations have a surface that is coated with NTCs. When compared to CNTs, NTCs often feature porous architectures, which results in a distinct retention and trapping mechanism that may be regulated by diffusion. Although these NTCs do not possess the good sorbent features of the CNTs, the purity of the CNTs is an essential component that impacts their overall efficacy as a sorbent [32, 33]. More sorption sites on the wall and interstitial gaps between the tubes of the pure CNTs make it simple to perform adsorption and desorption, which in turn improves preconcentration. The CNTs' availability is decreased by the NTC cover because the sorbate must diffuse through the NTC in order to reach the CNTs. Nevertheless, the porous structure of NTC slows down both the adsorption and desorption processes due to the mass transfer constraints introduced by the structure.

The most generally used procedures for removing NTCs and residual catalysts are acid treatment and gas phase oxidation [34–36]. Oxygen-containing functional groups, such as OH, C=O, and COOH, are produced during these processes, enhancing the polarity, hydrophilicity, and ion-exchange capacity of the CNTs. When activated carbon is oxidised in the gas phase, hydroxyl and carbonyl surface groups become more concentrated, whereas carboxylic acid content becomes more prominent when oxidised in the liquid phase [37]. The quantity of carboxyl and lactone groups on CNT treated with nitric acid was found to be greater than when similar procedure was carried out with H_2O_2 and $KMnO_4$ [38]. Nonetheless, CNT fragmentation (shortening) and defect creation in the graphitic network are two of the most significant problems associated with processes involving acid oxidation [39]. It has been discovered that the high ultrasonic power that is often used during the oxidation process in order to disperse CNT agglomerates also contributes to the process of fragmentation. This result follows hot on the heels of the one that found acid content and exposure length to be key factors to fragmentation. Others include gentler treatments, such as microwave treatment with dilute acids and a chelating agent [40–42]. Hence, it is vital to find a middle ground between the circumstances of the purifying process and the damage done to the CNTs.

As a natural byproduct of the manufacturing process, carbon nanotubes often include traces of amorphous carbon as well as metal catalyst residues. In addition, since the process creates carbon nanotubes of varying diameters, an additional step of purification is required before these nanotubes may be used in a wide range of applications. Thus, there is a need for analytical techniques in order for the nanotubes to be characterised and then purified. There are several methods for processing the raw material, techniques such as acid washing, followed by centrifugation, resuspension in surfactant or polymeric media, cross-flow filtration, and flocculation using surfactants are all included [43–45]. In spite of the purification process, the produced sample of carbon nanotubes still consists of a heterogeneous combination of SWNT and MWNT of varying lengths and diameters. The electrical properties of nanotubes may be significantly influenced by their geometrical features, which also have an impact on how well they disperse in organic or aqueous fluids and how they interact with other substances like surfactants, polymers, or aromatic chemicals. For the purpose of purifying fluorescent SWNT fragments, gel electrophoresis has been explicitly suggested as a preparative method. As a consequence, there are three major fractions. The process is time-consuming, and additional fractionation of previously isolated bands is necessary [46]. It has been suggested that capillary electrophoresis or liquid chromatography be used for this second stage. Despite carbon nanotubes' poor solubilization in water, capillary electrophoresis has shown promise for removing contaminants and fractionating SWNTs based on size and length [47]. In order to use capillary electrophoresis to determine nanotube purity, it is necessary to first solubilize the nanotubes [47]. Suspending SWNTs in polyvinylpyrrolidone or sodium dodecyl sulfate has been shown to be an effective way to overcome this limitation [48, 49]. The electrophoretic peaks observed were smaller, suggesting improved performance, with the latter. In both investigations, Raman technique was used for the detection. The use of atomic force microscopy allowed for the verification of the various fractions' levels of purity. Because of the heterogeneous nature of the nanotube suspension, this approach suffers from a poor level of repeatability across runs in terms of the numbers and locations of the peaks. This is the primary drawback associated with this technology [47].

Despite this, the CNTs that are generated by any of these techniques often include NTC in addition to remnants of previous catalysts here and there such as Fe, Co, and Ni. Metallic impurities may account for up to 40% of the total weight. The oxidation of CNTs, which may change their sorbent properties, and the generation of defects during purification are also factors to be taken into account [4].

Techniques for the dispersion and solubilization of carbon nanotubes may, in general, be divided up into three distinct categories. These categories are I dispersion following oxidative acid treatments, (ii) noncovalent stabilisation, and (iii) covalent stabilisation. Common solubilization methods include oxidative acid treatment stages like refluxing in diluted nitric acid or refluxing/sonicating in a concentrated

H_2SO_4/HNO_3 combination [50]. A black supernatant solution that contains carboxylated carbon nanotubes is formed after an oxidative acid treatment and several cycles of washing and centrifugation. On the other hand, this type of solubilization may cause surface flaws, and it can also cause tubes to become shorter than they were before. Because of this, several researchers have advised for the use of a noncovalent method to stabilize carbon nanotubes in solution. This is due to the fact that the structures and characteristics of carbon nanotubes are often maintained after they have been dispersed. While using this method, the carbon nanotubes are derivatized or complexed with micelles, polymers, or other aggregation systems in order to solubilize and disperse them in an aqueous environment. These aggregates contain a hydrophobic core, which is the location of the nanotube, and a hydrophilic surface, which allows them to dissolve in water. The nanotube is found in the core of the aggregates [51]. Sodium dodecyl sulphate (also known as SDS), Triton X-100, and sodium dodecylbenzene sulfonate (also known as SDBS) are examples of the types of surfactants that are often used for nanotube dispersion. In order to make nanotubes more soluble, an organic functionalization of the surface of the nanotubes is required. This results in the product having a high degree of solubility. Controllable processes for altering carbon nanotubes with biologic or bioactive species have been developed, which is a major step forward in the direction of using carbon nanotubes in the biological and biomedical domains [52]. These improvements, which are especially significant for the creation of sensors in analytical research, have been documented in detail elsewhere. You may find those reports here and here [47]. It is possible to modify the CNTs' physiochemical properties via the process of functionalization [53, 54]. Changes in polarity, hydrophilicity, and other features of the CNT surface may be brought about by the presence of a functional group bonded covalently to the surface. The presence of functional groups may also change the diffusional barrier of CNT surfaces, making them less accessible and lowering their affinity for certain analytes. Also, functionalization improves interaction with polymers and other materials, leading to the easier creation of composites that may serve as microtrapping substrates. Polymers and solgel immobilisation are the typical examples [55]. The specifics of CNT functionalization have been discussed in depth in a several works [56, 57].

For the functionalization of CNTs and the synthesis of hybrid structures with intriguing features, a range of covalent or noncovalent methods have been established [58–65]. For instance, covalent functionalization may be utilized to produce functional groups such as COOH or NH2, which can substantially modify the chemical characteristics of CNTs and boost their performance in SPE. This can be done in order to improve the performance of CNTs in the context of solid phase extraction. This may be especially significant to the study of polar pharmaceutical compounds that have a broad range of pKa, which is one of the situations in which pH changes are often required to achieve good extraction efficiency. As a result, it becomes necessary to do several extractions at various pH levels, which presents a significant practical challenge when analyzing mixtures [66–68].

4.2 Laser Ablation Method

Nanotubes made of a single wall of carbon, also known as SWCNT have been produced by arcing or laser ablating carbon rod in the presence of a metal catalyst in an environment containing inert gas [70]. Nitrogen gas has been utilised in the production of BN or B–C–N composite nanotubes [73]. Laser ablation is one of the methods used for synthesizing carbon nanotubes. In this process, a high-energy laser is directed at a carbon target, typically in the form of graphite or a carbon-containing substrate. The laser's energy vaporizes the target material, and the resulting vapor condenses to form nanotubes. Nitrogen-Containing Materials Nitrogen can be incorporated into the carbon lattice of nanotubes during their formation, potentially altering their properties. This can have both positive and negative effects, depending on the application. Nitrogen-doped carbon nanotubes can exhibit different electronic and catalytic properties compared to undoped nanotubes. The concern about nitrogen-containing materials making SWCNTs more prone to instability likely arises from the potential disruption of the carbon nanotube's hexagonal lattice structure. The arrangement of carbon atoms in a nanotube is crucial for its mechanical, electrical, and other properties [69]. Experiments involving laser ablation were carried out inside of an electrical tube furnace that had been heated to a temperature of 1200 °C [75]. Nitrogen was employed as the carrier gas, and the flow rate was around 300 standard cubic centimeters per minute. During the laser ablation process, the pressure inside the quartz reactor tube was maintained at approximately 500 Torr. On the surface of the target, the beam with the second harmonic of aNd-yttrium-aluminium-garnet (YAG) laser with a pulse width of 8 nanoseconds was focused to achieve a concentration of energy within approximately 3 J/cm^2 per pulse. For the synthesis of SWCNT, a graphite target containing 1.2% (Ni1Co) was utilized [72].

4.3 Chemical Vapour Deposition Method

The creation of carbon nanotubes can be accomplished by a wide variety of processes, including arc discharge, CVD, laser ablation, flame synthesis, HiPco, electrolysis, pyrolysis, and flame synthesis are some of the processes that are used to produce carbon nanotubes and many others [77]. On the other hand, the chemical vapour decomposition (CVD) approach is currently the standard procedure that is widely accepted in the synthesis of CNTs due to the fact that it is both straightforward and inexpensive [77]. This technology was first developed in the 1960s and 1970s, and it has since been utilised, with great success, in the manufacturing of carbon fibres as well as carbon nanofibres [80]. In 1996, chemical vapour decomposition (CVD) was identified as a potentially useful approach for the manufacturing and synthesis of CNTs on a wide scale [81]. When compared to the methods of arc discharge and laser ablation, the CVD technique is regarded as the superior choice for the manufacturing of carbon nanotubes on a large scale and with a high degree of purity [82].

4.4 *Arc Discharge Method*

Arc discharge is a method that can be utilized during the production of carbon nanotubes (CNTs) that involves the use of higher temperatures (over 1700 degrees Celsius). This method typically results in the creation of CNTs that have less structural faults when compared to other procedures [83]. When all of the necessary conditions for development are met, the synthesis of MWNTs through arc discharge is a very smooth process. The most common approach uses a DC arc discharge within a space that was pressurised and filled with helium lower than that of the atmosphere between two graphite electrodes, which are often cooled by water, and have diameters ranging from 6 to 12 millimetres. Despite this, some more works involving the utilization of hydrogen or methane atmosphere have also been described [84]. For the synthesis of SWNTs, the arc discharge method is still used, although typically in conjunction with an updated strategy. The FHarc discharge method was discovered by Chen and colleagues. In this process, the production of SWNTs involves a hydrogen DC arc discharge coupled with the evaporation of a carbon anode that has 1% iron catalyst in a mixture of hydrogen and argon gas [84].

5 Surface Modification

5.1 *Surface Physical Modification*

In general, the term "surface physical modification" is used to describe a form of surface modification of nanoscales powders that does not involve the use of surface modifiers. Surface physical modification techniques include radiation treatment, plasma treatment ultrasonic treatment n, thermal treatment, electrochemical treatment, and other similar techniques [84].

5.2 *Ultrasonic Treatment*

The aggregation of nanoparticles can be effectively reduced through the use of the ultrasonic dispersion method. Both the nano-effect energy that is present between nanoparticles and the agglomeration of nanoparticles can be significantly reduced, and the latter can be effectively prevented through the utilization of local regions of high temperatures and pressures, as well as powerful shock waves and micro jets that are generated through ultrasonic cavitation. And ensure that it is completely disseminated [85].

5.3 Radiation Treatment

Radiation technology is a relatively recent field of study that focuses on the investigation, development, and practical putting into practice the physical laws, chemical reactions, and biological processes that are triggered when ionizing radiation comes into contact with matter. Irradiation with high energy creates active spots on the surface of the nanoparticle, and then an A monomer that is active organically is grafted onto the surface of the polymer film. This process modifies the micro- and nanoparticles' surface characteristics, hence enhancing their compatibility with the polymer substrate [86].

6 Surface Chemical Modification

Surface chemical modification is a process that can be used to alter the nanoparticle's surface, with its structure and state. Surface modification aims to achieve this goal, and surface chemical modification is one way to do so. Surface chemical modification refers to the intentional alteration of a nanoparticle's surface characteristics. This can include changing its chemical composition, reactivity, and interactions with other substances. The main goal of surface modification is to tailor nanoparticles' properties to suit particular applications. This could involve enhancing stability, improving dispersibility, attaching functional groups, or enabling interactions with other materials [87].

6.1 Estherification Method

Esterification is the name given to the chemical reaction that takes place when metal oxides react with alcohols. The transformation of a nanoparticle's surface from one that is hydrophilic and oleophobic into one that is lipophilic and hydrophobic is the most crucial phase of the esterification process that is used to modify the surface of nanoparticles. It is of utmost significance in a variety of practical applications [88].

6.2 *Coupling Agent Method*

The "Coupling Agent Method" is a surface modification technique that involves using a chemical compound with an amphoteric structure, which means it has both acidic and basic functional groups. The primary purpose of this coupling agent is to facilitate the bonding and interaction between different materials, often involving polymers..A strong chemical bond can be formed between two parts of the group in the molecule by one part of the group reacting includes a variety of functional groups on the surface of the powder, and another part of the group that is capable of carrying out some chemical reaction or physical entanglement with the organic polymer. Both of these processes are possible. Because of this, the coupling agent is sometimes referred to as a "molecular bridge." Its goal is to enhance the interaction at the interface between the inorganic component and the organic component both together, which ultimately results in a significant increase in the efficiency of the composite material. Coupling agents that are commonly utilized include, but are not limited to, zirconium aluminate coupling agents, aluminate coupling agents, silane coupling agents, and titanate coupling agents. Other agents that are similar in nature also fall into this category [89].

6.3 *Surface Graft Modification*

Due to the strong sp^2 C–C bonds, carbon nanomaterials possess higher inert surfaces. Surface Graft Modification technique is used to graft functional groups onto their surfaces to generate new properties or to improve the interaction with different materials. Some of the common functional groups include amino ($-NH_2$), hydroxyl (–OH), Carboxyl (–COOH) etc., can be introduced using various surface graft modification methods (Fig 2). We can modify the properties of the polymer grafted particles by choosing appropriate grafting monomers and grafting conditions.

In the present day, there are many strategies for surface modification; nevertheless, there are still relatively few approaches to handle the problem on a fundamental level; hence, additional study is required. In addition, the usage of composite modifiers is something that needs to be taken into consideration when working on the surface modification of nanoparticles [90].

7 Conclusion

As the materials are micro nano-sized, particularly in its nano-state, have many special properties, which have attracted researchers. This resulted in a variety of nano- and nano-composites with advanced functions and improved performance for deployment in numerous fields. Further, surface engineering techniques makes

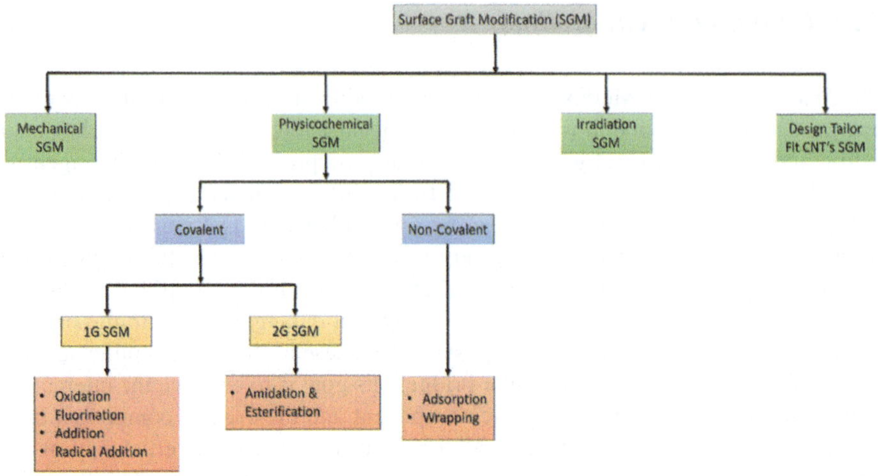

Fig. 2 Classification of surface graft modification methods

it applicable for automobile, aerospace, biomedical, quick prototyping industries. The nano-particles are accumulated with large specific surface area, high surface energy which makes them highly active, and easy to agglomerate. But this agglomerated secondary particle hinders to exert the nano-effects, making the material less than ideal. Therefore, in order to improve the dispersibility of material and improve their binding force with other components, surface modification of the nanoparticles is highly advisable. Nanoparticle surface modification has become even more effective recently as the field has witnessed a paradigm shift age-old electroplating technique to latest techniques such as vapor phase deposition, thermal spraying, plasma welding, laser welding, and welding with electron, microwave, solar beams, synchrotron radiation, etc... These control the internal stress, increasing the repulsive force, decreasing the gravitation, which gives the nanoparticle a new function to meet the current needs of material processing and its applications.

References

1. Klabunde, K.J., Richards, R.M. (eds.): Nanoscale Materials in Chemistry. Wiley (2009)
2. Wang, Z.L.: Characterizing the structure and properties of individual wire-like nanoentities. Adv. Mater. **12**(17), 1295–1298 (2000)
3. Kalfa, O.M., Yalçınkaya, Ö., Türker, A.R.: Synthesis of nano B_2O_3/TiO_2 composite material as a new solid phase extractor and its application to preconcentration and separation of cadmium. J. Hazard. Mater. **166**(1), 455–461 (2009)
4. Khajeh, M., Laurent, S., Dastafkan, K.: Nanoadsorbents: classification, preparation, and applications (with emphasis on aqueous media). Chem. Rev. **113**(10), 7728–7768 (2013)
5. Cui, Y., et al.: ICP-AES determination of trace elements after preconcentrated with p-dimethylaminobenzaldehyde-modified nanometer SiO_2 from sample solution. Microchem. J. **83**(1), 35–41 (2006)

6. Türker, A.R.: New sorbents for solid-phase extraction for metal enrichment. Clean-Soil, air, water **35**(6), 548–557 (2007)
7. Lemos, V.A., et al.: New materials for solid-phase extraction of trace elements. Appl. Spectroscopy Rev. **43**(4), 303–334 (2008)
8. Zhang, L., et al.: Studies on the capability and behavior of adsorption of thallium on nano-Al_2O_3. J. Hazard. Mater. **157**(2–3), 352–357 (2008)
9. Kim, B.-H., et al.: Preparation of TiO_2 thin film by liquid sprayed mist CVD method. Mater. Sci. Eng.: B **107**(3), 289–294 (2004)
10. Wu, X.-M., et al.: Preparation, characterization, and low-temperature heat capacities of nanocrystalline TiO_2 ultrafine powder. J. Solid State Chem. **156**(1), 220–224 (2001)
11. Thiruchitrambalam, M., Palkar, V.R., Gopinathan, V.: Hydrolysis of aluminium metal and sol–gel processing of nano alumina. Mater. Lett. **58**(24), 3063–3066 (2004)
12. Tagmatarchis, N., (ed.): Advances in Carbon Nanomaterials: Science and Applications. CRC Press (2012)
13. Liu, Y., Liang, P., Guo, L.: Nanometer titanium dioxide immobilized on silica gel as sorbent for preconcentration of metal ions prior to their determination by inductively coupled plasma atomic emission spectrometry. Talanta **68**(1), 25–30 (2005)
14. Mirkin, C.A., et al.: A DNA-based method for rationally assembling nanoparticles into macroscopic materials. Nature **382**(6592), 607–609 (1996)
15. Taleb, A., Petit, C., Pileni, M.P.: Synthesis of highly monodisperse silver nanoparticles from AOT reverse micelles: a way to 2D and 3D self-organization. Chem. Mater. **9**(4), 950–959 (1997)
16. Ascencio, J.A., et al.: Transmission electron microscopy and theoretical analysis of AuCu nanoparticles: atomic distribution and dynamic behavior. Microsc. Res. Tech. **69**(7), 522–530 (2006)
17. Osaka, T., et al.: Synthesis of magnetic nanoparticles and their application to bioassays. Anal. Bioanal. Chem. **384**, 593–600 (2006)
18. Zhong, W.: Nanomaterials in fluorescence-based biosensing. Anal. Bioanal. Chem. **394**, 47–59 (2009)
19. Howard, A.G., Statham, P.J.: Inorganic Trace Analysis: Philosophy and Practice. Wiley Incorporated (1993)
20. Wu, C.-S., Liu, F.-K., Ko, F.-H.: Potential role of gold nanoparticles for improved analytical methods: an introduction to characterizations and applications. Anal. Bioanal. Chem. **399**, 103–118 (2011)
21. Morozov, I.D., Trusov, L.I., Chizhik, S.P.: Ul'tradispersnyemetallicheskiesredy (Ultradispersed Metal Media). Atomizdat, Moscow (1977)
22. O'Brien, S., Brus, L., Murray, C.B.: Synthesis of monodisperse nanoparticles of barium titanate: toward a generalized strategy of oxide nanoparticle synthesis. J. Am. Chem. Soc. **123**(48), 12085–12086 (2001)
23. Hyeon, T., et al.: Synthesis of highly crystalline and monodisperse cobalt ferrite nanocrystals. The J. Phys. Chem. B **106**(27), 6831–6833 (2002)
24. Mao, Y., Banerjee, S., Wong, S.S.: Large-scale synthesis of single-crystalline perovskite nanostructures. J. Am. Chem. Soc. **125**(51), 15718–15719 (2003)
25. Song, Q., John Zhang, Z.: Shape control and associated magnetic properties of spinel cobalt ferrite nanocrystals. J. Am. Chem. Soc. **126**(19), 6164–6168 (2004)
26. Gubin, S.P.: What is nanoparticle? Development trends for nanochemistry and nanotechnology. Ross. Khim. Zh. **44**(6), 23 (2000)
27. Gubin, S.P., et al.: Magnetic nanoparticles: preparation, structure and properties. Russian Chem. Rev. **74**(6), 489 (2005)
28. Harris, P.J.F.: Carbon Nanotubes and Related Structures. Cambridge University Press, Cambridge, UK (1999)
29. Popov, V.N.: Carbon nanotubes: properties and application. Mater. Sci. Eng. R. Rep. **43**(3), 61–102 (2004)

30. Karwa, M., Iqbal, Z., Mitra, S.: Scaled-up self-assembly of carbon nanotubes inside long stainless steel tubing. Carbon **44**(7), 1235–1242 (2006)
31. Chung, J., Lee, J.: Nanoscale gap fabrication and integration of carbon nanotubes by micromachining. Sens. Actuators, A **104**(3), 229–235 (2003)
32. Brukh, R., Mitra, S.: Mechanism of carbon nanotube growth by CVD. Chem. Phys. Lett. **424**(1–3), 126–132 (2006)
33. Brukh, R., Sae-Khow, O., Mitra, S.: Stabilizing single-walled carbon nanotubes by removal of residual metal catalysts. Chem. Phys. Lett. **459**(1–6), 149–152 (2008)
34. MacKenzie, K., Dunens, O., Harris, A.T.: A review of carbon nanotube purification by microwave assisted acid digestion. Sep. Purif. Technol. **66**(2), 209–222 (2009)
35. Heras, A., et al.: Electrochemical purification of carbon nanotube electrodes. Electrochem. Commun. **11**(7), 1535–1538 (2009)
36. Hu, H., et al.: Nitric acid purification of single-walled carbon nanotubes. The J. Phys. Chem. B **107**(50), 13838–13842 (2003)
37. Wang, Y., Iqbal, Z., Malhotra, S.V.: Functionalization of carbon nanotubes with amines and enzymes. Chem. Phys. Lett. **402**(1–3), 96–101 (2005)
38. Aitchison, T.J., et al.: Purification, cutting, and sidewall functionalization of multiwalled carbon nanotubes using potassium permanganate solutions. The J. Phys. Chem. C **111**(6), 2440–2446 (2007)
39. Avilés, F., et al.: Evaluation of mild acid oxidation treatments for MWCNT functionalization. Carbon **47**(13), 2970–2975 (2009)
40. Wang, Y., Iqbal, Z., Mitra, S.: Microwave-induced rapid chemical functionalization of single-walled carbon nanotubes. Carbon **43**(5), 1015–1020 (2005)
41. Wang, Y., Iqbal, Z., Mitra, S.: Rapidly functionalized, water-dispersed carbon nanotubes at high concentration. J. Am. Chem. Soc. **128**(1), 95–99 (2006)
42. Wang, Y., Iqbal, Z., Mitra, S.: Rapid, low temperature microwave synthesis of novel carbon nanotube–silicon carbide composite. Carbon **44**(13), 2804–2808 (2006)
43. Bandow, S., et al.: Purification of single-wall carbon nanotubes by microfiltration. The J. Phys. Chem. B **101**(44), 8839–8842 (1997)
44. Dillon, A.C., et al.: A simple and complete purification of single-walled carbon nanotube materials. Adv. Mater. **11**(16), 1354–1358 (1999)
45. Rinzler, A.G., et al.: Large-scale purification of single-wall carbon nanotubes: process, product, and characterization. Appl. Phys. A: Mater. Sci. Process. **67**(1) (1998)
46. Xu, X., et al.: Electrophoretic analysis and purification of fluorescent single-walled carbon nanotube fragments. J. Am. Chem. Soc. **126**(40), 12736–12737 (2004)
47. Valcarcel, M., et al.: Present and future applications of carbon nanotubes to analytical science. Anal. Bioanal. Chem. **382**, 1783–1790 (2005)
48. Doorn, S.K., et al.: Capillary electrophoresis separations of bundled and individual carbon nanotubes. The J. Phys. Chem. B **107**(25), 6063–6069 (2003)
49. Doorn, S.K., et al.: High resolution capillary electrophoresis of carbon nanotubes. J. Am. Chem. Soc. **124**(12), 3169–3174 (2002)
50. Liu, J., et al.: Fullerene pipes. Science **280**(5367), 1253–1256 (1998)
51. Matarredona, O., et al.: Dispersion of single-walled carbon nanotubes in aqueous solutions of the anionic surfactant NaDDBS. The J. Phys. Chem. B **107**(48), 13357–13367 (2003)
52. Lin, Y., et al.: Advances toward bioapplications of carbon nanotubes. J. Mater. Chem. **14**(4), 527–541 (2004)
53. Dyke, C.A., Tour, J.M.: Covalent functionalization of single-walled carbon nanotubes for materials applications. J. Phys. Chem. A **108**(51), 11151–11159 (2004)
54. Vazquez, E., Prato, M.: Carbon nanotubes and microwaves: interactions, responses, and applications. Acsnano **3**(12), 3819–3824 (2009)
55. Fang, X.-L., et al.: From self-assembled microspheres to self-templated nanotubes: morphologies and properties of sulfur-bridged fluoranthene-based organic materials. Chem. Mater. **21**(24), 5763–5771 (2009)

56. Meng, L., Fu, C., Lu, Q.: Advanced technology for functionalization of carbon nanotubes. Progr. Nat. Sci. **19**(7), 801–810 (2009)
57. Wang, S.: Optimum degree of functionalization for carbon nanotubes. Curr. Appl. Phys. **9**(5), 1146–1150 (2009)
58. Niyogi, S., et al.: Chemistry of single-walled carbon nanotubes. Acc. Chem. Res. **35**(12), 1105–1113 (2002)
59. Zhang, L., et al.: Sidewall functionalization of single-walled carbon nanotubes with hydroxyl group-terminated moieties. Chem. Mater. **16**(11), 2055–2061 (2004)
60. Hirsch, A.: Functionalization of single-walled carbon nanotubes. AngewandteChemie Int. Edition **41**(11), 1853–1859 (2002)
61. Banerjee, S., Kahn, M.G.C., Wong, S.S.: Rational chemical strategies for carbon nanotube functionalization. Chem.–A Eur. J. **9**(9), 1898–1908 (2003)
62. Liu, Y., Li, Y., Yan, X.-P.: Preparation, characterization, and application of L-cysteine functionalized multiwalled carbon nanotubes as a selective sorbent for separation and preconcentration of heavy metals. Adv. Func. Mater. **18**(10), 1536–1543 (2008)
63. Lu, C., Chiu, H.: Chemical modification of multiwalled carbon nanotubes for sorption of Zn2+ from aqueous solution. Chem. Eng. J. **139**(3), 462–468 (2008)
64. Xiao, Y., et al.: Anti-HER2 IgY antibody-functionalized single-walled carbon nanotubes for detection and selective destruction of breast cancer cells. BMC Cancer **9**, 1–11 (2009)
65. Ntim, S.A., et al.: Effects of polymer wrapping and covalent functionalization on the stability of MWCNT in aqueous dispersions. J. Colloid Interface Sci. **355**(2), 383–388 (2011)
66. Hylton, K., Sangwan, M., Mitra, S.: Microscale membrane extraction of diverse antibiotics from water. Analyticachimicaacta **653**(1), 116–120 (2009)
67. Bae, C., et al.: Template directed oxide nanotubes: synthesis, characterization, and applications. Chem. Mater **20**, 756–767 (2008)
68. Moynihan, S., et al.: Template synthesis of highly oriented polyfluorene nanotube arrays. Chem. Mater. **20**(3): 996–1003 (2008)
69. Awasthi, K., Srivastava, A., Srivastava, O.N.: Synthesis of carbon nanotubes. J. Nanosci. Nanotechnol. **5**(10), 1616–1636 (2005)
70. Iijima, S., Ichihashi, T.: Single-shell carbon nanotubes of 1-nm diameter. Nature **363**(6430), 603–605 (1993)
71. Bethune, D.S., et al.: Cobalt-catalysed growth of carbon nanotubes with single-atomic-layer walls. Nature **363**(6430), 605–607 (1993)
72. Raidongia, K.: Synthesis and characterization of inorganic nanorods and nanotubes and Kirkendall effect-induced transformations of metal nanowires of oxide or chalcogenide nanotubes. Diss. Jawaharlal Nehru Centre for Advanced Scientific Research (2007)
73. Loiseau, A., Pascard, H.: Synthesis of long carbon nanotubes filled with Se, S, Sb and Ge by the arc method. Chem. Phys. Lett. **256**(3), 246–252 (1996)
74. Niu, C., Lu, Y.Z., Lieber, C.M.: Experimental realization of the covalent solid carbon nitride. Science **261**(5119), 334–337 (1993)
75. Thess, A., et al.: Crystalline ropes of metallic carbon nanotubes. Science **273**(5274), 483–487 (1996)
76. Shah, K.A., Tali, B.A.: Synthesis of carbon nanotubes by catalytic chemical vapour deposition: a review on carbon sources, catalysts and substrates. Mater. Sci. Semicond. Process. **41**, 67–82 (2016)
77. Yahya, N., et al.: Synthesis of carbon nanostructures by CVD method. In: Carbon and Oxide Nanostructures: Synthesis, Characterisation and Applications, pp. 23–49 (2011)
78. Baker, R.T.K.: Catalytic growth of carbon filaments. Carbon **27**(3), 315–323 (1989)
79. Cassell, A.M., et al.: Large scale CVD synthesis of single-walled carbon nanotubes. The J. Phys. Chem. B **103**(31), 6484–6492 (1999)
80. Su, M., Zheng, B., Liu, J.: A scalable CVD method for the synthesis of single-walled carbon nanotubes with high catalyst productivity. Chem. Phys. Lett. **322**(5), 321–326 (2000)
81. Palizdar, M., et al.: Investigation of Fe/MgO catalyst support precursors for the chemical vapour deposition growth of carbon nanotubes. J. Nanoscience and nanotechnology **11**(6), 5345–5351 (2011)

82. Prasek, J., et al.: Methods for carbon nanotubes synthesis. J. Mater. Chem. **21**(40), 15872–15884 (2011)
83. Parkansky, N., et al.: Single-pulse arc production of carbon nanotubes in ambient air. J. Phys. D Appl. Phys. **37**(19), 2715 (2004)
84. Wu, Y., et al.: In situ synthesis of graphene/single-walled carbon nanotube hybrid material by arc-discharge and its application in supercapacitors. Nano Energy **1**(6), 820–827 (2012)
85. Zhong, J., Isayev, A.I., Zhang, X.: Ultrasonic twin screw compounding of polypropylene with carbon nanotubes, graphenenanoplates and carbon black. Eur. Polymer J. **80**, 16–39 (2016)
86. Tolvanen, A., et al.: Modifying the electronic structure of semiconducting single-walled carbon nanotubes by Ar+ ion irradiation. Phys. Rev. B **79**(12), 125430 (2009)
87. Rehman, A., Park, M., Park, S.-J.: Current progress on the surface chemical modification of carbonaceous materials. Coatings **9**(2), 103 (2019)
88. Abuilaiwi, F.A., et al.: Modification and functionalization of multiwalled carbon nanotube (MWCNT) via Fischer esterification. The Arab. J. Sci. Eng. **35**(1), 37–48 (2010)
89. Ma, P.C., Kim, J.-K., Tang, B.Z.: Functionalization of carbon nanotubes using a silane coupling agent. Carbon **44**(15), 3232–3238 (2006)
90. Kim, S.W., et al.: Surface modifications for the effective dispersion of carbon nanotubes in solvents and polymers. Carbon **50**(1), 3–33 (2012)

Application and Research Progress of Nanomaterials as Adsorbents in Environment Field

Hailan Qin, Yunkang Liu, Huan Liu, Siyuan Di, and Shukui Zhu

Abstract Nanoscale materials represent a new class of materials with extremely broad application prospects. Their development and utilization are intricately intertwined with societal progress. Nanomaterial adsorbents possess distinctive attributes, including a large specific surface area, a quantum size effect, and special surface properties. Therefore, it exhibits excellent adsorption performance. Compared with other traditional adsorbents, the application prospects of nanomaterial adsorbents in the environmental field are more attractive. This chapter elucidates the types, synthesis and, characterization of nanomaterial adsorbents according to their properties. Combined with the application of relevant examples in the environment, the corresponding nanoscale adsorbents are briefly introduced. Furthermore, the application of nanomaterial adsorbents in environment treatment was prospected.

Keywords Nanoscale materials · Adsorbents · Environmental pollution · Pollutants

1 Introduction

In recent years, research on nanoscale materials and nanoscale technology has received extensive attention [1, 2]. The late 1980s marked the ascent and rapid evolution of nanotechnology, playing a pivotal role in societal advancement, economic rejuvenation, and national progress. Amidst these developments, nanoscale materials emerged as key protagonists. Currently, nanometer materials are widely used in diverse technical domains, including aerospace, medicine, energy, environment, and sensors. At the same time, because of their large chemical activity and surface energy, nanoscale materials readily interact with foreign atoms to form stable structures, rendering them valuable as adsorbents [3]. Therefore, this chapter will comprehensively explore various facets of nanomaterial adsorbents.

H. Qin · Y. Liu · H. Liu · S. Di · S. Zhu (✉)
State Key Laboratory of Biogeology and Environmental Geology, China University of Geosciences, Wuhan 430074, China
e-mail: shukuizhu@126.com; shukuizhu@cug.edu.cn

J. Tharini and S. Thomas (eds.), *Carbon Nanomaterials and their Composites as Adsorbents*, Carbon Nanostructures, https://doi.org/10.1007/978-3-031-48719-4_7

The notion of nanomaterials and nanotechnology was proposed by Richard Feynman in his 1959 lecture "There's Plenty of Room at the Bottom". Over time, as these fields have evolved, scientists have come to recognize the significant role that nanomaterials and technologies will play in the future. In the 1970s, scientists came up with different guesses about the technology of nanomaterials from different angles. In 1974, scientist Donnie Gooch first used the term nanotechnology to describe precision machining. In 1982, scientists invented the scanning tunneling microscope (STM), a crucial instrument for studying the nanoworld. It revealed a visible world of atoms and molecules, contributing positively to the advancement of nanotechnology. The inaugural International Conference on Nanoscience and Technology was held in Baltimore, USA, 1990. This marked the birth of nanoscience and technology. In 1991, due to the discovery of carbon nanotubes, it became a hot spot for nanotechnology research. In 1993, the Vacuum Physics Laboratory of the Chinese Academy of Sciences in Beijing successfully wrote out the word "China", marking China's position in the international field of nanotechnology. In 1997, American scientists succeeded for the first time in moving a single electron with a single electron. In 1999, Brazilian and American scientists invented the world's smallest "scale" in an experiment with carbon nanotubes. Nanotechnology has swiftly emerged as a new and rapidly developing field of technology. Recently, several countries have formulated relevant strategies or plans, and the development pattern of industrial applications of nanomaterials and technologies has been gradually formed.

Nanometer is a physical unit of measure. Nanomaterials are ultramicroscopic particles in the size range of 1–100 nm and their dense aggregates, as well as materials made of nanoscale microcrystals [4–6]. This category encompasses a wide range of materials, including metallic, nonmetallic, inorganic, organic, and biological substances. Nanomaterials exhibit strong adsorption properties due to their relatively large specific surface area and high number of surface atoms [7]. They find extensive utility across various environmental applications, such as water pollution [8], soil pollution [9] and air pollution [10]. Nanoscale sorbents are widely employed across diverse fields, including information, energy, environment, biology, agriculture, and defense.

An adsorbent is a solid substance capable of effectively adsorbing some components from gases or liquids. It has a substantial specific surface area, an appropriate pore structure, and surface composition, it has a strong adsorption capacity for adsorbed substances [11]. Generally, adsorbents do not undergo chemical reactions with adsorbed components or the surrounding medium [12]. Moreover, it is easy to manufacture and regenerate and has excellent adsorption and mechanical properties. Nonetheless, traditional adsorption materials have limitations in adsorption performance, including slow adsorption rate, small adsorption capacity, difficult regeneration, and short effective life [13]. It's widely recognized that the adsorption performance of adsorbents is primarily influenced by the specific surface area and adsorption capacity of the materials. Consequently, the essential criteria for a material to serve as an effective adsorbent revolve around possessing a substantial specific surface area and impressive adsorption capacity. As a brand-new material, nanomaterials can conform to the characteristics of adsorbents and have excellent adsorption properties. This quality renders them highly promising for crafting high-performance adsorbents [14]. Both domestic and international scientists have extensively researched nanomaterials and implemented them across a variety of fields, spanning ceramics, chemical production, environment, bioengineering and medicine, military, and energy, among others.

2 Properties of Nanoscale Material Adsorbents

Compared to ordinary bulk materials, nanoscale materials demonstrate vastly distinct properties. At the nanoscale level, the wave nature of electrons and interatomic interactions in matter will be affected. Thus, nanomaterial adsorbents feature distinctive attributes absent in conventional counterparts. This section will describe the properties of nanomaterials.

2.1 Surface Effect

The surface effect is a phenomenon where alterations in particle size lead to a dramatic shift in the ratio of surface atomic numbers of nanoparticles to the total atomic numbers, subsequently resulting in changes in their nature [15]. When the particle diameter becomes smaller, the specific surface area of spherical particles will increase significantly. This will cause a relative increase in the number of atoms situated on the particle surface. Consequently, these surface atoms exhibit elevated activity and extreme instability, ultimately manifesting in different particle characteristics. The surface effect can be neglected when the particle diameter is greater than 100 nm, but as the particle diameter decreases, the percentage of surface atoms will increase significantly. When the particle diameter decreases to approximately 1 nm, the percentage of surface atoms can surpass 90%, implying that a significant majority of atoms are concentrated on the nanoparticle surface. Due to the surge in the count of surface atoms, accompanied by a deficiency in surface atomic coordination and heightened surface energy, these atoms are inherently inclined to bond with other atoms for stabilization [16]. Therefore, they possess high chemical activity.

2.2 Small Size Effect

When the size of a substance is processed to 1–100 nm, its physical and chemical properties change compared to the larger size. The change in macroscopic properties caused by smaller particle size is referred to as the "small size effect" [17–19]. For ultramicro particles, their decreasing size and increase in specific surface area bring about modifications in some of their original properties. (1) Special optical properties [20]: Nanometals exhibit significantly enhanced light absorption. The smaller size of the nanomaterial, the lower the light reflectivity. All metals appear black in the nanoparticle state. The reflectance of metal ultramicro particles to light is usually less than 1%, and they can be completely extinct when the thickness is only a few microns. Conversely, some non-metallic materials will appear reflective as they approach the nanoscale. (2) Special magnetic properties [21, 22]: The magnetic properties of small-sized superfine particles are significantly different from those of large materials. For instance, the coercivity of pure iron in large pieces is about 80 A/m. However, when the particle size drops below 20 nm, coercivity surges by a factor of 1,000. When the particle size is less than 6 nm, its coercivity decreases to 0, showing superparamagnetic properties. (3) Special thermal properties [23, 24]: The reduction of solid material to ultra-fine dimensions leads to a significant decrease in its melting point. This effect becomes most pronounced when particles are less than 10 nm. For example, the conventional melting point of gold is 1064 °C. When the particle size is reduced to 2 nm, the melting point is only about 327 °C. (4) Special mechanical properties: Nanomaterials demonstrate notably enhanced strength, hardness, and toughness. For instance, nanoceramic materials exhibit commendable toughness. Due to their vast interfaces, nanomaterials facilitate the migration of atoms under external forces, given the relatively chaotic atomic arrangement at these interfaces.

2.3 *Quantum Size Effect*

The quantum size effect refers to the effect of decreasing the number of atoms inside a particle. Due to volume reduction when the particle size drops to an extreme value. Kubo, a Japanese scientist, defines the quantum size effect, when the particle size drops to a minimum value, as a phenomenon in which the electron energy levels near the Fermi energy level appear to change from quasi-continuous to discontinuous discrete distribution [25]. This effect triggers significant alterations in the magnetic, optical, acoustic, thermal, and electrical properties of nanoparticles compared to macroscopic materials. Notably, nanoparticles display a pronounced quantum size effect in infrared absorption [26]. The peaks of resonance absorption are sharper than those of ordinary materials, and the specific heat capacity is nonlinearly related to temperature. The wave of electrons in the discrete quantized energy level of nanoparticles also brings a series of special properties to the nanoparticles. These encompass high nonlinear optical properties, special chemical and photocatalytic properties, strong oxidation and reducing properties [27–29].

2.4 *Macroscopic Quantum Tunneling Effect*

The tunneling effect refers to the ability of microscopic particles to cross the potential barriers [30]. Recently, it has been found that some macroscopic quantities have tunneling effects, encompassing aspects like the magnetization intensity of super-particles, magnetic flux in quantum coherent devices, and electric charge. They undergo changes as they surmount the barriers within macroscopic systems, giving rise to the term "macroscopic quantum tunneling". The scale of nanomaterials is in the transition domain at the interface of atomic clusters and macroscopic objects. It is a transitional metastable substance between macroscopic matter and a microscopic atom or molecule. The quantum size effect, surface effect, small size effect, and macroscopic quantum tunneling effect constitute the properties inherent to nanoparticles and nanosolids. These factors bestow nanoparticles and nanosolids with a multitude of distinct mechanical, magnetic, optical, thermal, and chemical attributes. In comparison to conventional materials (such as single crystals, polycrystalline, and amorphous materials), they exhibit numerous exceptional characteristics. Exploring these properties in the context of nanomaterial adsorbents holds vital theoretical and practical significance, notably for advancing environmental catalyst development.

3 Classification of Nanomaterial Adsorbents

The role of nanomaterial adsorbents in adsorption applications mainly depends on the surface properties and external functionalization of nanomaterial. In broad terms, nanoscale adsorbents can be classified into nanoparticles, carbon nanomaterials, and other nanomaterials [31]. The surface atoms of nanometals, particularly those situated at the edges and corners, exhibit heightened chemical activity.

3.1 Nanoparticles

3.1.1 Metal Nanoparticles

At the nanoscale, several fundamental properties of metals deviate significantly from those of bulk materials. These disparities encompass factors such as melting point, color, magnetism, electrochemical properties, and optical properties. These atoms are both the active site of the adsorbent and the active center of the catalyst. Nanoscale metal atomic clusters exist as a metastable state characterized by high surface energy, thereby rendering them prone to aggregation. The following two methods are commonly used to change their stability: (1) Utilizing porous materials such as Al_2O_3, SiO_2, activated carbon and polymers as carriers, and immersing these carriers in a metal salt solution. This process involves fixing the metal ions onto the carrier by reducing them to zero-valent metal atoms. (2) Modifying the surfaces of nanoscale metals to alter their characteristics.

3.1.2 Metal Oxide Nanoparticles

Nanoscale metal oxides are widely used in the adsorption field of heavy metal ions. The γ-alumina nanoscale sorbents were synthesized for the removal of nickel ions from wastewater [32]. Nanoscale adsorbents with low and high specific surface areas (LSSA and HSSA) were prepared by solvothermal method. Under optimal conditions, HSSA and LSSA nano adsorbents achieved nickel removal rates of 99.6% and 96.9% respectively. Lounsbury et al. [33] utilized nano-hematite (nα-Fe_2O_3) as a selenium adsorbent to study the effect of specific surface area on adsorption. Liu et al. [34] examined the effect of pore-forming agents (MnO_2, Fe_2O_3, Co_2O_3, NiO, and CuO) content (12–34 wt%) and type on the dynamic pore compensation effect and adsorption performance of the adsorbents. The corresponding synthesis path is illustrated in Fig. 1. It can be observed from the figure that $CaCO_3$ is thermally decomposed in a pure N_2 atmosphere at 800 °C to obtain sintered CaO during regeneration. The pore dynamic compensation is realized by releasing the vapor generated from the reduction reaction of transition metal oxide MO_x and H_2. The adsorption process is carried out by changing the atmosphere. Two chemical reactions occurred in the whole pore forming process, including the carbonation of CaO and oxidization of the reduced product MO_{x-1}. Compared with the nano-CaO/Al_2O_3 adsorbent, the nano-CaO-NiO/Al_2O_3 adsorbent showed the best dynamic pore compensation effect at a doped NiO content of 21wt%. A nano CaO-CaTiO adsorbent exhibiting high CO adsorption capacity and durability was synthesized [35]. The grain size of CaTiO can be reduced by the adsorption reaction method, and the doping content, temperature and time of thermal pretreatment can be reduced appropriately. Enhanced adsorption capacity is observed with smaller particle sizes of CaTiO, whereas greater durability is observed with larger particle sizes of CaTiO.

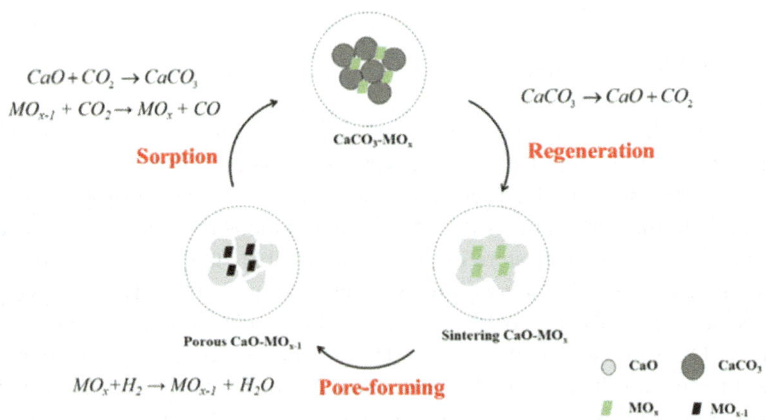

Fig. 1 Schematic diagram of CaL process of nano-CaO-based adsorbents with pore forming agent [34]. Reproduced with the permission of Springer Nature

3.1.3 Magnetic Nanoparticles

In recent years, as a means to address the limitations of traditional adsorbents, magnetic nanoscale sorbents have garnered significant attention in the field of industrial wastewater treatment [36]. These magnetic nano sorbents, obtained through magnetization and nano sorption by introducing magnets, exhibit exceptional solid–liquid separation ability. Under the action of an applied magnetic field, it can be reused after washing and regeneration through desorption. Currently, composite magnetic nanoscale sorbents are considered highly promising and efficient agents for treating heavy metals [37], dyes [38], and other organic pollutants [39] in wastewater. Studies have demonstrated that mesoporous magnetic activated carbon nanoparticles with good magnetic field response possess a huge specific surface area (671.2 m^2/g). This attribute enables them to provide numerous active surface sites for the removal of cationic dye methylene blue and anionic dye reactive red 198 [40].

3.1.4 Metal Sulfide Nanoparticles

Metal sulfide nanomaterials as adsorbents have excellent characteristics, including a fast adsorption rate, high adsorption capacity, and stable adsorption products, which have potential research and application value [41]. In a study, K/Zn/Sn/S layered metal sulfide nanosheets (KZTS-NS) were employed as adsorbents to remove Co^{2+} in the aqueous solution [42]. The adsorption mechanism is depicted in Fig. 2. The results reveal two mechanisms for Co^{2+} removal by KZTS-NS. At low concentrations, KZTS-NS mainly removes Co^{2+} through the ion exchange mechanism with K^+ ions in Fig. 2a. As shown in Fig. 2b, when the initial concentration of Co^{2+} is high (20–100 mg/L), the adsorption mechanism includes both the ion exchange with K^+ and the surface adsorption to form $Zn_{0.76}Co_{0.24}S$ composite. The formation of $Zn_{0.76}Co_{0.24}S$ destroys the layered structure. Furthermore, tubular nano copper sulfide (CuS) was prepared by hydrothermal method. It served as an adsorbent to remove 17α-ethynyl estradiol (EE2), demonstrating exceptional adsorption performance [43]. The nanomaterial adsorbents can remove 90% EE2 (5 mg/L). In another study, a nano-ZnS gel network was employed by Irina [44] as a cation exchange material to eliminate Pb^{2+} and Hg^{2+} from aqueous solutions. Nano-sized zinc sulfide (ZnS) crystals with a specific surface area of 360 m^2/g were prepared by thermal decomposition using single zinc ethylxanthate as precursor, which was used to remove As^{5+}, As^{3+} and Pb^{2+} [45]. Additionally, cauliflower-like SnS nanoparticles exhibited rapid adsorption of all MB dye in aqueous solutions within a few seconds [46]. The unique cauliflower structure plays an important role in the adsorption process.

3.1.5 Nanostructured Mixed Oxides

Incorporating magnetic oxides with other functional materials has been shown to reduce the aggregation of magnetic nanoparticles in wastewater, significantly

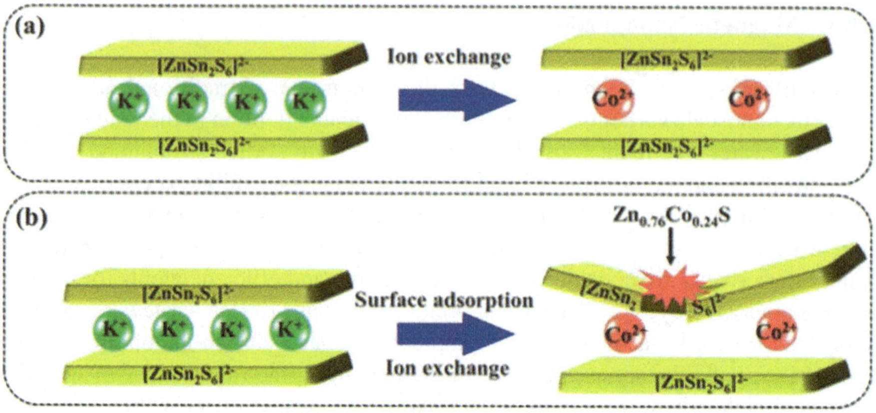

Fig. 2 Schematic diagram of adsorption mechanism: **a** Mainly ion exchange in low Co^{2+} concentration solutions (10–20 mg/L); **b** both ion exchange and surface adsorption by forming $Zn_{0.76}Co_{0.24}S$ composites when Co^{2+} concentration higher than 20 mg/L. The formation of $Zn_{0.76}Co_{0.24}S$ composite could destroy the layered structure of KZTS-NS [42]. Reproduced with the permission of Elsevier

enhancing the adsorption of heavy metals [47]. The adsorption performance of Pb^{2+} and Cu^{2+} in aqueous solution is improved obviously by the magnetic nano-adsorbent coated with silica and zeolite. This adsorbent not only retains the high adsorption property of zeolite but also reduces the agglomeration phenomenon of nanoparticles [48]. It has been revealed that bentonite combined with magnetic nanoparticles can effectively increase the specific surface area of the adsorbent. The abundant hydroxyl active sites on its surface can coordinate with As and Cu, making it move quickly from aqueous solution to adsorbent in a short time [49]. Finally, achieve the purpose of removing heavy metals. Furthermore, functional groups such as amino, carboxyl, and thiol groups have been shown to significantly impact adsorption properties. It is well known that electrostatic attraction and electrostatic coordination is considered to be the two most important interactions between metal ions and adsorbents. Typically, the coordination between $-NH_2$ or -SH and the bivalent metal ion predominates. And the prepared $Fe_3O_4@SiO_2$-PEI-SH surface contains $-NH_2$, -NH, -SH groups, both intra-group and inter-group coordination contribute to heavy metal ion adsorption, as illustrated in Fig. 3 [47]. Fe_3O_4 nanoparticles coated with polymeric humic acid (HA) showed favorable stability in natural water, acidic, and alkaline solutions. HA has an alkyl and aromatic unit backbone linked to carboxylic acid, phenolic hydroxyl, and quinone functional groups, which can ligand with heavy metals through oxygen and nitrogen-containing functional groups, effectively facilitating the removal of a wide range of heavy metals [37].

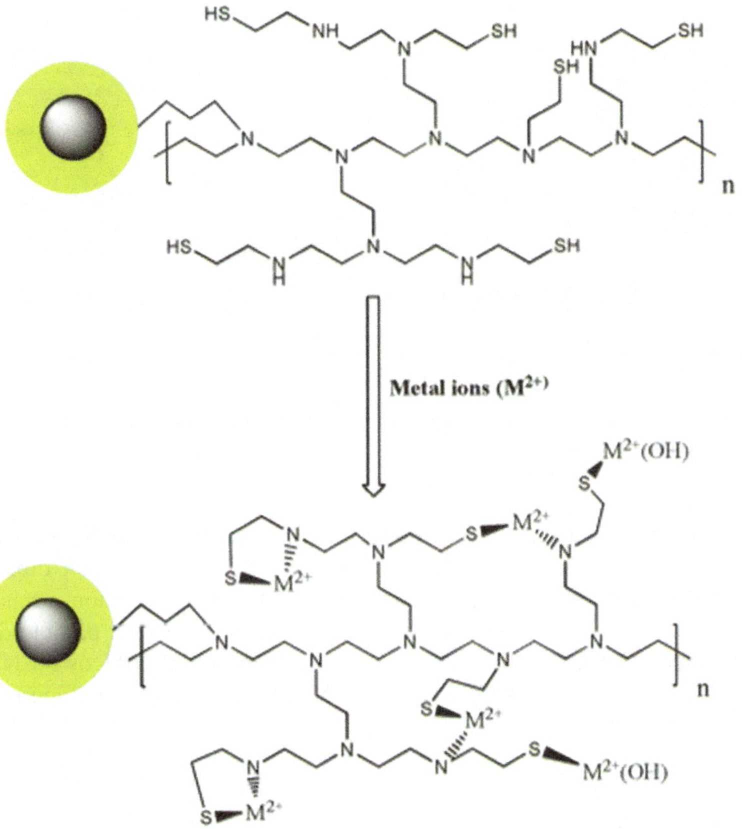

Fig. 3 A possible mechanism adsorption of metal ions (M^{2+}) on Fe_3O_4@SiO_2-PEI-SH [47]. Reproduced with the permission of Elsevier

3.2 *Carbon Nanoscale Materials*

3.2.1 Carbon Nanotubes

Research has discovered the presence of multi-walled carbon nanotubes and single-walled carbon nanotubes on the surfaces of all graphite electrodes during discharge [50]. This discovery opens up a promising direction in the field of carbon materials research. Hydrogen storage is a crucial facet of gas adsorption research involving carbon nanotubes. Some studies have revealed that single-walled carbon nanotubes with an average diameter of 1.85 nm can absorb hydrogen with an adsorption capacity equivalent to 4.2% of the weight of the carbon nanotubes [51]. In the presence of Cu^{2+}, Pb^{2+}, and Cd^{2+} in an aqueous solution simultaneously, competitive adsorption occurs when nitrated carbon nanotubes are regarded as adsorbents [52]. Compared with activated carbon, carbon nanotubes are very effective for the adsorption of

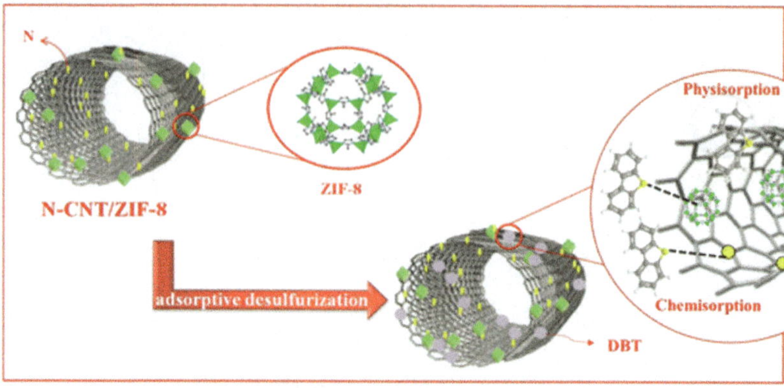

Fig. 4 The DBT adsorption mechanism of the NCNT/ZIF-8 nanosorbent [53]. Reproduced with the permission of Elsevier

these heavy metal ions. A nitrogen-doped carbon nanotube/ZiF-8 (NCNT/ZiF-8) nanocomposite was prepared, and its performance in adsorption of dibenzothiophene (DBT) from the liquid phase was investigated. Its adsorption mechanism is shown in Fig. 4, the adsorption phenomenon is associated with particle diffusion into the internal surface of the adsorbent. The adsorption of DBT by NCNT/ZIF-8 reached 81.2 mg/g, significantly surpassing the adsorption capacity of pure ZIF-8 at 69.1 mg/g [53].

3.2.2 Graphene

Graphene is an innovative material wherein carbon atoms, linked via sp2 hybridization, form a tightly arranged single layer within a two-dimensional honeycomb lattice structure. The common methods of graphene production in powder are mechanical exfoliation, redox, and SiC epitaxial growth, and the method of thin film production is chemical vapor deposition. Benefitting from high optical transmittance and a substantial specific surface area, graphene exhibits immense potential for applications within the adsorption domain. Notably, Graphene has outstanding photocatalytic properties, proving effective in the degradation of diverse pollutants. The large specific surface area gives graphene excellent adsorption properties, thus allowing it to adsorb heavy metal ions from water bodies. A nitrogen-rich porous N-GO doped chitosan double composite N-GO@HTCS (Fig. 5), serves as an efficient adsorbent for chromium removal in water, with its adsorption capacity reaching 42.64 mg/g within 50 min [54].

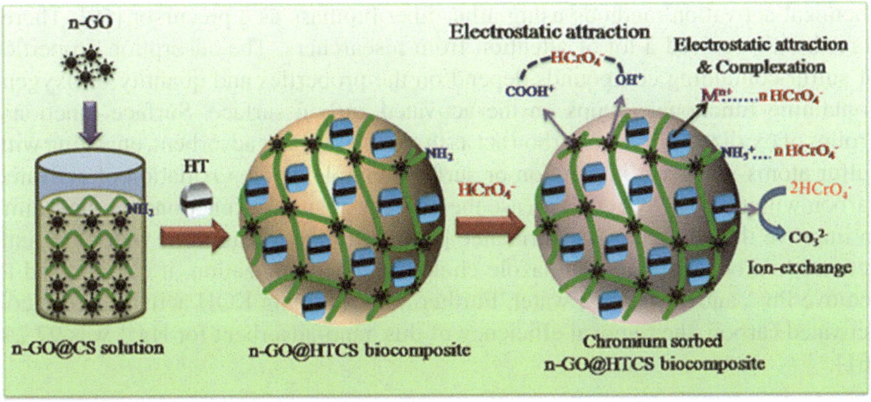

Fig. 5 The feasible chromium sorption mechanism of n-GO@HTCS biocomposite [54]. Reproduced with the permission of Elsevier

3.2.3 Fullerene

Fullerene is a hollow spherical molecule composed of 60 carbon atoms. The discovery of fullerenes has devoted more attention to carbon nanostructures. It has high chemical stability, a large specific surface area, and good electrical conductivity [55]. By preparation $Fe_3O_4@SiO_2@C_{60}$ magnetic material was exploited to extract many pesticide residues from chrysanthemum. Notably, the nanoscale adsorbent demonstrates favorable stability and reusability [56]. Using fullerene to modify magnetic molecularly imprinted polymers (MIPs) increases the specific surface area of magnetic particles, forms a delocalized π conjugate structure, and enhances the hydrophobicity of magnetic particles. Eventually, the adsorption performance of methylene blue (MB) is enhanced [57]. Besides, the most interesting aspect of fullerene as an adsorbent is its selectivity. Many metal ions can be enriched with fullerenes.

3.2.4 Nano-activated Carbon

Nano-activated carbon possesses a relatively developed and rich pore structure, large specific surface area, rich surface functional groups, and excellent adsorption energy. Additionally, it also boasts robust mechanical strength and chemical stability, heat resistance, acid resistance, and alkali resistance. This material is insoluble in both water and organic solvents, rendering activated carbon-based catalysts easily regenerable following deactivation. In general, the production of nano-activated carbon involves grinding clean waste into fine particles, followed by heat treatment (pyrolysis) of these particles, and subsequent physical and/or chemical activation [58]. Recently, nano-activated carbon was synthesized via two stages of pyrolysis and chemical activation methods using arhar fiber biomass as a precursor [59]. Therefore, it has received a lot of attention from researchers. The adsorption properties of sulfur-containing compounds depend on the properties and quantity of oxygen-containing functional groups on the activated carbon surface. Surface functional groups of oxidized activated carbon act as the active sites for adsorbent, engaging with sulfur atoms via direct interaction or surface complexation. Oxidation of activated carbon with $(NH_4)_2S_2O_8$, to increase the number of surface functional groups, aims to improve the adsorption performance [60]. After modification of water hyacinth activated carbon by 2-aminothiazole chelation functionalization, it can be used to remove Pb^{2+} and Hg^{2+} from water. Furthermore, utilizing KOH activated corn cob activated carbon, the removal efficiency of this nano-adsorbent for Hg^{2+} was 97.2% [61].

3.3 *Other Nanoscale Materials*

3.3.1 Silicon Dioxide Nanoparticles

Nano SiO_2 is a non-toxic, odorless, and non-polluting non-metallic material. Due to its extensive specific surface area and chemical reactivity, silica finds broad application in the field of adsorption. The adsorbents, prepared using 3-aminopropyltrimethoxysilane modified silica nanoparticles, successfully achieved the adsorption of Cu^{2+} and Pb^{2+} up to 2.18 mmol/g and 4.74 mmol/g [62], respectively. The adsorption capacity of Pb^{2+} was 47.8 mg/g by using N-(2-amino-ethyl)-2, 3-dihydroxybenzaldehyde imide modified SiO_2 [63]. SiO_2-CNT nano-adsorbent successfully achieved the removal of Hg^{2+}, its removal amount reached 164 mg/g [64]. Using CTAB-functionalized nanoscale SiO_2 (Fig. 6), the maximum adsorption capacity of β-lactam amoxicillin reached approximately 25 mg/g, surpassing the capacities of other adsorbents [65].

3.3.2 Nanoclay

Clay is a natural resource endowed with numerous superior properties resulting from prolonged exposure to specific environmental conditions. For example, it has an extensive specific surface area, good adsorption properties, high adsorption capacity, and ion exchange capacity. The magnetic bentonite@$MnFe_2O_4$ (BCMFC) composite nanoscale adsorbent exhibited remarkable adsorption efficiency for Cr^{3+} and Cr^{4+} with 97.37% and 98.65% removal rates [66], respectively. The adsorption amounts of KG-g-PMETAC/MMT nanocomposite adsorbent prepared by microwave-assisted method were 155.85 mg/g, 149.64 mg/g and 137.77 mg/g for MB, toluidine blue, and crystalline violet [67], respectively.

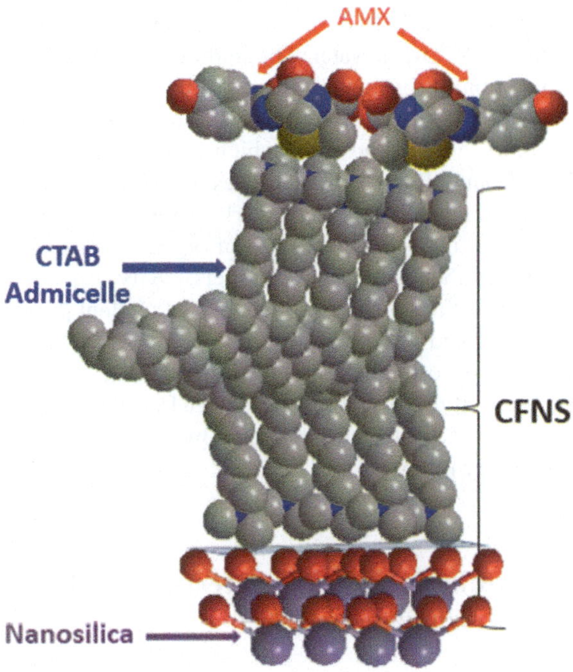

Fig. 6 Cartoon representation of AMX uptake onto CTAB functionalized nanosilica (CFNS) [65]. Reproduced with the permission of Elsevier

3.3.3 Polymer-Based Nanoscale Adsorbents

The adsorption of metal and organic pollutants through polymer-based nanoscale adsorbents largely depends on the chemical properties of the adsorbents' surface. Conventional magnetic adsorbents can be modified by polymers to increase the adsorption capacity. Polymer chemical modifications can lead to the selective adsorption of different metal ions by magnetic nanoparticles. Moreover, the regeneration and reusability of hybrid adsorbents are typically more straightforward compared to single nanoparticle adsorbents. Employing polymer anion exchange resin D201 as the carrier, nano-cerium oxide (NCO) was loaded into the D201 pore to prepare the composite nano-adsorbent NCO@201. This innovative material was employed to extract fluoride from acidic wastewater. Impressively, the maximum adsorption capacity for fluoride removal reached 17.67 mg/g [68].

3.3.4 Organic Nanoscale Materials

Metal–organic frameworks (MOFs) are crystalline porous materials formed by bridging structures of metal ions or organic ligands. MOFs have a rich crystal structure, high porosity, homogeneous pore size, and considerable specific surface area. These attributes have propelled MOFs to the forefront of interest in the catalytic adsorption field. BUC-17 with 2D structure was used for the adsorption treatment of Cr^{6+} wastewater. The findings revealed that BUC-17 powder had excellent adsorption and removal effect on Cr^{6+}, and the adsorption amount could reach 121 mg/g at pH = 4 [69]. A columnar layered co-organic skeleton TMU-74 was designed and synthesized, which can rapidly, effectively, and selectively remove Pb^{2+} from polluted water samples, boasting an adsorption capacity of 385.71 mg/g [70]. Covalent organic frameworks (COFs) materials are emerging crystalline porous polymeric materials composed of organic molecules connected by covalent bonds. There are abundant porous channels, cavity structures, and particle gaps on the inner or outer surface of the material. It has excellent prospect in adsorption field. COF-S-SH, produced using treating COF-V with 1,2-ethanedithiol, showcased its potential for Hg^{2+} adsorption and removal from wastewater, reaching an exceptional adsorption capacity of up to 1350 mg/g [71]. Furthermore, the ionic covalent organic framework material BT-DGCl, prepared by introducing guanidinium groups with Cl^-, can effectively remove Cr^{6+} [72].

4 Synthesis and Characterization of Common Nanoscale Adsorbents

The chemical composition and structure of nanomaterial adsorbents are the key factors that determine their properties and applications. Up to now, a variety of nanoscale materials with various active adsorption sites and crystal structures have been developed and characterized. However, because of the different microstructures of nanomaterial adsorbents, their adsorption behaviors often vary significantly [73]. Characterization techniques of nanomaterial adsorbents play a pivotal role in theoretical research and application of nanoscale adsorbents, which is a cross technology between nanomaterial research and multiple applications. The chemical composition and structure of nanomaterial materials are key factors in their performance and applications. Therefore, it is necessary to characterize their microscopic features to explore their adsorption properties. This promotes the application of nanoscale adsorbents in multiple fields.

4.1 Synthesis of Nanoscale Adsorbents

4.1.1 Physical Synthesis Method

Researchers have continuously developed synthetic methods for nanoscale adsorbents with small particle sizes, excellent stability, and difficult agglomeration in recent years. Common physical synthesis methods of nanomaterial adsorbents include ball milling [74], spraying, photolithography [75], and more. These techniques are mainly based on mechanical processes, as well as the method of vaporizing precursors in vacuum or inert gas using optical and electrical technologies to form nanomaterials [76].

The physical crushing method, also known as the mechanical alloy (ball milling) method, pulverizes raw materials directly into ultrafine powder particles through mechanical crushing. In 1970, a fresh technology called "mechanical alloying" was first reported for synthesizing oxide dispersion strengthened alloys capable of withstanding high pressures and temperatures [77]. The crushing process involves the deformation and rupture of solid materials or particles under the action of crushing forces. When the crushing force becomes sufficiently high, the instantaneous force between the solid material or particles greatly exceeds the mechanical strength of material. As a result, the material is crushed, and its size is effectively reduced [78]. Therefore, this method is especially suitable for preparing nano-powder particles of brittle materials. The specific surface area, surface functional groups and phase composition of materials are affected by grinding time, rotating speed, size of grinding balls, and atmosphere (air or vacuum, oxygen or no oxygen) [79]. The method has the advantages of inexpensive equipment, simple operation process, and no need for solvents. Therefore, it is considered an environmentally friendly, efficient, time-saving and cost-effective technique for material fabrication. However, the variety of materials that can be prepared using this method is relatively limited. Improved the synthesis process of nitrogen-containing bone biochar by ball milling technology, which greatly increased the pore area and specific surface area of nanoscale adsorbents, and significantly increased the adsorption capacity of nitrogen-containing bone biochar to heavy metal ions [80]. Environmental friendly thiol-modified biochar (BMS-biochar) was synthesized from the original biochar and 3-mercaptopropyltrimethoxysilane (3-MPTS) by the ball milling process [81]. The results showed that the specific surface area of thiol-modified biochar materials could be improved by the ball milling method. And then more negatively charged surfaces and more functional group adsorption sites are provided. Ultimately, the adsorption performance is greatly improved. The potential mechanism for Hg^{2+} and CH_3Hg^+ adsorption through BMS-biochar was proposed, as shown in Fig. 7.

Physical vapor deposition (PVD), also referred to as the evaporation and condensation method, is commonly conducted within a vacuum chamber. This method utilizes a heat source to promote the evaporation and gasification of materials, allowing them to undergo physical reactions in a gaseous state. Eventually, they condense and grow into nanoparticles or coatings during the cooling process. The corrosion resistance, wear resistance, mechanical, thermal, and optical properties of the precursor were changed during the synthesis process [82]. The heating methods commonly used in the evaporation and gasification process include resistance heating, plasma jet heating, high frequency induction heating, electron beam heating, laser heating, arc heating, and microwave heating. The synthesis of ultrafine particles by different heating methods has different structures, particle sizes and distributions, and quantities [83]. For instance, the synthesis of 2, 4, 5-triphenylimidazole nanowires is achieved through adsorbent-assisted physical vapor deposition. In short, the

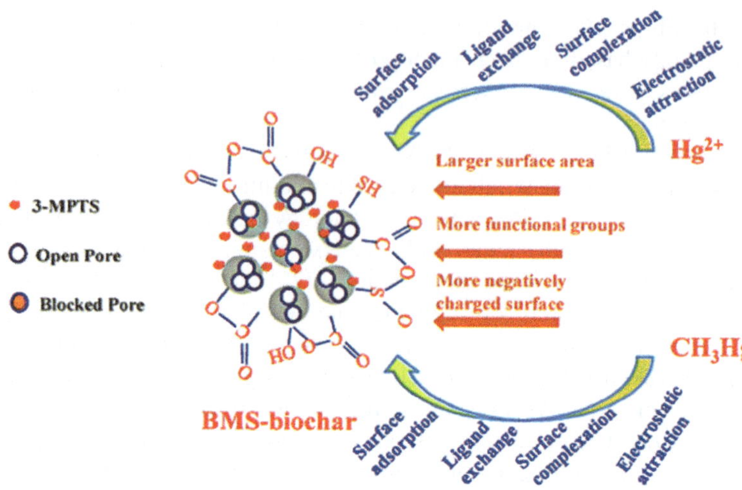

Fig. 7 Schematics of potential adsorption mechanisms of Hg^{2+} and CH_3Hg^+ on BMS-biochar [81]. Reproduced with the permission of Elsevier

precursor is heated and evaporated in a tube furnace with inert gas, and the product is collected downstream of flowing gas [84]. Studies have revealed that the unique size of nanowires is primarily dependent on optical properties. When the diameter of nanowires decreases, its absorption spectrum will shift blue, and the vibration fine structure of the emission spectrum will become more obvious.

4.1.2 Chemical Synthesis Method

Common chemical preparation methods include the sol–gel method, solvothermal method, chemical vapor deposition method, chemical precipitation method, and so on. The sol–gel method, a wet chemical technique, finds extensive application in nanomaterial development. This method is employed to develop a variety of high-quality metal oxide-based nanomaterials. In the synthesis of metal oxide nanoparticles, liquid precursors are converted into sols, which subsequently evolve into network structures known as gels [85]. In a typical sol–gel process, inorganic salt or metal alkoxide are employed as precursor. They are dissolved in water or organic solvent to form a homogeneous solution. The solute undergoes a hydrolysis reaction, resulting in the formation of nano-sized particles and sols. These sols undergo transformation into gels under certain conditions. By utilizing low-temperature chemical processes, the microstructure of the material can be tailored and controlled within a relatively small size range, achieving uniformity at the submicron, nanometer, and even molecular levels. Subsequently, impurities are removed, and heat treatment is applied to form oxides or other composite nanomaterials. A mesoporous silica-magnesium oxide ceramic fiber with excellent adsorption performance was

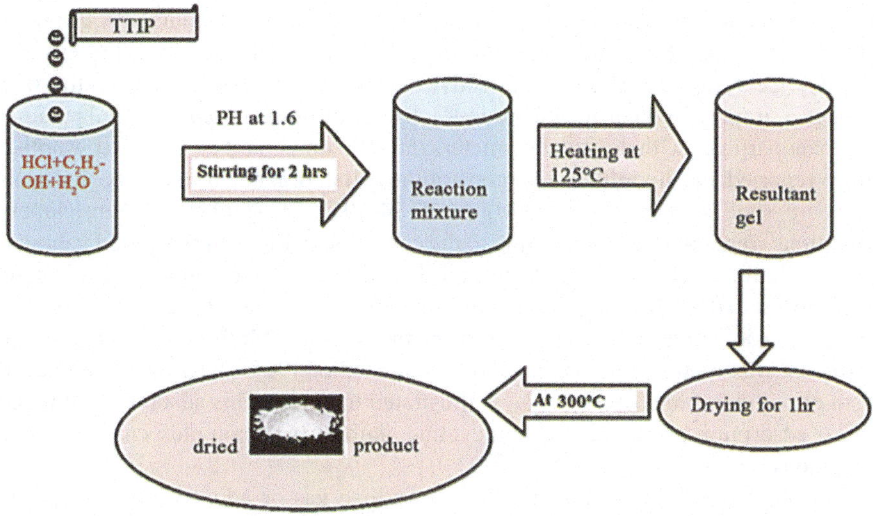

Fig. 8 Schematic diagram of the synthesis method of TiO_2 nanoparticles [87]. Reproduced with the permission of Wiley

synthesized by sol–gel method using citric acid as a surfactant [86]. It has a diameter of 1.23 μm and a specific surface area of 142.61 m^2/g and has a strong adsorption capacity for Pb (II), Cu (II), methylene blue, and fulvic acid. Another instance demonstrates a method for preparing titanium dioxide nanoparticles through the sol–gel approach, as depicted in Fig. 8 [87].

The hydrothermal method is one of the most famous and widely used methods for producing nanostructured materials. In the hydrothermal method, nanoscale materials are obtained by heterogeneous reactions in a high pressure and temperature water medium near the critical point in a sealed container [88]. It is considered as a simple and environmentally friendly preparation method of nanomaterials. It can be divided into hydrothermal crystallization method, hydrothermal synthesis method, hydrothermal decomposition method, hydrothermal oxidation method, hydrothermal reduction method, microwave hydrothermal method, supercritical hydrothermal synthesis method, and reactive electrode submerged arc method [89]. Of this amount, the solvothermal method is a method developed on the basis of hydrothermal reaction approach. The solvothermal method utilizes organic solvents instead of an aqueous solutions medium in the hydrothermal method, and reacts in a closed high-pressure reactor. At present, this method was used to prepare metal oxide nanoparticles [90], metal organic framework materials [91], covalent organic framework materials [92], and so on. The characteristics of nanomaterials prepared by the solvothermal method are closely related to the reaction conditions [93]. Compared with the sol–gel method, this technique does not require high temperature combustion and avoids the mixing of impurities and the formation of hard agglomeration of particles. At the same time, it is easy to adjust the reaction conditions, but also can control the crystallization, particle size, and porosity of nanoparticles. Moreover, it is advantageous to obtain suitable stoichiometric substances and the advantages of crystal morphology. For instance, CeO_2 nanoparticles with effective adsorption capabilities were synthesized through a simple solvothermal method [94]. By controlling the experimental parameters, nanoparticles with different diameters (50–800 nm) could be prepared. Another study reported the hydrothermal preparation metal–organic framework materials with core–shell structure (Fe_3O_4@Mg-MOF-74) [95]. The synthesis and enrichment conditions were optimized according to the control variables. The prepared nanomaterials achieved efficient enrichment and detection of glycopeptides via magnetic solid phase extraction, maintaining outstanding adsorption effect after 6 cycles. Likewise, a defective, carboxyl-exposed and highly stable MOF-808_{def} nanomaterial adsorbent was synthesized using solvothermal method [96]. The adsorption mechanism of pollutants and MOF-808_{def} is illustrated in Fig. 9. This adsorbent exhibited robust adsorption properties for tetracycline antibiotics in complex environmental conditions.

Chemical vapor deposition (CVD) is to utilize gas or adopt various means to change substances into gas, and make them react chemically in the gas state. Subsequently, they condense and grow into nanoparticles in the cold treatment process. According to evaporation methods, CVD can be divided into plasma enhanced chemical vapor deposition (PECVD) and laser induced chemical vapor deposition (LICVD). CVD is recognized as a solvent-free, one-step "green" technology. It has

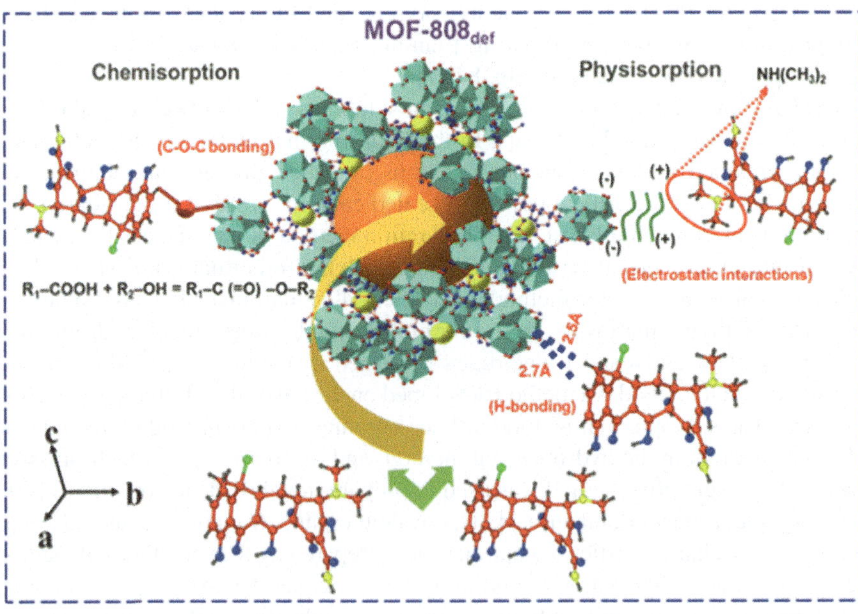

Fig. 9 A schematic diagram of as-proposed interaction mechanisms between TC-HCl and MOF-808_{def} [96]. Reproduced with the permission of Elsevier

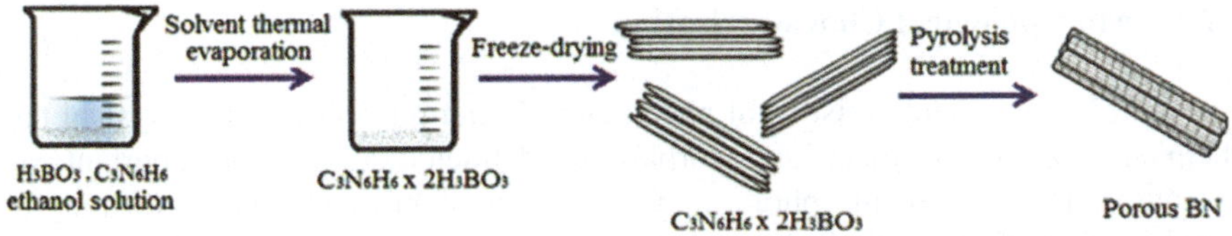

Fig. 10 Schematic illustration of the synthetic procedure of oxygen-doped bundlelike porous boron nitride [101]. Reproduced with the permission of American Chemical Society

been widely employed for the synthesis of diverse nanomaterials, such as graphene [97], carbon nanotubes [53], transition metal disulfides [98], metal nanoparticles [99], and more. Utilizing a metal–organic chemical vapor deposition method, metal–organic precursors are sublimated into the microporous structure of MIL-101 [100], and then metal nanoparticles were obtained after deposition. Oxygen-doped bundle-like porous boron nitride was synthesized through the thermal evaporation and high temperature pyrolysis of solvent [101]. The synthesis schematic is shown in Fig. 10. The research shows that the resulting nanoscale materials possess a high specific surface area (474.3 m^2/g) and high porosity (0.332 m^3/g). In the meantime, under the synergistic effect of oxidizing groups and lattice defects, bulk porous boron nitride can selectively adsorb and extract copper ions and lead ions.

The chemical precipitation method involves introducing a precipitant into a solution of soluble salts containing one or more ions at a certain temperature. Subsequently, the solution is directly hydrolyzed or precipitated to form insoluble hydroxides, oxides, or inorganic salts. Ultimately, the required nanoparticles can be obtained by pyrolysis or thermal removal. It is usually used to prepare metal oxide nanoparticles and functionalize them to obtain excellent properties [102]. A novel magnetic nano-adsorbent was developed by covalently binding polyacrylic acid to the surface of Fe_3O_4 nanoparticles, followed by amino-functionalization using diethylenetriamine [103]. The amino-functionalized adsorbent material can quickly adsorb heavy metal ions and anions through chelation and ion exchange. Generally speaking, the general chemical precipitation method is difficult to control the size of nanoparticles [104]. In order to solve this limitation, an inverse coprecipitation method was proposed by adding a cationic solution to the precipitation solution, which can produce more uniform nucleation precursors [105]. In addition, adding surfactants is also instrumental to obtain small-sized nanomaterials.

4.2 Characterization

To investigate the relationship between the structure and properties of nanomaterial adsorbents, it is necessary to characterize them on the atomic scale and nanoscale. The main characterization encompasses aspects such as morphology, structure, composition analysis and property assessment.

4.2.1 Morphological Characterization

The morphology characteristics of nanomaterials can influence many physical and chemical properties of materials. Morphology characterization is an important part of material analysis, and morphology analysis is one of the main contents of studying nanomaterials. The morphology characteristics of nanomaterials can influence many physical and chemical properties of materials. Commonly employed methods for morphology characterization of nanomaterials encompass the scanning electron microscope (SEM) [106], transmission electron microscope (TEM) [107], STM [108], and atomic force microscope (AFM) [109]. SEM and TEM are the most widely employed characterizations, which can provide information such as geometric morphology, microstructure, dispersion state, particle size, and crystal facet information. SEM can provide nanoscale morphology images, while TEM has high spatial resolution, enabling the acquisition of element composition and phase structure information within specific regions. Notably, STM is only suitable for morphology analysis and surface atomic structure distribution analysis of conductive thin film materials, but not for nano-powder materials. Meanwhile, AFM can analyze conductive and nonconductive materials, but its resolution is worse than that of STM. Among these techniques, STM and AFM mainly applied in in-situ morphology analysis.

4.2.2 Structural Characterization

X-ray diffraction (XRD) phase analysis is based on the X-ray diffraction effect of a polycrystalline sample [110]. This method that can analyze and determine the existing form and phase structure of each component in a sample. The content of the analysis includes the crystallization of each component, the crystal phase, the crystal structure, the valence state of each element in the crystal, the bonding state, etc. It can also be used to identify unknown components. In the meantime, the specific surface area is a critical physical characteristic of nanoscale adsorbent. Typically, the nitrogen adsorption–desorption technique is employed to determine the BET (Brunauer Emmett Teller) specific surface area, and the BJH (Barret Joyner Helena) method is wielded to investigate the pore size distribution [110]. This technology is widely utilized and plays an important role. Infrared spectroscopy can be utilized to analyze defects, interstitial atoms, dislocations, grain boundaries, organic functional groups, and other information on nanomaterials. Among them, fourier transform infrared spectroscopy (FTIR) has been effectively applied [107]. Raman spectra can be used to probe various excitation states in nanoscale materials [110]. It exercised the distinct vibrational and rotational energy levels of different substances to produce discernible raman frequency shifts, thus revealing information regarding the molecular structure and bond state characteristics of samples.

4.2.3 Composition Analysis

The physical properties of nanomaterials are closely related to their chemical compositions. Therefore, it is very important to determine the element composition, type, and amounts of impurities in nanomaterials. According to different analysis purposes, the composition analysis of nano adsorbent mainly involves spectral analysis, mass spectrometry analysis and energy spectrum analysis. Inductively coupled plasma (ICP) is a method to analyze the elements to be measured according to the characteristic spectral lines emitted when excited atoms of the elements to be measured return to the ground state, using an inductively coupled plasma as the excitation source [111]. Inductively coupled plasma-mass spectrometry (ICP-MS) is a kind of elemental mass spectrum analysis method using inductively coupled plasma as an ion source. The ion source produces sample ions by mass analyzer and detector after mass spectrometry to obtain mass spectra. X-ray photoelectron spectroscopy (XPS) is a highly sensitive surface analysis technique [111]. Its detection depth is generally 4–15 nm of surface layer information, which is an effective means to study the physical properties such as element composition and chemical state on sample surface. Electron probe technology is a practical method for micro-area composition analysis. It combines electron beam scanning imaging with X-ray emission spectrum analysis, which can not only observe the microstructure of samples but also analyze the chemical composition of samples in selected areas. Additionally, energy dispersive spectrometer (EDS) is also a common micro-area composition analysis method, which exploits the different X-ray photon characteristic energies of different elements to analyze the composition.

4.2.4 Characterization of Properties

The common properties of nanoscale adsorbent materials include thermogravimetric analysis, zeta potential, magnetic characterization, and more. Thermogravimetric (TG) analysis is a technique employed to measure the change of material mass with temperature or time at a set temperature. The composition, thermal stability, thermal decomposition, and samples of nanoscale adsorbent materials and possible intermediates related to product quality can be understood by analyzing the TG curve. Zeta potential is the potential between the ionic layer attached to the surface of particles and the aqueous solution, which comprehends the electrostatic properties of materials and measures their stability [107]. The magnetism of nanoscale materials is usually characterized adopt a vibrating sample magnetometer [91].

5 Application of Nanoscale Adsorbents in Environment Field

In the environmental field, water pollution, soil pollution, and air pollution will bring extremely harmful to the survival and development of humans and other organisms. However, conventional prevention measures and solutions can merely mitigate the extent of pollution, and many of these processes are cumbersome. At this time, the creation and application of nanomaterial adsorbents can effectively address these issues. Due to its high specific surface area, the nanoscale adsorbent exhibits superior adsorption performance. Therefore, it shows great potential in preparing high-performance adsorbents. Eventually, the rational utilization of nanomaterial adsorbents can provide more possibilities for environmental treatment.

5.1 Wastewater Treatment

Recently, the leap development of nanoscale adsorbents provides a golden opportunity for innovation in water treatment technology. This progress has sparked extensive research among numerous scholars. With the improvement of nanomaterial adsorbent preparation technology, nanoscale adsorbents have also been developed in wastewater treatment. Nanoscale adsorbent with a large specific surface area can provide a large number of active sites, resulting in a significant adsorption capacity. Therefore, it has an extremely strong ability to capture both oxygen radicals and alkane molecules, thus effectively removing pollutants from water environment. β-Cyclodextrin (β-CD) is loaded onto the surfaces of zeolite and vermiculite to adsorb lead and cadmium ions from water. Research reveals that following the loading of β-CD, the adsorption saturation of zeolite for Cd^{2+} and Pb^{2+} are 93.06 and 175.25 mg/g, respectively [112]. The MgFe-LDH hollow nanospheres prepared by one-step thermal method, with large specific surface area and hollow pore structure, showed excellent adsorption performance for both As(V) and Cr(VI), and the removal rate was 99% within 5 min [8]. In addition, numerous studies have been conducted on the utilization of nanoscale adsorbents for the effective removal of antibiotic pollutants from water. Fe_3O_4@MoS_2 nanomaterials are utilized as adsorbents for extracting sulfonamide antibiotics (SAs) from water samples [113]. Researchers have developed a variety of methods to eliminate permanganate ions (MnO_4^-) from wastewater, encompassing chemical precipitation, electrochemical methods, biological methods, and adsorption methods [114]. The oxygen functional groups on the surface of graphene oxide are used as the preferential sites for rapid heterogeneous nucleation. Therfore, a nanofiltration membrane made of ultrafine metal oxide/reduced graphene oxide nanocomposites was synthesized, which can remove 98% methylene blue [115]. Furthermore, a magnetic graphene oxide (GO) gel was devised for the adsorption of Cu(II), Cd(II) and Pb(II) from wastewater. The material's maximum adsorption capacity reached 55.96, 86.28, and 189.04 mg/g, respectively. Moreover, it has better selective adsorption for Pb(II) [116]. This also shows that the application of nanotechnology in water treatment has a promising future.

5.2 Soil Remediation

Traditional methods for remediating soil contaminated with organic compounds often encounter challenges such as low efficiency, high cost and susceptibility to secondary contamination. In contrast, the development and application of nanotechnology hold promising solutions to these issues. With the rapid development of materials science and nanotechnology, nanomaterials have found increasing applications in the realm of contaminated soil remediation. Compared with traditional remediation materials, nanomaterials have the advantages of large specific surface area, strong adsorption capacity, and high reactivity, which make nanomaterial remediation technology have excellent application prospects in contaminated soil remediation. As we all know, herbicides are extensively employed to manage and eradicate weeds across various crops. Consequently, atrazine is frequently detected in soil, and using Fe_3O_4@PS/DVB-MNPs as nanoscale adsorbent can effectively extract pesticides from the soil [9]. CMC-stabilized MnO_2 nanoparticles was used for degrading estradiol present in soil [117]. In response to the enrichment of heavy metals in soil, the preparation of nanoscale sorbents has enabled the preconcentration and determination of trace heavy metal ions. This achievement has been realized through the utilization of cobalt magnetic nanoparticles (MNPs) [118], which can effectively adsorb lead ions within the soil environment. The use of 50 mg/L multi-walled carbon nanotubes (MWCNTs) as carriers has significantly improved the migration of 0.1 mg/L phenanthrene in soil [119]. These examples underscore the superior attributes of nanoscale adsorbents that surpass those of traditional remediation materials. In the field of contaminated soil treatment, nanomaterials are irreplaceable by traditional remediation technologies in terms of their ability to adsorb and degrade heavy metals and organic pollutants as well as redox reactions. Consequently, nanomaterials have established themselves as indispensable tools in advancing contaminated soil treatment practices.

5.3 Air Purification

Atmospheric air contamination profoundly impacts human health in industrial cities. The adsorption process stands as the most prevalent and optimal technique for removing volatile organic compounds (VOCs). A novel application involving magnetic nano-adsorbents, employing Fe_3O_4 nanoparticles and zeolite Y, has been developed for air decontamination [120]. This concept was used to purify toluene from the air. Zhao et al. prepared a layer of cellulose nanofibers (CNF)/copper nitrate on a stainless-steel hollow structure to remove target air pollutants. The adsorption efficiency of PM2.5 remained consistently above 95% for over 24 h, and the adsorption amount of formaldehyde was 47.71 mg/g [10]. Toluene was also adsorbed using a CNC-PA6 composite nanofiber filter. This nano-adsorbent not only enhances the toluene adsorption capacity but also improves the fiber morphology and mechanical strength characteristics [121]. Nanotechnology has emerged as an advanced treatment strategy, harnessing various nanomaterials (NMs) to effectively remediate air pollution. Previous studies have shown that magnetic nanoparticles $Mg_{0.25}Fe_{2.75}O_4$ effectively reduce carbon monoxide, particulate mass, and hydrocarbon emissions [122]. Compared to conventional strategies, nano-sorbents display significant potential for air pollution remediation, owing to their high reactivity, efficiency and lower. Thus, nanotechnology presents a promising avenue for advancing air pollution remediation.

6 Conclusions and Future Perspectives

Given the escalating environmental challenges, society is exhibiting a heightened focus on nanoscale adsorbents. Nanomaterial adsorbents, characterized by distinctive structures and exceptional adsorption capabilities, are garnering increased attention within the realm of environmental concerns. In the meantime, the types of nanoscale adsorbents have also become diverse due to their different properties. Currently, the development and application of nanoscale materials are still in the stage of research and breakthrough. And it is hoped that the rational development and application of nanomaterial adsorbents can achieve low cost, high efficiency, and large-scale application.

References

1. Liu, S., Xia, T.: Continued efforts on nano-EHS is critical to maintain sustainable growth of nano-industry. Small **16**(21), 2000603 (2020)
2. Qian, J., Gao, X., Pan, B.: Nanoconfinement-mediated water treatment: from fundamental to application. Environ. Sci. Technol. **54**(14), 8509–8526 (2020)
3. Emenike, E., Adeniyi, A., Omuku, P., Okwu, K., Iwuozor, K.: Recent advances in nano-adsorbents for the sequestration of copper from water. J. Water Process. Eng. **47**, 102715 (2022)
4. Ogunfowora, L., Iwuozor, K., Ighalo, J., Igwegbe, C.: Trends in the treatment of aquaculture effluents using nanotechnology. Clean. Mater. **2**, 100024 (2021)
5. Iwuozor, K., Ogunfowora, L., Oyekunle, I.: Review on sugarcane-mediated nanoparticle synthesis: a green approach. SugarTech **24**, 1186–1197 (2022)
6. Alao, I., Oyekunle, I., Iwuozor, K., Emenike, E.: Green synthesis of copper nanoparticles and investigation of its anti-microbial properties. Adv. J. Chem. Sec. B **4**, 39–52 (2022)
7. Kyzas, G., Matis, K.: Nanoadsorbents for pollutants removal: a review. J. Mol. Liq. **203**, 159–168 (2015)
8. Gao, Y., Li, Y., Zhang, L., Huang, H., Hu, J., Shah, S., Su, X.: Adsorption and removal of tetracycline antibiotics from aqueous solution by graphene oxide. J. Colloid Interface Sci. **368**(1), 540–546 (2012)

9. Kermani, M., Sereshti, H., Nikfarjam, N.: Application of a magnetic nanocomposite of cross-linked poly(styrene/divinylbenzene) as an adsorbent for the magnetic dispersive solid phase extraction-dispersive liquid-liquid microextraction of atrazine in soil and aqueous samples. Anal. Methods **12**, 1834–1844 (2020)
10. Zhao, X., Chen, L., Guo, Y., Ma, X., Li, Z., Ying, W., Peng, X.: Porous cellulose nanofiber stringed HKUST-1 polyhedron membrane for air purification. Appl. Mater. Today **14**, 96–101 (2019)
11. Aguilar-Pérez, K., Avilés-Castrillo, J., Ruiz-Pulido, G.: Nano-sorbent materials for pharmaceutical-based wastewater effluents-an overview. Case Stud. Chem. Environ. Eng. **2**, 100028 (2020)
12. Wang, C., Zeng, W., Jiang, T., Chen, X., Zhang, X.: Incorporating attapulgite nanorods into graphene oxide nanofiltration membranes for efficient dyes wastewater treatment. Separ. Purif. Technol. **214**, 21–30 (2019)
13. Cincinelli, A., Martellini, T., Coppini, E., Fibbi, D., Katsoyiannis, A.: Nanotechnologies for removal of pharmaceuticals and personal care products from water and wastewater. J. Nanosci. NanotechnolnoL. **15**(5), 3333–3347 (2015)
14. Mo, L., Pang, H., Tan, Y., Zhang, S., Li, J.: 3D multi-wall perforated nanocellulose-based polyethylenimine aerogels for ultrahigh efficient and reversible removal of Cu (II) ions from water. Chem. Eng. J. **378**, 122157 (2019)
15. Fan, T.: Nano porous piezoelectric energy harvester by surface effect model. Mech. Adv. Mater. Struct. **27**(9), 754–760 (2020)
16. Xiang, G., Tang, Y., Liu, Z., Zhu, W., Liu, H., Wang, J., Zhong, G., Li, J., Wang, X.: Probing ligand-induced cooperative orbital redistribution that dominates nanoscale molecule-surface interactions with one-unit-thin TiO_2 nanosheets. Nano Lett. **18**(12), 7809 7815 (2018)
17. Wang, J., Lian, J., Greer, J., Nix, W., Kim, K.: Size effect in contact compression of nano- and microscale pyramid structures. Acta Mater. **54**(15), 3973–3982 (2006)
18. Liu, G., Xiao, F., Qin, T.: Rock mechanics characteristics and acoustic emission rule under small-size effect. Chin. J. Rock Mech. Eng. **37**, 3905–3917 (2018)
19. Varnavski, O., Ramakrishna, G., Kim, J., Lee, D., Goodson, T.: Optically excited acoustic vibrations in quantum-sized monolayer-protected gold clusters. ACS Nano **4**(6), 3406–3412 (2010)
20. Schwartzberg, A., Zhang, J.: Novel optical properties and emerging applications of metal nanostructures. J. Phys. Chem. C **112**, 10323–10337 (2008)
21. Jun, Y., Seo, J., Cheon, J.: Nanoscaling laws of magnetic nanoparticles and their applicabilities in biomedical sciences. Acc. Chem. Res. **41**(2), 179–189 (2008)
22. Cheon, J., Kang, N., Lee, S., Yoon, J., Oh, S.: Shape evolution of single-crystalline iron oxide nanocrystals. J. Am. Chem. Soc. **126**(7), 1950–1951 (2004)
23. Beck, R., John, P., Homer, M., Whetten, R.: Impact-induced cleaving and melting of alkali-halide nanocrystals. Science **253**(5022), 879–883 (1991)
24. Castro, T., Reifenberger, R., Choi, E., Andres, R.: Size-dependent melting temperature of individual nanometer-sized metallic clusters. Phys. Rev. B **42**(13), 8548–8556 (1990)
25. Halperin, W.: Quantum size effects in metal particles. Rev. Mod. Phys. **58**(3), 533–606 (1986)
26. Ekimov, A., Efros, A., Onushchenko, A.: Quantum size effect in semiconductor microcrystals. Solid State Commun. **56**(11), 921–924 (1985)
27. Fujishima, A., Honda, K.: Electrochemical photolysis of water at a semiconductor electrode. Nature **238**, 37–38 (1972)
28. Tsolekile, N., Parani, S., Matoetoe, M., Songca, S., Oluwafemi, O.: Evolution of ternary I-III-VI QDs: synthesis, characterization and application. Nano-Structures and Nano-Objects **12**, 46–56 (2017)
29. Zhang, Y., Zhang, N., Tang, Z., Xu, Y.: Graphene transforms wide band gap ZnS to a visible light photocatalyst. the new role of graphene as a macromolecular photosensitizer. ACS Nano. **6**(11), 9777–9789 (2012)
30. Thomas, L., Lionti, F., Ballou, R., Gatteschi, D., Sessoli, R., Barbara, B.: Macroscopic quantum tunneling of magnetization in a single crystal of nanomagnets. Nature **383**, 145–147 (1996)

31. Gubin, S., Koksharov, Y., Khomutov, G., Yurkov, G.: Magnetic nanoparticles: preparation, structure and properties. Russ. Chem. Rev. **74**, 489 (2005)
32. Keshtkar, Z., Tamjidi, S., Vaferi, B.: Intensifying nickel (II) uptake from wastewater using the synthesized γ-alumina: an experimental investigation of the effect of nano-adsorbent properties and operating conditions. Environ. Technol. Innov. **22**, 101439 (2021)
33. Lounsbury, A., Yamani, J., Johnston, C., Casanova, P., Zimmerman, J.: The role of counter ions in nano-hematite synthesis: implications for surface area and selenium adsorption capacity. J. Hazard. Mater. **310**, 117–124 (2016)
34. Liu, H., Wu, S.: Improvement of durability of nano-CaO-based adsorbents by chemical looping in situ dynamic pore compensation. Int. J. Environ. Sci. Technol. (2022)
35. Liu, H., Wu, S.: In situ preparation of $CaTiO_3$ and its effect on CO_2 sorption performance of nano-CaO-$CaTiO_3$ adsorbent. Environ. Technol. **43**, 673–683 (2022)
36. Melnyk, I., Pogorilyi, R., Zub, Y., Vaclavikova, M., Gdula, K., Dąbrowski, A., Seisenbaeva, G., Kessler, V.: Protection of thiol groups on the surface of magnetic adsorbents and their application for wastewater treatment. Sci. Rep. **8**, 8592 (2018)
37. Liu, J., Zhao, Z., Jiang, G.: Coating Fe_3O_4 magnetic nanoparticles with humic acid for high efficient removal of heavy metals in water. Environ. Sci. Technol. **42**(18), 6949–6954 (2008)
38. Ma, H., Pu, S., Hou, Y., Zhu, R., Zinchenko, A., Chu, W.: A highly efficient magnetic chitosan "fluid" adsorbent with a high capacity and fast adsorption kinetics for dyeing wastewater purification. Chem. Eng. J. **345**, 556–565 (2018)
39. Wang, W., Xu, Z., Zhang, X., Wimmer, A., Shi, E., Qin, Y., Zhao, X., Zhou, B., Li, L.: Rapid and efficient removal of organic micropollutants from environmental water using a magnetic nanoparticles-attached fluorographene-based sorbent. Chem. Eng. J. **343**, 61–68 (2018)
40. Abuzerr, S., Darwish, M., Mahvi, A.: Simultaneous removal of cationic methylene blue and anionic reactive red 198 dyes using magnetic activated carbon nanoparticles: equilibrium, and kinetics analysis. Water Sci. Technol. **2017**(2), 534–545 (2018)
41. Huang, X., Kong, L., Huang, S., Li, L.: Metal sulfide nanomaterials based adsorbents. Prog. Chem. **29**(1), 83–92 (2017)
42. Zhang, M., Gu, P., Yan, S., Liu, Y., Zhang, G.: Effective removal of radioactive cobalt from aqueous solution by a layered metal sulfide adsorbent: Mechanism, adsorption performance, and practical application. Sep. Purif. Rev. **256**(1), 117775 (2021)
43. Zhang, S., Meng, W., Wang, L., Li, L., Long, Y., Hei, Y., Zhou, L., Wu, S., Zheng, Z., Luo, L., Jiang, F.Z.: Preparation of nano-copper sulfide and its adsorption properties for 17α-Ethynyl estradiol. Nanoscale Res. Lett. **15**, 48 (2020)
44. Pala, I., Brock, S.: ZnS nanoparticle gels for remediation of Pb^{2+} and Hg^{2+} polluted water. ACS Appl. Mater. Interfaces **4**(4), 2160–2167 (2012)
45. Piquette, A., Cannon, C., Apblett, A.: Remediation of arsenic and lead with nanocrystalline zinc sulfide. Nanotechnology **23**(29), 294014 (2012)
46. Mohebbi, S., Mikaeily, S., Azar, A.: Cauliflower-like tin sulphide nanoparticles: an amazing material for adsorption of methylene blue dye. Micro. Nano Lett. **10**, 220–223 (2015)
47. Ji, J., Chen, G., Zhao, J.: Preparation and characterization of amino/thiol bifunctionalized-magnetic nanoadsorbent and its application in rapid removal of Pb (II) from aqueous system. J. Hazard. Mater. **368**, 255–263 (2019)
48. Khodadadi, M., Malekpour, A., Ansaritabar, M.: Removal of Pb (II) and Cu (II) from aqueous solutions by NaA zeolite coated magnetic nanoparticles and optimization of method using experimental design. Microporous Mesoporous Mater. **248**, 256–265 (2017)
49. Shabani, E., Salimi, F., Jahangiri, A.: Removal of arsenic and copper from water solution using magnetic iron/bentonite nanoparticles (Fe_3O_4/bentonite). SILICON **11**, 961–971 (2018)
50. Iijima, S.: Helical microtubules of graphitic carbon. Nature **354**, 56–58 (1991)
51. Liu, C., Fan, Y., Liu, M., Cong, H., Cheng, H., Dresselhaus, M.: Hydrogen storage in single-walled carbon nanotubes at room temperature. Science (New York, N.Y.) **286**(5442), 1127–1129 (1999)
52. Li, Y., Ding, J., Luan, Z., Di, Z., Zhu, Y., Xu, C., Wu, D., Wei, B.: Competitive adsorption of Pb^{2+}, Cu^{2+} and Cd^{2+} ions from aqueous solutions by multiwalled carbon nanotubes. Carbon **41**(14), 2787–2792 (2003)

53. Daraee, M., Saeedirad, R., Ghasemy, E., Rashidi, A.: N-CNT/ZIF-8 nano-adsorbent for adsorptive desulfurization of the liquid streams: experimental and. J. Environ. Chem. Eng. **9**(1), 104806 (2021)
54. Periyasamy, S., Manivasakan, P., Jeyaprabha, C., Meenakshi, S., Viswanathan, N.: Fabrication of nano-graphene oxide assisted hydrotalcite/chitosan biocomposite: an efficient adsorbent for chromium removal from water. Int. J. Biol. Macromol. **132**, 1068–1078 (2019)
55. Pan, Y., Liu, X., Zhang, W., Liu, Z., Zeng, G., Shao, B., Liang, Q., He, Q., Yuan, X., Huang, D., Chen, M.: Advances in photocatalysis based on fullerene C_{60} and its derivatives: properties, mechanism, synthesis, and applications. Appl. Catal. B Environ. **265**, 118579 (2020)
56. Song, W., Zhou, D., Guo, C., Han, F., Ding, L., Lyu, Y., Jia, X., Zheng, P., Deng. X.: Determination of multiresidue of pesticide in chrysanthemum by magnetic fullerene dispersive solid phase extraction with liquid chromatography-tandem mass spectrometry. Chin. J. Anal. Chem. **49**(9), 1587–1596 (2021)
57. Xu, H., Zhang, P., Zhou, S., Jia. Q.: Fullerene functionalized magnetic molecularly imprinted polymer: synthesis, characterization and application for efficient adsorption of methylene blue. Chin. J. Anal. Chem. **48**(9), 20107–20113 (2020)
58. Muhammad, S., Abdul Khalil, H., Hamid, S., Albadn, Y., Suriani, A., Kamaruzzaman, S., Mohamed, A., Allaq, A., Yahya. E.: Insights into agricultural-waste-based nano-activated carbon fabrication and modifications for wastewater treatment application. Agriculture **12**, 1737 (2022)
59. Prakash, M., Gujjala, R., Panchal, M., Ojha. S.: Mechanical characterization of arhar biomass based porous nano activated carbon polymer composites. Polym. Compos. **41**, 3113–3123 (2020)
60. Han, X., Lin, H., Zheng, Y.: Adsorptive denitrogenation and desulfurization of diesel using activated carbons oxidized by $(NH_4)_2S_2O_8$ under mild conditions. Can. J. Chem. **93**, 538–548 (2015)
61. Liu, Z., Sun, Y., Xu, X., Meng, X., Qu, J., Wang, Z., Liu, C., Qu. B.: Preparation, characterization and application of activated carbon from corn cob by KOH activation for removal of Hg(II) from aqueous solution. Bioresour. Technol. **306**, 123154 (2020)
62. Hao, S., Zhong, Y., Pepe, F., Zhu, W.: Adsorption of Pb^{2+} and Cu^{2+} on anionic surfactant-templated amino-functionalized mesoporous silicas. Chem. Eng. J. **189–190**, 160–167 (2012)
63. Hadavifar, M., Bahramifar, N., Younesi, H., Li, Q.: Adsorption of mercury ions from synthetic and real wastewater aqueous solution by functionalized multi-walled carbon nanotube with both amino and thiolated groups. Chem. Eng. J. **237**, 217–228 (2014)
64. Saleh, T.: Isotherm, kinetic, and thermodynamic studies on Hg(II) adsorption from aqueous solution by silica- multiwall carbon nanotubes. Environ. Sci. Pollut. Res. **22**, 16721–16731 (2015)
65. Dinh, T., Phan, M., Nguyen, D., Le, T., Nadda, A., Srivastav, A., Pham, T., Pham, T.: Removal of beta-lactam antibiotic in water environment by adsorption technique using cationic surfactant functionalized nanosilica rice husk. Environ. Res. **210**, 112943 (2022)
66. Ahmadi, A., Foroutan, R., Esmaeili, H., Tamjidi, S.: The role of bentonite clay and bentonite clay@$MnFe_2O_4$ composite and their physico-chemical properties on the removal of Cr(III) and Cr(VI) from aqueous media. Environ. Sci. Pollut. Res. **27**, 14044–14057 (2020)
67. Krishnappa, P., Badalamoole, V.: Microwave assisted synthesis of karaya gum based montmorillonite nanocomposite: characterisation, swelling and dye adsorption studies. Int. J. Biol. Macromol. **154**, 739–750 (2020)
68. Tang, H., Xu, H., Shi, X., Mao, Y., Xie, H., Yang, W.: Fluorine removal from acidic wastewater by a polymer-based nanosized cerium oxides composite adsorbent. Mod. Chem. Ind. **42**(9), 141–145 (2022)
69. Guo, J., Li, J., Wang, C.: Adsorptive removal of Cr(VI) from simulated wastewater in MOF BUC-17 ultrafine powder. J. Environ. Chem. Eng. **7**(1), 102909 (2019)
70. Zarekarizi, F., Morsali, A., Buyukgungor, O.: Rapid and selective water remediation through a functionalized pillar's core of a novel metal-organic framework. Cryst. Growth Des. **20**, 6109–6116 (2020)

71. Sun, Q., Aguila, B., Perman, J., Earl, L., Abney, C., Cheng, Y., Wei, H., Nguyen, N., Wojtas, L., Ma, S.: Post synthetically modified covalent organic frameworks for efficient and effective mercury removal. J. Am. Chem. Soc. **139**(7), 2786–2793 (2017)
72. Jansone-Popova, S., Moinel, A., Schott, J., Mahurin, S., Popovs, I., Veith, G., Moyer, B.: Guanidinium-based ionic covalent organic framework for rapid and selective removal of toxic Cr(VI) oxoanions from water. Environ. Sci. Technol. **53**(2), 878–883 (2019)
73. Kolluru, S., Agarwal, S., Sireesha, S., Sreedhar, I., Kale, S.: Heavy metal removal from wastewater using nanomaterials-process and engineering aspects. Process. Saf. Environ. Prot. **150**, 323–355 (2021)
74. Zhuang, S., Lee, E., Lei, L., Nunna, B., Kuang, L., Zhang, W.: Synthesis of nitrogen-doped graphene catalyst by high-energy wet ball milling for electrochemical systems. Energy Res. **40**, 2136–2149 (2016)
75. Xu, K., Chen, J.: High-resolution scanning probe lithography technology: a review. Appl. Nanosci. **10**, 1013–1022 (2020)
76. Nam, J., Jang, M., Jang, H., Park, W., Wang, X., Choi, S., Byungjin, C.: Room-temperature sputtered electrocatalyst WSe_2 nanomaterials for hydrogen evolution reaction. J. Energy Chem. **47**, 107–111 (2020)
77. Benjamin, J.: Dispersion strengthened superalloys by mechanical alloying. Metall. Mater. Trans. B **1**, 2943–2951 (1970)
78. Amusat, S., Kebede, T., Dube, S., Nind, M.: Ball-milling synthesis of biochar and biochar-based nanocomposites and prospects for removal of emerging contaminants: a review. J. Water Process. Eng. **41**, 101993 (2021)
79. Cai, H., Xu, L., Chen, G., Peng, C., Ke, F., Liu, Z., Li, D., Zhang, Z., Wan, X.: Removal of fluoride from drinking water using modified ultrafine tea powder processed using a ball-mill. Appl. Surf. Sci. **375**, 74–84 (2016)
80. Xiao, J., Hu, R., Chen, G.: Micro-nano-engineered nitrogenous bone biochar developed with a ball-milling technique for high-efficiency removal of aquatic Cd(II), Cu(II) and Pb(II). J. Hazard. Mater. **387**, 121980 (2019)
81. Lyu, H., Xia, S., Tang, J., Zhang, Y., Gao, B., Shen, B.: Thiol-modified biochar synthesized by a facile ball-milling method for enhanced sorption of inorganic Hg^{2+} and organic CH_3Hg^+. J. Hazard. Mater. **384**(15), 121357 (2019)
82. Kumar, A.: Different methods used for the synthesis of TiO_2 based nanomaterials: a review. Am. J. Nano Res. Appl. **6**(1), 1–10 (2018)
83. Zia, A., Birkett, M., Badshah, M., Iqbal, M.: Progress in-situ synthesis of graphitic carbon nanoparticles with physical vapour deposition. Prog. Cryst. Growth Charact. Mater. **67**(3), 100534 (2021)
84. Zhao, Y., Xiao, D., Yang, W., Peng, A., Yao, J.: 2,4,5-triphenylimidazole nanowires with fluorescence narrowing spectra prepared through the adsorbent-assisted physical vapor deposition method. Chem. Mater. **18**(9), 2302–2306 (2006)
85. Baig, N., Kammakakam, I., Falath, W.: Nanomaterials: a review of synthesis methods, properties, recent progress, and challenges. Mater. Adv. **2**, 1821–1871 (2021)
86. Xu, C., Shi, S., Dong, Q., Zhu, S., Wang, Y., Zhou, H., Wang, X., Zhu, L., Zhang, G., Xu, D.: Citric-acid-assisted sol-gel synthesis of mesoporous silicon-magnesium oxide ceramic fibers and their adsorption characteristics. Ceram. Int. **46**(8), 10105–10114 (2020)
87. Sharma, R., Sarkar, A., Jha, R., Sharma, A., Sharma, D.: Sol-gel-mediated synthesis of TiO_2 nanocrystals: structural, optical, and electrochemical properties. Int. J. Appl. Ceram. Technol. **17**(3), 1400–1409 (2020)
88. Salem, S., Hammad, E., Mohamed, A., El-Dougdoug, W.: A comprehensive review of nanomaterials: types, synthesis, characterization, and applications. Biointerface Res. Appl. Chem. **13**, 41 (2022)
89. Saleh, T.: Protocols for synthesis of nanomaterials, polymers, and green materials as adsorbents for water treatment technologies. Environ. Technol. Innov. **24**, 101821 (2021)
90. Liu, J., Sun, Z., Deng, Y., Zou, Y., Li, C., Guo, X., Xiong, L., Gao, Y., Li, F., Zhao, D.: Highly water-dispersible biocompatible magnetite particles with low cytotoxicity stabilized by citrate groups. Angew. Chem. Int. Ed. **48**(32), 5875–5879 (2009)

91. Li, X., Wei, L., Nie, R., Wang, Z., Huang, W., Liu, J., Zhang, X., Chen, Y.: Integrating magnetic metal-organic frameworks-based sample preparation with microchannel resistance biosensor for rapid and quantitative detection of aflatoxin B1. J. Hazard. Mater. **438**(15), 129425 (2022)
92. Maschita, J., Banerjee, T., Lotsch, B.: Direct and linker-exchange alcohol-assisted hydrothermal synthesis of imide-linked covalent organic frameworks. Chem. Mater. **34**(5), 2249–2258 (2022)
93. Chen, X., Geng, K., Liu, R., Tan, K., Gong, Y., Li, Z., Tao, S., Jiang, Q., Jiang, D.: Covalent organic frameworks: chemical approaches to designer structures and built-in functions. Angew. Chem. Int. Ed. **59**(13), 5050–5091 (2020)
94. Sun, J., Wang, C., Zeng, L., Xu, P., Yang, X., Chen, J., Xing, X., Jin, Q., Yu, R.: Controllable assembly of CeO_2 micro/nanospheres with adjustable size and their application in Cr (VI) adsorption. Mater. Res. Bull. **75**, 110–114 (2016)
95. Li, J., Wang, J., Ling, Y., Chen, Z., Gao, M., Zhang, X., Zhou, Y.: Unprecedented highly efficient capture of glycopeptides by Fe_3O_4@Mg-MOF-74 core-shell nanoparticles. Chem. Comm. **53**, 4018–4021 (2017)
96. Fu, M., Deng, X., Wang, S., Yang, F., Lin, L., Zaworotko, M., Dong, Y.: Scalable robust nano-porous Zr-based MOF adsorbent with high-capacity for sustainable water purification. Sep. Purif. Technol. **288**(1), 120620 (2022)
97. Cheng, Y., Wang, K., Qi, Y., Liu, Z.F.: Chemical vapor deposition method for graphene fiber materials. Acta. Phys. -Chim. Sin. **38**, 2006046 (2022)
98. Wang, Q., Lei, Y., Wang, Y., Liu, Y., Song, C., Zeng, J., Song, Y., Duan, X., Wang, D., Li, Y.D.: Atomic-scale engineering of chemical-vapor-deposition-grown 2D transition metal dichalcogenides for electrocatalysis. Energy Environ. Sci. **13**, 1593–1616 (2020)
99. Xiao, B., Rutherford, G., Sharma, A., Pradhan, S., Bonner, C., Bahoura, M.: Surface modification and charge injection in a nanocomposite of metal nanoparticles and semiconductor oxide nanostructures. Sci. Rep. **10**, 1–7 (2020)
100. Hermannsdörfer, J., Friedrich, M., Miyajima, N., Albuquerque, R., Kümmel, S., Kempe, R.: Ni/Pd@ MIL-101: synergistic catalysis with cavity-conform Ni/Pd nanoparticles. Angew. Chem. Int. Ed. **51**(46), 11473–11477 (2012)
101. Liu, F., Li, S., Yu, D., Su, Y., Shao, N., Zhang, Z.: Template-free synthesis of oxygen-doped bundlelike porous boron nitride for highly efficient removal of heavy metals from wastewater. ACS Sustain. Chem. Eng. **6**(12), 16011–16020 (2018)
102. Ma, J., Xiong, Y., Dai, X., Yu, F.: Zinc spinel ferrite nanoparticles as a pseudocapacitive electrode with ultrahigh desalination capacity and long-term stability. Environ. Sci. Technol. Lett. **7**(2), 118–125 (2020)
103. Huang, S., Chen, D.: Rapid removal of heavy metal cations and anions from aqueous solutions by an amino-functionalized magnetic nano-adsorbent. J. Hazard. Mater. **163**(1), 174–179 (2009)
104. Fan, H., Zhou, S., Jiao, W., Qi, G., Liu, Y.: Removal of heavy metal ions by magnetic chitosan nanoparticles prepared continuously via high-gravity reactive precipitation method. Carbohydr. Polym. **174**(15), 1192–1200 (2017)
105. Hassanzadeh-Tabrizi, S., Taheri-Nassaj, E.: Synthesis of high surface area Al_2O_3-CeO_2 composite nanopowder via inverse co-precipitation method. Ceram. Int. **37**(4), 1251–1257 (2011)
106. Subbotina, E., Ram, F., Dvinskikh, S., Berglund, L., Olsén, P.: Aqueous synthesis of highly functional, hydrophobic, and chemically recyclable cellulose nanomaterials through oxime ligation. Nat. Commun. **13**, 1–12 (2022)
107. Helal, S., Abdel-Aziz, H., El-Zayat, M., Hasaneen, M.: Preparation, characterization and properties of three different nanomaterials either alone or loaded with nystatin or fluconazole antifungals. Sci. Rep. **12**, 22110 (2022)
108. Zhang, X., Gärisch, F., Chen, Z., Hu, Y., Wang, Z., Wang, Y., Xie, L., Chen, J., Li, J., Barth, J., Narita, A., List-Kratochvil, E., Müllen, K., Palma, C.: Self-assembly and photoinduced fabrication of conductive nanographene wires on boron nitride. Nat. Commun. **13**, 1–8 (2022)

109. Zhang, X., Liu, Y., Zheng, B., Zang, J., Lv, C., Zhang, T., Wang, H., Zhao, G.: Protein interface redesign facilitates the transformation of nanocage building blocks to 1D and 2D nanomaterials. Nat. Commun. **12**, 1–11 (2021)
110. Sharifpour, E., Arabkhani, P., Sadegh, F., Mousavizadeh, A., Asfaram, A.: In-situ hydrothermal synthesis of CNT decorated by nano ZnS/CuO for simultaneous removal of acid food dyes from binary water samples. Sci. Rep. **12**, 1–20 (2022)
111. Wu, H., Wu, Q., Zhang, J., Gu, Q., Guo, W., Rong, S., Zhang, Y., Wei, X., Wei, L., Sun, M., Li, A., Jing, X.: Highly efficient removal of Sb(V) from water by franklinite-containing nano-FeZn composites. Sci. Rep. **11**, 1–17 (2021)
112. Zheng, S., Xia, S., Han, S., Yao, F., Zhao, H., Huang, M.: β-Cyclodextrin-loaded minerals as novel sorbents for enhanced adsorption of Cd^{2+} and Pb^{2+} from aqueous solutions. Sci. Total. Environ. **693**(25), 133676 (2019)
113. Zhao, Y., Wu, R., Yu, H., Li, J., Liu, L., Wang, S., Chen, X., Chan, T.: Magnetic solid-phase extraction of sulfonamide antibiotics in water and animal-derived food samples using core-shell magnetite and molybdenum disulfide nanocomposite adsorbent. J. Chromatogr. A **1610**(11), 460543 (2020)
114. Mubarak, M., Jeon, H., Islam, M., Yoon, C., Bae, J., Hwang, S., Choi, W., Lee, H.: One-pot synthesis of layered double hydroxide hollow nanospheres with ultrafast removal efficiency for heavy metal ions and organic contaminants. Chemosphere **201**, 676–686 (2018)
115. Zhang, W., Xu, H., Xie, F., Ma, X., Niu, B., Chen, M., Zhang, H., Zhang, Y., Long, D.: General synthesis of ultrafine metal oxide/reduced graphene oxide nanocomposites for ultrahigh-flux nanofiltration membrane. Nat. Commun. **13**, 471 (2022)
116. Wu, Z., Deng, W., Zhou, W., Luo, J.: Novel magnetic polysaccharide/graphene oxide @Fe_3O_4 gel beads for adsorbing heavy metal ions. Carbohyd. Polym. **216**(15), 119–128 (2019)
117. Han, B., Zhang, M., Zhao, D., Feng, Y.: Degradation of aqueous and soil-sorbed estradiol using a new class of stabilized manganese oxide nanoparticles. Water Res. **70**(1), 288–299 (2015)
118. Roushani, M., Baghelani, Y., Mavaei, M., Abbasi, S., Mohammadi, S.: Preparation of modified magnetic cobalt nanoparticles as a new magnetic sorbent for the preconcentration and determination of trace amounts of lead ions in environmental water and soil (air-dust) samples. Commun. Soil Sci. Plant Anal. **49**(6), 645–657 (2018)
119. Fang, J., Shan, X., Wen, B., Huang, R.: Mobility of TX100 suspended multiwalled carbon nanotubes (MWCNTs) and the facilitated transport of phenanthrene in real soil columns. Geoderma **207**, 1–7 (2013)
120. Zendehdel, R., Ansari, S., Sedghi, R., Jafari, M.: Magnetic nano-zeolite Y as a novel fluidized bed for air decontamination. J. Environ. Sci. Technol. **16**, 1261–1268 (2019)
121. Buyukada-Kesici, E., Gezmis-Yavuz, E., Aydin, D., Cansoy, C., Alp, K., Koseoglu-Imer, D.: Design and fabrication of nano-engineered electrospun filter media with cellulose nanocrystal for toluene adsorption from indoor air. Mater. Sci. Eng. B **264**, 114953 (2021)
122. Sarvestani, N., Tabasizadeh, M., Abbaspour-Fard, M., Nayebzadeh, H., KarimiMaleh, H., Van, T., Jafari, M., Ristovski, Z., Brown, R.: Influence of doping Mg cation in Fe_3O_4 lattice on its oxygen storage capacity to use as a catalyst for reducing emissions of a compression ignition engine. Fuel **272**(15), 117728 (2020)

Adsorption Isotherms and Kinetic Models

M. Kanagalakshmi, S. Gopika Devi, P. Ananthi, and Anitha Pius

Abstract Adsorption is a significant phenomenon that underlies a variety of crucial technical and environmental processes. There is no question about the importance of adsorption in protecting the environment and the industrial sector. Additionally, the first stage in many catalytic processes is adsorption of reactants over the catalyst. Therefore, numerous attempts have been made to explore the various adsorption process elements. However, a vital stage in the design and operation of adsorption equipment is having the working grasp of adsorption equilibrium and kinetics. To comprehend the adsorption equilibrium and kinetics, several different isotherms and kinetics models have been created. This chapter makes an effort to present an overview of a particular set of theories, isotherms and kinetic models used to describe adsorption events at the gas–solid interface. This chapter is split into two sections: the first section covers the theories and models (kinetics). The models that are used to analyze the isotherms of adsorption are the main topic of the second section.

Keywords Adsorption · Isotherms · Kinetics · Pseudo-first-order · BET

1 Introduction

One or more adsorbates may be used in an adsorption process and they may be fixed to an adsorbent by chemical or physical bonding. Due to its simplicity, effectiveness, viability from an economic standpoint and social acceptability, adsorption is frequently employed in the wastewater treatment process [1]. Molecular sieves, polymeric adsorbents, activated carbon, and various other inexpensive substances are the examples of common adsorbents. Adsorption should be studied from a thermodynamic and kinetic perspective to learn more about its functioning and processes. In addition to adsorption capacity, a chosen adsorbent's kinetic performance is the crucial for the pilot application [2].

M. Kanagalakshmi · S. G. Devi · P. Ananthi · A. Pius (✉)
Department of Chemistry, The Gandhigram Rural Institute – Deemed to be University, Gandhigram, Dindigul, Tamil Nadu 624 302, India
e-mail: dranithapius@gmail.com

J. Tharini and S. Thomas (eds.), *Carbon Nanomaterials and their Composites as Adsorbents*, Carbon Nanostructures, https://doi.org/10.1007/978-3-031-48719-4_8

The kind and type of adsorption is determined by the degree of contact between the molecules of the adsorbates and the adsorbents. The mechanism can also be categorized as either physisorption (physical adsorption) or chemisorption (chemical adsorption). Van der Waals forces regulate the molecular interactions during physisorption, whereas valence forces are involved during chemisorptions [3]. The physisorption (low enthalpy of adsorption, formation of multilayers and reversibility) or chemisorption (electron transfer, formation of ionic or covalent bonds, occurring on monolayers only and irreversibility) of the adsorption process is determined by the thermodynamic parameters, whose estimation methodology is discussed [4].

The three basic phases of mass-transfer-governed process of adsorption are: (i) moving the adsorbate from the bulk to the exterior (ii) moving into the pores and (iii) binding the adsorbate to the active sites of the adsorbent. As a result, chemical interactions as well as the contribution of adsorbent and intraparticle diffusion processes affect the rate of adsorption. Kinetic models seek to explain dynamic behavior, comprehend adsorbent-adsorbate interactions, and offer an understanding of the intricate adsorption mechanism. The pseudo-first-order and pseudo-second-order models are the most often utilized equations, even though many sophisticated equations such as those by Elovich, Avrami, Bangham or film diffusion models have been produced [5].

For environmental remediation and wastewater treatment facilities in particular, understanding the design and operation of adsorption systems and processes is crucial. Understanding the adsorption process of contaminants on adsorbents requires a thorough grasp of adsorption isotherm models. The equilibrium of the sorption of adsorbate on the surface of adsorbents is also demonstrated by adsorption isotherms, which gives essential information for optimizing adsorbents for the removal of contaminants from the environment [6].

2 Adsorption Kinetics

Adsorption is a complicated phenomenon that usually takes an approach that combines surface adsorption and diffusion into the pores. Adsorption kinetics are mainly controlled by numerous steps, including diffusion and reaction processes (Fig. 1).

Step1. Molecule and/or ion transfer from the bulk solution to the boundary film bordering the adsorbent surface.

Step 2. Ion transport from boundary film to the surface of the adsorbent.

Step 3. Transfer of ion from the surface to the intraparticular active sites and porous structure.

Step 4. Chemical reaction, i.e., adsorption reaction on adsorption sites via chelating, ion exchange or complexation.

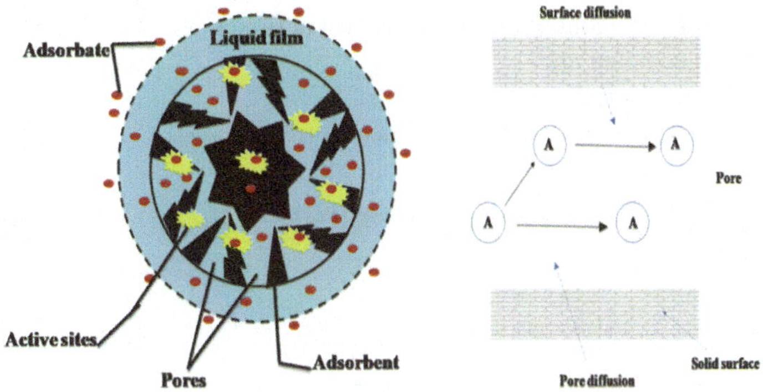

Fig. 1 Schematic expression of mass transfer resistances in the pathway of species adsorption by porous adsorbents

2.1 *Kinetic Model of Avrami*

According to this concept, the reaction occurs at the solid support's active surface sites, and its equation has the following two forms: exponential and double logarithmic, respectively represented by Eqs. (1) and (2)

$$Q_1 = Q_{av}(1 - e^{(k_{av}t)n_{av}}) \tag{1}$$

$$\ln(-\ln(1 - Q_t)) = \ln(K_{av}) + n_{av}\ln(t) \tag{2}$$

where Q_{av}: Avrami theoretical value of the amount of the adsorption (mg g^{-1})

K_{av}: Avrami rate constant

t: time of contact (min)

n_{av}: Avrami order model and

Q_t: the amount of adsorbate in the adsorbent at time t (mg g^{-1}).

For the exponential model, it is easier to graph Q_t as a function of contact time t, whereas, for the double logarithmic model, it is more convenient to graph ln (-ln (1-Q_t) as a function of contact time t), noting the coefficient of determination R^2 [7].

2.2 *Elovich's Equation*

Zeldowitsch (1934) developed the so-called Elovich equation, a kinetic equation of chemisorption, to characterize the rate of carbon monoxide adsorption on manganese dioxide, which falls exponentially with increasing gas adsorbed:

$$\frac{dq}{dt} = ae^{-aq} \tag{3}$$

where q represents the amount of gas adsorbed at time t
a desorption constant and
α is the initial adsorption rate.
Equation (3) can be rearranged to a linear form:

$$q = \left(\frac{2.3}{\alpha}\right)\log(t + t_o) - \left(\frac{2.3}{\alpha}\right)logt_o \tag{4}$$

with

$$t_o = \frac{1}{\alpha a} \tag{5}$$

A straight line with the appropriate value of t_0 should result from plotting q vs $\log(t + t_0)$. The Elovich equation is used to calculate the chemisorption kinetics of gases onto heterogeneous solids.

With the assumption of aαt >> 1 Eq. (4) was integrated by using the boundary conditions of $q = 0$ at $t = 0$ and $q = q$ at $t = t$ to yield:

$$q = \alpha\ln(\alpha a) + \alpha lnt \tag{6}$$

The adsorption of gas onto solid materials is best described by Elovich's equation. I has seen used to explain the adsorption of contaminants from aqueous solutions, such as the removal of Cadmium from effluents using bone char and the adsorption of Cr(VI) and Cu(II) by chitin, chitosan and Rhizopus arrhizus [8].

2.3 Lagergren's Pseudo-First-Order

The liquid–solid adsorption system based on the pseudo-first-order equation is the most frequently used with the Lagergren kinetic. The following ordinary first-order differential equation characterizes the adsorption kinetics of a species in an adsorbent particle in this model Eq. (7)

$$\frac{dQ}{dt} = K_1(Q_e - Q_t) \tag{7}$$

The differential equation Eq. (7) demonstrates that the adsorption capacity is inversely correlated with the "distance to equilibrium," which is measured by the difference between the average concentration of the species in the adsorbed phase and the final concentration of the phase adsorbed in equilibrium with the fluid phase: For a clean particle, the average species concentration in the adsorbed phase is zero at time $t = 0$. As time goes on, the distance at equilibrium decreases and is neutralized when equilibrium is attained, being equal to the ultimate concentration of the adsorbed phase in equilibrium with the fluid phase. The average concentration of species in the adsorbed phase is equal to the concentration of the final adsorbed phase in equilibrium with the fluid phase, which is equal to zero.

By integration of this differential equation Eq. (7) for the boundary conditions: qt = 0 to t = 0 and qt = qt to t = t, the following pseudo-first-order Lagergren equation was obtained Eq. (8)

$$ln(Q_e - Q_t) = ln(Q_e) - K_1 t \tag{8}$$

where Q_e: the amount of adsorbate in the adsorbent at equilibrium (mg g^{-1})

Q_t: the amount of adsorbate in the adsorbent at time t (mg g^{-1});

K_1: constant rate of Lagergren's first order and

t: time of contact (min).

To test this isotherm, it is convenient to represent ln (Q_e-Q_t) graphically as a function of the contact time t and to note the coefficient of determination R^2 [9].

2.4 *Pseudo-Second-Order*

The plots of t/q_t against t from Eq. (9) and the pseudo-second-order parameters, q_e and k_2 were obtained from the pseudo-second-order plot. The correlation coefficients, R^2 values, at different concentrations are much higher (>0.99) than the pseudo-first-order. These results indicate that pseudo-second-order kinetics is applicable.

$$\frac{t}{q_t} = \frac{1}{k_2 q_e^2} + \frac{1}{q_e} \tag{9}$$

where the slope and intercept of a plot of t/q_t vs t can be used to empirically calculate the second-order constants k_2 (g/mg.min) [8].

2.5 *Intraparticle Diffusion Kinetic Models*

Weber–Morris Diffusion Kinetic Model

In this model, intraparticle diffusion significantly affects the kinetics of the adsorption process and the plots of q_t vs $t_{1/2}$ based on Eq. (10) produce straight lines that pass through the origin which yields the rate, constant k_{id}. Additionally, the contribution of surface adsorption in the rate-controlling phase increases with the size of the intercept, c.

$$q_1 = k_{id} t^{1/2} + C \tag{10}$$

where k_{id} is the intraparticle diffusion rate constant (g/mg.min$^{1/2}$) and c is the intercept in the plot of q_t against $t_{1/2}$ that is related to the thickness of the boundary layer [2].

3 Adsorption Isotherms

The amount of material adsorbed by a substrate is often expressed as a function of the equilibrium concentration at the constant temperature. The process of representing this function is known as adsorption isotherms.

3.1 One Parameter Isotherm Model

Henry's Isotherm Model

In Henry's Isotherm model, partial pressure of the adsorptive fluid is exactly proportional to the amount of adsorbate. Henry's adsorption constant, which appears in linear adsorption isotherms and is analogous to Henry's gas law. It is based on Gibbs adsorption and it is used to identify the equilibrium adsorption state for adsorbates at the constant temperature. As the result, the partial pressure of the adsorptive fluid and the equilibrium quantity of adsorbate in the fluid are connected in Henry's adsorption isotherm by

$$Q_e = K_H C_e \tag{11}$$

where Q_e is the amount of adsorbate at equilibrium condition (mg/g)

K_H is Henry's adsorption constant and

C_e is the equilibrium concentration of the adsorbate on the adsorbent.

Henry's isotherm is not suitable for high temperature and pressure experiments due to the violent molecular motion based on the strong contact force and potential energy [10].

3.2 Two Parameter Isotherm Models

Dubinin-Radushkevic (D-R) Isotherm

This isotherm assumes that the features of the adsorption curve are linked to the porosity of the adsorbent. It helps us to identify the type of adsorption that take place such as chemisorption or physisorption, and also to calculate the apparent adsorption energy. D-R does not presume that the adsorbent's surface is uniform. The model's linear form is written as:

$$lnq_e = \ln Q_D - B_D + \left[RTln\left(1 + \frac{1}{C_e}\right)\right]^2 \quad (12)$$

where Q_D (mol/g) represents the maximum adsorption capacity and B_D (mol^2 /KJ^2) are D-R constants derived from the slope and intercept of the plot of ln q_e versus RT ln (1 + 1/C_e). It is possible to calculate the mean free energy of adsorption, E (KJ/mol), which is the amount of energy needed to move one mole of adsorbate from solution to the surface of an adsorbent:

$$E = \frac{1}{\sqrt{2BD}} \quad (13)$$

B_D (mol^2 /KJ^2) is the free energy of sorption per mole of the adsorbate as it moves towards the surface of the adsorbent and Q_D (mol/g) is related to the extent of adsorption of adsorbate [11].

Elovich Isotherm

Multilayer adsorption was implied by the equation defining this model, which is based on a kinetic principle that states that adsorption sites grow exponentially with adsorption. The kinetics of gas chemisorption onto solids were the initial target of the equation's derivation. The following expressions represent the Elovich model:

$$\frac{q_e}{q_m} = K_E C_e e \frac{q_e}{q_m} \quad (14)$$

but the linear form is expressed as follows:

$$\ln\frac{q_e}{C_e} = \ln K_e q_m - \frac{q_e}{q_m} \quad (15)$$

Elovich maximum adsorption capacity and Elovich constant can be calculated from the slope and intercept of the plot of ln(q_e/C_e) versus q_e [12].

Flory–Huggins Isotherm

It is possible to predict the thermodynamic viability of an adsorption process using the Flory–Huggins isotherm model, which depends on how much the adsorbent surface is covered by the substance being adsorbed. The relation represents the Flory–Huggins isotherm model.

$$\frac{\theta}{C_o} = K_{FH}(1 - \theta)^{n_{FH}} \quad (16)$$

$$\theta = 1 - \frac{C_e}{C_o} \quad (17)$$

where K_{FH} is the Flory–Huggins' constant [L mg^{-1}]. The n_{FH} parameter represents the number of adsorbate ions occupying sorption sites. Further, the equilibrium constant, K_{FH}, can be used to inspect the spontaneity of the reaction by calculating Gibbs free energy using the relation [13].

$$\Delta_G = -RTlnK_{FH} \tag{18}$$

Fowler-Gunggenheim Isotherm

During the adsorption process, the Fowler–Guggenheim method takes into account the lateral interactions of the adsorbed molecules. One of the most fundamental correlations, the model enables the prediction of the lateral interaction between adsorbates. This model is discussed in detail:

$$K_{FG}.C_e = \frac{\theta}{1-\theta}\exp\left(\frac{2.\theta.W}{R.T}\right) \tag{19}$$

K_{FG} stands for the L/mg Fowler–Guggenheim equilibrium constant. W is the interaction energy (kJ/mol) between molecules that have been adsorbed, and its value is correlated with the adsorption of heat. The temperature changes linearly as more molecules are adsorbed onto the surface of the adsorbent.

(i) $W > 0$ kJ/mol denotes both the presence of exothermic reactions and the attraction between adsorbed molecules.
(ii) $W < 0$ kJ/mol is the result of endothermic reactions and the attraction of adsorbed molecules.
(iii) when $W = 0$ kJ/mol the adsorbed molecules do not interact [14].

Freundlich Isotherm

A physical kind of adsorption known as a multilayer weakly bonded Freundlich isotherm characterizes this sort of adsorption. The heterogeneity of the adsorption sites is another presumption made by the Freundlich isotherm. The empirical relationship for expressing Freundlich isotherm is given in the equation:

$$lnQ_e = lnK_f + \frac{1}{n}lnC_e \tag{20}$$

where K_f is the Freundlich constant

C_e is the concentration of adsorbate under equilibrium conditions (mg/L)

Q_e is the amount of adsorbate absorbed per unit of adsorbent (mg/g) and

n is the value indicating the degree of linearity between the adsorbate solution and the adsorption process.

The value of n is described as follows:

(i) $n = 1$, linear adsorption.
(ii) $n < 1$, adsorption process with chemical interaction.

(iii) $n > 1$, adsorption process with physical interaction.
(iv) Favorable adsorption process is declared when $0 < 1/n < 1$, and a cooperative adsorption process occurs when $1/n > 1$ [15].

Halsey Isotherm

This equation is appropriate for assessing the multilayer adsorption system for adsorbate ions adsorption at the significant distance from the surface. The Halsey adsorption isotherm may be expressed as Eq. (21)

$$q_e = \exp\left(\frac{lnK_H - \ln C_e}{m_H}\right) \tag{21}$$

The fact that the experimental findings coincide with this equation, especially at the higher concentrations which supports the adsorbent's heterogeneous pore distribution. When it comes to isotherms of type II that occur in heterosporous substances, this equation provides a fair description of the adsorption data. Comparing multilayer adsorption isotherm models to the literature, the Halsey model best matches the experimental results [16].

Harkin-Jura Isotherm

The Harkin-Jura isotherm model, which applies to solid–gas systems, is predicated based on the concept that multilayer adsorption may take place on the surface of absorbents with heterogeneous pore distribution. This Harkin-Jura isotherm model is expressed as follows:

$$\frac{1}{Q_e^2} = \frac{B}{A} - \left(\frac{1}{A}\right)\log C_e \tag{22}$$

where A and B are Harkin-Jura constants [10].

Hill Isotherm

Hill's equation was proposed as a theory to explain how different species attach to the homogenous substrates. According to the model, adsorption is a cooperative process in which the capacity of a macromolecule to bind with the ligand at one site could affect other binding sites at the same macromolecule. The results of the Hill model's calculations and Hill isotherm's linear form are displayed in

$$q_e = \frac{q_H C_e^{nH}}{K_D + C_e^{nH}} \tag{23}$$

where K_D, n_H, and q_H are constants [17].

Hill-Deboer Isotherm

The case where there is mobile adsorption, as well as lateral contact between adsorbed molecules, is described by the Hill-Deboer isotherm model. This isotherm equation's linearized form is as follows

$$\ln\left[\frac{C_e(1-\theta)}{\theta}\right]-\frac{\theta}{1-\theta}=-lnK_1-\frac{K_2\theta}{RT} \tag{24}$$

where K_1 is Hill-Deboer constant (Lmg^{-1}) and K_2 is the energetic constant of the interaction between adsorbed molecules ($KJmol^{-1}$). Equilibrium data from adsorption experiments can be analyzed by plotting $\ln[Ce(1-\theta)/\theta]-\theta/(1-\theta)$ versus θ [14].

Jovanovic Isotherm

The Langmuir model and the Jovanovic isotherm model both make the same hypothesis. But it also takes into account the potential of mechanical interactions between the molecules that are adsorbing in addition to the Langmuir model. The Jovanovic model is also valid for localized adsorption, in contrast to the Langmuir model, which is only valid for mobile adsorption. The general Jovanovic equation for the homogeneous surface is given by the expression:

$$Q_e(P,T,e)=A[1-\exp(-bP)] \tag{25}$$

where A is the adsorption capacity at equilibrium and b is the Jovanovic parameter defined by:

$$b=b_o\exp\left(\frac{q}{RT}\right) \tag{26}$$

b_0 is the corresponding limiting value of b as the temperature approaches infinity, q is the isosteric heat of adsorption. The linearized form of the Jovanovic isotherm model is shown below:

$$\ln Q_e=\ln Q_{max}-K_jC_e \tag{27}$$

where Q_e is the amount of adsorbate in the adsorbent at equilibrium ($mg\ g^{-1}$), Q_{max} is the maximum uptake of adsorbate obtainable when ln Q_e is plotted against C_e, K_J is the Jovanovic constant.

A modified form of the Jovanovic isotherm model known as the Jovanovic-Freundlich isotherm model has proven to better fit the optimization model for L-Lysine imprinted Polymer [10].

Kiselev Isotherm

The Kiselev adsorption isotherm equation is a localized monomolecular layer model. Its linearized expression is as follows Eq. (28):

$$\ln\left[\frac{1}{C_e(1-\theta)}\right] = \frac{K_1}{\theta} + K_1 K_n \tag{28}$$

where Ki is Kiselev equilibrium constant (Lmg^{-1}) and K_n is the equilibrium constant of the formation of a complex between adsorbed molecules determined from the linear plot of 1/C_e (1-q) versus 1/q [18].

Langmuir Isotherm

The Langmuir isotherm was primarily created for the interaction between gases and solids, but it is also utilized for other adsorbents. The surface rates of adsorption and desorption are equal with zero accumulation under equilibrium circumstances since it is an empirical model based on kinetic principles. The Langmuir isotherm may be expressed as follows under the following conditions: (a) monolayer adsorption; (b) homogenous sites; (c) constant adsorption energy; and (d) absence of lateral contact between the adsorbed molecules.

$$q_e = \frac{q_o K_L C_e}{1 + K_L C_e} \tag{29}$$

where q_o is the maximum amount of adsorbed surfactant in mg/ g and

K_L is the Langmuir constant in L/mg.

The linearized version is

$$\frac{C_e}{q_e} = \frac{1}{K_L q_o} + \frac{C_e}{q_o} \tag{30}$$

A plot between C_e/q_e versus C_e will generate a straight line with a slope of $1/q_o$ and an intercept equal to $1/K_L q_o$.

Only one molecule can bind to each of the identical adsorption sites required by the monolayer assumption. Once a surfactant molecule has occupied a site, there is no longer any adsorption taking place there. At very low concentrations ($K_L C_e$ 1), this model transforms into Henry's model. The Langmuir isotherm model seems to have a single plateau according to the L-shaped curve.

The separation factor, also known as the equilibrium parameter (R_L) is a significant parameter associated with the Langmuir model that is used to determine whether surfactant adsorption is advantageous or disadvantageous. It may be demonstrated mathematically as

$$R_L = \frac{1}{1 + K_L C_o} \tag{31}$$

where K_L and C_o are the Langmuir constants and highest initial concentration of surfactant, respectively.

In general, $R_L < 1$ indicates that adsorption is favorable

$R_L \sim 0$ indicates that adsorption is irreversible

$R_L = 1$ indicates that the adsorption isotherm is linear and

$R_L > 1$ corresponds to unfavorable adsorption [19].

Temkin Isotherm

This isotherm has a component that explicitly accounts for the interactions between the adsorbent and adsorbate. The model assumes that the heat of adsorption (a function of temperature) of all molecules in the layer will diminish linearly rather than logarithmically with coverage by disregarding the extremely low and high concentration values. By plotting the amount sorbed q_e versus ln C_e and calculating the constants from the slope and intercept, it was possible to demonstrate that, as predicted by the equation, the derivation is characterized by a uniform distribution of binding energies (up to some maximum binding energy). The following equation represtents the model:

$$q_e = \frac{RT}{b} \ln(A_T C_e) \tag{32}$$

$$q_e = \frac{RT}{b_T} \ln A_T + \left(\frac{RT}{b}\right) \ln C_e \tag{33}$$

$$B = \frac{RT}{b_T} \tag{34}$$

$$q_e = B \ln A_T + B \ln C_e \tag{35}$$

A_T = Temkin isotherm equilibrium binding constant (L/g).
b_T = Temkin isotherm constant
R = universal gas constant (8.314 J/mol/K)
T = Temperature at 298 K
B = Constant related to heat of sorption(J/mol) [12].

3.3 Three-Parameter Isotherms

Brunauer–Emmett–Teller Isotherm

In gas–solid equilibrium systems, the Brunauer–Emmett–Teller (BET) isotherm is a theoretical model that is often employed. BET is an expanded variant of the Langmuir isotherm that is used to create multilayer adsorption systems. From data on nitrogen adsorption, the BET model is frequently used to calculate the surface area of an adsorbent. The extension of this model to a liquid–solid interface is described by Eq. (36)

$$q_e = \frac{q_{mBET} C_{BET} C_e}{(C_e - C_s)[1 + (C_{BET} - 1)\frac{C_e}{C_s}]} \tag{36}$$

The energy of surface contact is correlated with the C_{BET} parameter. This idea is valid for relative concentrations between 0.02 and 0.4. The adsorption energy does not vary with the rate of adsorption in the same layer, and there is no interaction between adsorbed molecules. This isotherm is based on the same assumptions as the Langmuir isotherm: surface and distribution of sites are uniform, and the surface is energetically homogenous. The rate of desorption from each layer is also equal to the rate of adsorption on that layer. Other simplifying assumptions were added to this model by multilayer adsorption of the BET model: the second, third and higher layers have the same adsorption energy, which equals the heat of fusion and is not directly affected by adsorbent-adsorbate interactions. However, compared to the second or subsequent levels, the energy for the top layer is different. In addition, the number of layers tends to be infinite as the concentration approaches the saturation concentration [16].

Jossens Isotherm

The distribution of energy of adsorbate-adsorbent interactions on adsorption sites is the basis of the model developed by Jossens et al. It considers that the activated carbon surface is heterogeneous, concerning the interactions which it engages with the adsorbate:

$$C_e = \frac{q_e}{H}\exp(Fq_e^p) \tag{37}$$

where q_e is the adsorbed amount at equilibrium (mg g^{-1})

C_e is the equilibrium concentration of the adsorbate (mg L^{-1}) and

H, F, and p are the parameters of the equation of Jossens. H and F depend only on temperature.

This equation can be reduced to Henry's law at low capacities [20].

Kahn Isotherm

The Kahn isotherm model represents the adsorption potential to be temperature independent. It suggests that the nature of the adsorbent and adsorbate is dependent on the adsorption procedure. Kahn isotherm equation is given as follows:

$$q_e = \frac{q_{max} b_K C_e}{(1 + b_K C_e) a_K} \tag{38}$$

where a_k is the exponent value of Kahn isotherm and

b_k signifies the Kahn isotherm constant [21].

Koble–Corrigan Isotherm

The Sips isotherm model is similar to the Koble-Carrigan isotherm model. This model combines the Freundlich isotherm and the Langmuir isotherm. Koble–Carrigan isotherm model is given in the following equation:

$$q_{eq} = \frac{A_{KC} C_{eq}^{nKC}}{1 + B_{KC} C_{eq}^{nKC}} \tag{39}$$

The Freundlich isotherm is reached by this model at high adsorbate concentrations. The constant n must be bigger than or equal to 1 for it to be true. When n is less than 1, it means that the model, although having a high concentration coefficient or a low error value, is unable to define the experimental data [22].

Langmuir–Freundlich Isotherm

The Langmuir and Freundlich isotherms are the most often utilized analytical isotherms for simulating the adsorption of arsenic on iron oxides. The Langmuir–Freundlich isotherm, also referred to as Sip's equation, is a flexible isotherm expression that can simulate both Langmuir and Freundlich behaviors. Langmuir–Freundlich isotherm, which can also be used to simulate pH-dependent sorption effects.

A general form of the Langmuir–Freundlich isotherm equation can be written as:

$$q = \frac{Q_m (K_a C_{eq})^n}{(K_a C_{eq})^n + 1} \tag{40}$$

where q is the amount of As(V) adsorbed on the sand at equilibrium (mg As(V)/ g sand)

Q_m is the adsorption capacity of the system (mg of sorbate/ g sorbent), which can also be expressed as N_t, which is a measure of the total number of binding sites available per gram of sorbent

C_{eq} is the aqueous phase concentration at equilibrium (mg/ L).

K_a is the affinity constant for adsorption (L/mg) n is the index of heterogeneity [23].

Radke-Prausnitz Isotherm

Radke-Prausnitz model is the favored option for the majority of adsorption systems with low adsorbate concentrations due to several important characteristics. The isotherm model becomes a linear isotherm at low adsorbate concentrations. It approaches the Freundlich isotherm at large adsorbate concentrations, and when $n_{RP} = 0$, it transforms into a Langmuir isotherm. This isotherm's ability to provide an accurate match over a wide range of adsorbate concentrations is another essential characteristic. Radke-Prausnitz equation can be expressed as:

$$q = \frac{q_{mRP} b_{RP} p}{(1 + b_{RP} p)^{n_{RP}}} \tag{41}$$

where q_{mRP} is the maximum adsorption capacity [mmol/g]

b_{RP} is the Radke-Prausnitz constant [bar^{-1}].

n_{RP} is Radke-Prausnitz model exponent [24].

Redlich-Peterson Isotherm

The linear form of R–P isotherm equation.

The R–P isotherm equation is expressed as

$$q_e = \frac{q_{mon'} b_{RP} C_e}{1 + b_{RP} C_e^{\alpha}} \tag{42}$$

where q'_{mon} and b_{RP} are parameters of the R–P isotherm equation.

Two linear forms of Eq. (42) can be obtained by transformation as

$$\ln\left(b_{RP} q_{mon'} \frac{C_e}{q_e} - 1 \right) = ln b_{RP} + \alpha ln C_e \tag{43}$$

and

$$\frac{C_e}{q_e} = \frac{1}{b_{RP} q_{mom'}} + \left(\frac{1}{q_{mom'}} \right) C_e^{\alpha} \tag{44}$$

The logarithmic linear form of Eq. (43) was adopted by many researchers. For fitting Eq. (43) to the experimental data to obtain a linear plot of ln b_{RP} q'_{mon} (C_e/q_e)−1) vs ln C_e, various constant (b_{RP} q'_{mon}) values must be tried before the optimum line is obtained. After obtaining the optimum line the constants of Eq. (43) are to be calculated. The range of b_{RP} q'_{mon} values is wide from 0.01 to several hundred, so it is not easy to obtain the correct value. Equation (44) is the exponential linear form obtained by plotting C_e/q_e vs C_e^{α} [25].

Sips Isotherm

Sips or Langmuir–Freundlich isotherm has the following form:

$$Q = \frac{K_S C_f^{\beta_S}}{1 + a_S C_f^{\beta_S}} \tag{45}$$

where K_S is the Sips model isotherm constant (L/g)

a_S is the Sips model constant (L/mg) and

β_S the Sips model exponent.

It efficiently decreases to the Freundlich isotherm at low sorbate concentrations, opposing Henry's rule. It predicts a monolayer sorption capacity like that of the Langmuir isotherm at high sorbate concentrations.

Sips model constants showed a similar pattern to Redlich-Peterson model constants [26].

Toth Isotherm

The Toth isotherm is another empirical version of the Langmuir equation to minimize the discrepancy between experimental data and equilibrium data predicted by the equation. The most effective use of this model is in the description of heterogeneous adsorption systems that fulfill the low and high-end boundary of adsorbate concentration. The following is how the Toth isotherm model is expressed:

$$\frac{q_e}{q_m} = \theta = \frac{K_e C_e}{[1 + (K_L C_e)^n]^{1/n}} \tag{46}$$

where K_L is Toth isotherm constant (mg g^{-1}) and
n is Toth isotherm constant (mg g^{-1}).

This equation simplifies to the Langmuir isotherm equation when n = 1. As a result, the parameter n describes the adsorption system's heterogeneity, and if it deviates from unity (1) further, the system is considered heterogeneous. The Toth isotherm may be rearranged to give a linear form as follows:

$$ln \frac{q_e^n}{q_m^n - q_e^n} = nlnK_L + n\ln C_e \tag{47}$$

The values of parameters of the Toth model can be evaluated by the nonlinear curve fitting method using sigma plot software.

This isotherm model has been applied for the modeling of several multilayer and heterogeneous adsorption systems [27].

3.4 Four-Parameter Isotherms

Baudu Isotherm

According to Baudu, the Langmuir coefficients, b and q_{mL}, are not constants across a wide concentration range when calculated by measuring tangents at various equilibrium concentrations. Their variations can be written in the following forms:

$$b = b_o C_e^x \tag{48}$$

$$q_{m_L} = q_{m0} C_e^y \tag{49}$$

Graphical study of ln b = f(lnC_e) and ln q_{mL} = f (lnC_e) gives access to b_0, q_{m0}, x, and y. Baudu has transformed the Langmuir equation into the following expression:

$$q_e = \frac{q_{m0} b_0 C_e^{(1+x+y)}}{1 + b_0 C_e^{(1+x)}} \tag{50}$$

with (1+x+y) and (1+x) <1

where q_e is the adsorbed amount at equilibrium (mg g^{-1}).

C_e is the equilibrium concentration of the adsorbate (mg L^{-1})
q_{m0} is the Baudu maximum adsorption capacity (mg g^{-1})
b_0 the equilibrium constant, and x and y are the Baudu parameters.
For lower surface coverage, Eq. (50) reduces to the Freundlich equation, i.e.: [28]

$$q_e = \frac{q_{m0} b_0 C_e^{(1+x+y)}}{1 + b_0} \tag{51}$$

Fritz-Schlunder Isotherm

Another four-parameter equation of the Langmuir–Freundlich type was developed empirically by Fritz and Schlunder. It is expressed by the equation:

$$q_e = \frac{A C_e^{\alpha}}{1 + B C_e^{\beta}} \tag{52}$$

with α and $\beta \leq 1$

where q_e is the adsorbed amount at equilibrium (mg g^{-1})

C_e is the equilibrium concentration of the adsorbate (mg L^{-1})

A and B are the Fritz–Schlunder parameters and

α and β are the Fritz–Schlunder equation exponents.

At high liquid-phase concentrations of the adsorbate, Eq. (52) reduces to the Freundlich equation, i.e.:

$$q_e = \frac{1_m F S_s K_1 C_e^{\alpha FS}}{1 + K_2 C_e^{\beta FS}} \tag{53}$$

where q_e signifies the monolayer adsorption capacity of the adsorbent (mg g^{-1}) [29].

3.5 Five-Parameter Model

Fritz–Schlunder-V Isotherm Model

Fritz and Schlunder developed a five-parameter empirical model that is capable of simulating the model variations more precisely for application over a wide range of equilibrium data. The isotherm equation is

$$q_e = \frac{1_m F S_s K_1 C_e^{\alpha_{FS}}}{1 + K_2 C_e^{\beta_{FS}}} \tag{54}$$

where q_{mFS5} is Fritz-Schlunder maximum adsorption capacity (mg g^{-1}) and K_1, K_2, α_{FS}, and β_{FS} are Fritz-Schlunder parameters.

This isotherm is valid only in the range of L_{FS} value less than or equal to 1.

This model approaches the Langmuir model while the value of both exponents α_{FS} and β_{FS} equals 1 and for higher adsorbate concentrations it reduces to the Freundlich model [30].

4 Conclusion

This article examine the adsorption kinetics, adsorption isotherms and their mathematical expressions, and their important parameters. Adsorption kinetic parameters offer information on the adsorbent's adsorption capacity, rate of adsorption, and amount of adsorbate removed at equilibrium and at any given time. Adsorption reveals the sort of layer created during the adsorption as well as whether the process is physical or chemical. Adsorption kinetics and isotherms could be used in the adsorption processes, particularly those involving fluctuations in contact time and adsorbate concentrations.

References

1. Largitte, L., Pasquier, R.: A review of the kinetics adsorption models and their application to the adsorption of lead by an activated carbon. Chem. Eng. Res. Des. **109**, 495–504 (2016). https://doi.org/10.1016/j.cherd.2016.02.006
2. Qiu, H., Lv, L., Pan, B.C., Zhang, Q.J., Zhang, W.M., Zhang, Q.X.: Critical review in adsorption kinetic models. J. Zhejiang Univ. Sci. A **10**(5), 716–724 (2009). https://doi.org/10.1631/jzus.A0820524
3. Agboola, O.D., Benson, N.U.: Physisorption and chemisorption mechanisms influencing micro (Nano) plastics-organic chemical contaminants interactions: a review. Front. Environ. Sci. **9**, 1–27 (2021). https://doi.org/10.3389/fenvs.2021.678574
4. Vareda, J.P.: On validity, physical meaning, mechanism insights and regression of adsorption kinetic models. J. Mol. Liq. **376**, 121416 (2023). https://doi.org/10.1016/j.molliq.2023.121416
5. González-López, M.E., Laureano-Anzaldo, C.M., Pérez-Fonseca, A.A., Arellano, M., Robledo-Ortíz, J.R.: A critical overview of adsorption models linearization: methodological and statistical inconsistencies. Sep. Purif. Rev. **51**(3), 358–372 (2022). https://doi.org/10.1080/15422119.2021.1951757
6. Heinrich, P., Hanslik, L., Kämmer, N., Braunbeck, T.: The tox is in the detail: technical fundamentals for designing, performing, and interpreting experiments on toxicity of microplastics and associated substances. Environ. Sci. Pollut. Res. **27**(18), 22292–22318 (2020). https://doi.org/10.1007/s11356-020-08859-1
7. Benjelloun, M., Miyah, Y., Akdemir Evrendilek, G., Zerrouq, F., Lairini, S.: Recent advances in adsorption kinetic models: their application to dye types. Arab. J. Chem. **14**(4), 103031 (2021). https://doi.org/10.1016/j.arabjc.2021.103031
8. Lim, S.F., Lee, A.Y.W.: Kinetic study on removal of heavy metal ions from aqueous solution by using soil. Environ. Sci. Pollut. Res. **22**(13), 10144–10158 (2015). https://doi.org/10.1007/s11356-015-4203-6

9. Hameed, B.H., Mahmoud, D.K., Ahmad, A.L.: Equilibrium modeling and kinetic studies on the adsorption of basic dye by a low-cost adsorbent: Coconut (Cocos nucifera) bunch waste. J. Hazard. Mater. **158**(1), 65–72 (2008). https://doi.org/10.1016/j.jhazmat.2008.01.034
10. Ehiomogue, P., Ahuchaogu, I.I., Ahaneku, I.E.: Review of adsorption isotherms models. Acta Tech. Corviniensis **14**(4), 87–96 (2022), [Online]. Available: https://www.researchgate.net/publication/358271705
11. Musah, M., Azeh, Y., Mathew, J., Umar, M., Abdulhamid, Z., Muhammad, A.: Adsorption kinetics and isotherm models: a review. Caliphate J. Sci. Technol. **4**(1), 20–26 (2022). https://doi.org/10.4314/cajost.v4i1.3
12. Ayawei, N., Ebelegi, A.N., Wankasi, D.: Modelling and interpretation of adsorption isotherms. J. Chem. (2017). https://doi.org/10.1155/2017/3039817
13. Shikuku, V.O., Mishra, T.: Adsorption isotherm modeling for methylene blue removal onto magnetic kaolinite clay: a comparison of two-parameter isotherms. Appl. Water Sci. **11**(6), 1–9 (2021). https://doi.org/10.1007/s13201-021-01440-2
14. Nandiyanto, A.B.D. et al.: Curcumin adsorption on carbon microparticles: synthesis from soursop (annonamuricata l.) peel waste, adsorption isotherms and thermodynamic and adsorption mechanism. Int. J. Nanoelectron. Mater. **13**(Special issue), 173–192 (2020)
15. Ragadhita, R., Nandiyanto, A.B.D.: How to calculate adsorption isotherms of particles using two-parameter monolayer adsorption models and equations. Indones. J. Sci. Technol. **6**(1), 205–234 (2021). https://doi.org/10.17509/ijost.v6i1.32354
16. Saadi, R., Saadi, Z., Fazaeli, R., Fard, N.E.: Monolayer and multilayer adsorption isotherm models for sorption from aqueous media. Korean J. Chem. Eng. **32**(5), 787–799 (2015). https://doi.org/10.1007/s11814-015-0053-7
17. Shahbeig, H., Bagheri, N., Ghorbanian, S.A., Hallajisani, A., Poorkarimi, S.: A new adsorption isotherm model of aqueous solutions on granular activated carbon. World J. Model. Simul. **9**(4), 243–254 (2013)
18. Dada, A.O.: Modeling of Biosorption of Pb (Ii) and Zn (Ii) Ions Onto Pamrh : Langmuir, Huggins, Fowler-Guggenheim and Kiselev Comparative Isotherm Studies **10**(2), 1048–1058 (2019)
19. Kalam, S., Abu-Khamsin, S.A., Kamal, M.S., Patil, S.: Surfactant adsorption isotherms: a review. ACS Omega **6**(48), 32342–32348 (2021). https://doi.org/10.1021/acsomega.1c04661
20. Hamdaoui, O., Naffrechoux, E.: Modeling of adsorption isotherms of phenol and chlorophenols onto granular activated carbon. Part II. Models with more than two parameters. J. Hazard. Mater. **147**(1–2), 401–411 (2007). https://doi.org/10.1016/j.jhazmat.2007.01.023
21. Sundararaman, T.R. et al.: Adsorptive removal of malachite green dye onto coal-associated soil and conditions optimization. Adsorpt. Sci. Technol. (2021). https://doi.org/10.1155/2021/5545683
22. Ramadoss, R., Subramaniam, D.: Removal of divalent nickel from aqueous solution using blue-green marine algae: adsorption modeling and applicability of various isotherm models. Sep. Sci. Technol. **54**(6), 943–961 (2019). https://doi.org/10.1080/01496395.2018.1526194
23. Jeppu, G.P., Clement, T.P.: A modified Langmuir-Freundlich isotherm model for simulating pH-dependent adsorption effects. J. Contam. Hydrol. **129–130**, 46–53 (2012). https://doi.org/10.1016/j.jconhyd.2011.12.001
24. Kiełbasa, K., Kamińska, A., Niedoba, O., Michalkiewicz, B.: Co2 adsorption on activated carbons prepared from molasses: a comparison of two and three parametric models. Materials (Basel) **14**(23) (2021). https://doi.org/10.3390/ma14237458
25. Wu, F.C., Liu, B.L., Wu, K.T., Tseng, R.L.: A new linear form analysis of Redlich-Peterson isotherm equation for the adsorptions of dyes. Chem. Eng. J. **162**(1), 21–27 (2010). https://doi.org/10.1016/j.cej.2010.03.006
26. Vijayaraghavan, K., Padmesh, T.V.N., Palanivelu, K., Velan, M.: Biosorption of nickel(II) ions onto Sargassum wightii: application of two-parameter and three-parameter isotherm models. J. Hazard. Mater. **133**(1–3), 304–308 (2006). https://doi.org/10.1016/j.jhazmat.2005.10.016
27. Koble, R.A., Corrigan, T.E.: Adsorption isotherms for pure hydrocarbons. Ind. Eng. Chem. **44**(2), 383–387 (1952). https://doi.org/10.1021/ie50506a049

28. McKay, G., Mesdaghinia, A., Nasseri, S., Hadi, M., Solaimany Aminabad, M.: Optimum isotherms of dyes sorption by activated carbon: fractional theoretical capacity and error analysis. Chem. Eng. J. **251**, 236–247 (2014)
29. Yaneva, Z.L., Koumanova, B.K., Georgieva, N.V.: Linear regression and nonlinear regression methods for equilibrium modelling of p-nitrophenol biosorption by Rhyzopus oryzen: comparison of error analysic criteria. J. Chem. (2013), Article ID 517631, 10 p.
30. Parker, G.R., Jr.: Optimum isotherm equation and thermodynamic interpretation for aqueous 1,1,2-trichloroethene adsorption isotherms on three adsorbents. Adsorption **1**(2), 113–132 (1995)

Adsorption and Desorption of Adulterants in the Food Industry

Manisha Manimuthu, Sridhar Jayavel, Johni Rexliene Michael, Balaji Viswanathan, and Ashwini Karuppaswamy

Abstract Food products may be adulterated by the use of low-cost look-alike materials, low-quality materials, dilution of the original product, incorrect labeling of the material's age and origin, and low-quality materials. The majority of dishonest traders base their decision to illegally adulterate their goods to increase sales on these four main criteria. Adsorption technology is one of these strategies that is effective and inexpensive. A special diagnostic technique for determining the chemical makeup of food contaminants is DESI mass spectrometry (MS). Adsorption, a powerful technology that can separate complex mixtures under delicate working conditions, is frequently required in the food processing industry. Among these techniques, the adsorption method emerged as an efficient and cost-effective way of eliminating from the food sector both organic and inorganic allergens such as synthetic colorants and organic contaminants like methylene blue, benzene, phenols, and methyl orange. The study also examines the adsorption process's kinetics, isotherms, and mechanism, which can be used to effectively extract and concentrate additives or adulterants from extracts or the food products or wastewater produced by agricultural operations and the food industry.

Keywords Adulterant · Adsorption · Desorption · Methyl orange · Phenol

1 Introduction

Adsorption is the process of transferring an adsorbate in bulk from the aqueous stage to the adsorbent surfaces until a concentration-based thermodynamic equilibrium is attained and net adsorption is discontinued. The production of enzymes, fractionation of proteins and sugar, purification of penicillin, proteins, and isomers have all been accomplished using large-scale batch adsorption procedures. Nevertheless, batch adsorption has some disadvantages, such as a high stationary phase, low productivity, and high buffer consumption costs since not all of the adsorbent in

M. Manimuthu · S. Jayavel (✉) · J. R. Michael · B. Viswanathan · A. Karuppaswamy
Department of Biotechnology (DDE), Madurai Kamaraj University, Madurai 625021, India
e-mail: jsridharbiotech@mkuniversity.ac.in

J. Tharini and S. Thomas (eds.), *Carbon Nanomaterials and their Composites as Adsorbents*, Carbon Nanostructures, https://doi.org/10.1007/978-3-031-48719-4_9

the column is successfully used. A substantial quantity of side streams are produced during the manufacturing of food goods. These side streams are intricate mixtures that frequently contain valuable items that can be recovered and used once again in the food chain to improve their value [1, 2]. For example, grains include complex carbohydrates including glucans, arabinoxylans, and hemicellulose while fruits and vegetables contain polyphenols. Additionally, proteins can be extracted from agricultural waste materials like canola meal. With the use of conventional techniques including precipitation, adsorption, filtering, and extraction, among others, these chemicals have been valued and recovered utilizing tried-and-true methods, such as the universal five-step method offered by [3, 4].

Color is a critical measure in quality assurance and one of the main sensory aspects that influences consumers' food preferences and acceptability. Natural and synthetic dyes are used during processing with the intention of either adding color to otherwise colorless meals or recapturing the lost original color [5]. Synthetic dyes are widely used in the food industry because of their durability and capacity to create brilliant hues. According to some studies, however, consuming specific food additives may have adverse effects, such as the development of different malignancies, hypersensitivity reactions, the induction of overactivity in youngsters, and behavioral issue [6]. Due to increasing concerns about their safety, a number of these coloration are no longer allowed to be used in food. Frequently, artificial coloring chemicals are used to color foods. However, there is now concern for their safety. The industry is forced to replace artificial colorants as a result of this restriction and the global trend toward eating healthier meals, which has increased interest in colorants made from natural sources. Thus, the moisture adsorption isotherms of the three naturally occurring food colors betalains, curcumin, and anthocyanins were investigated, and there in vitro hygroscopicity, antioxidant capacity, and color capacity were assessed. Natural colorants can be used to effectively color food items, and they may also have additional beneficial health effects.

(Mildner-Szkudlarz and Jelen; 2008) state that there are numerous ways that rice might become polluted. Other cases include diluting milk and adding melamineadulterating mung beans in pistachio etc. [7, 8]. Customers are being taken advantage of, and it can be bad for their health. Food adulteration detection is essential to protect customers' legal interests. Due to their high adsorption capacities, purity in recovering the adsorbed molecules, affordability, and simplicity of regeneration, adsorption methods using nonpolar styrene–divinylbenzene (SDVB) copolymers or a few aqueous acrylic polymers are also known to be applicable for manufacturing applications. Foods rich in phenolic compounds can reduce a person's risk of contracting chronic diseases such arthritis, diabetes, atherosclerosis, cancer, and other age-related ailments [9, 10]. The scientific field has recently become interested in the possibility of non-ionic polymeric macroporous synthetic resins, particularly SDVB and acrylic resins, for the separation and/or purification of phenolic chemicals. They have been used to purify and concentrate phenolic compounds from Inga edulis leaves (total phenolics and flavonoids), treat citrus processing wastewater (hesperidin), lessen the bitterness of juice (naringin), purify substances (cyanidin-3-glucoside), and adsorb tangeritin, hesperidin, naringin, quercetin, and naringenin [11–13].

On the other hand, adsorption techniques provide a strong alternative for the removal of colors from industrial effluent together with other toxins, pollutants, and impurities. Traditional methods for evaluating foodstuffs to identify adulterants, such as high-performance liquid chromatography (HPLC), ion chromatography, gas chromatography-mass spectrometry, capillary electrophoresis, etc., are used to investigate the essential chemical compositions [14–16]. Artificial food coloring is used to cover up manufacturing errors and elevate inferior food to a higher standard. Artificial food dyes ingest into the body where they convert into poisons and cause cancer, mutations, and other adverse effects. These colors frequently cause allergic reactions in humans. Many agricultural and forestry products, including wood sawdust, sunflower seed hulls, and maize cob waste, among others, hold a great deal of potential for use as adsorbents despite the fact that they include carbohydrates and peptides with various chemical groups like hydroxyl, carboxyl, and phosphates. Because of its abundance, low cost in comparison to other adsorbents such activated carbon or inorganic materials used in adsorption methods, and limited industrial use, it was chosen as an adsorbent for the removal of organic compounds from aqueous solutions, odors, and colors [17, 18]. The various adsorbent kinds and their procedures for extracting both organic and inorganic pollutants will be demonstrated in the current study. Future adsorption process needs, the effects of multiple ion removal, various operational moods, kinetics, and isotherms will also be discussed (Figs. 1 and 2).

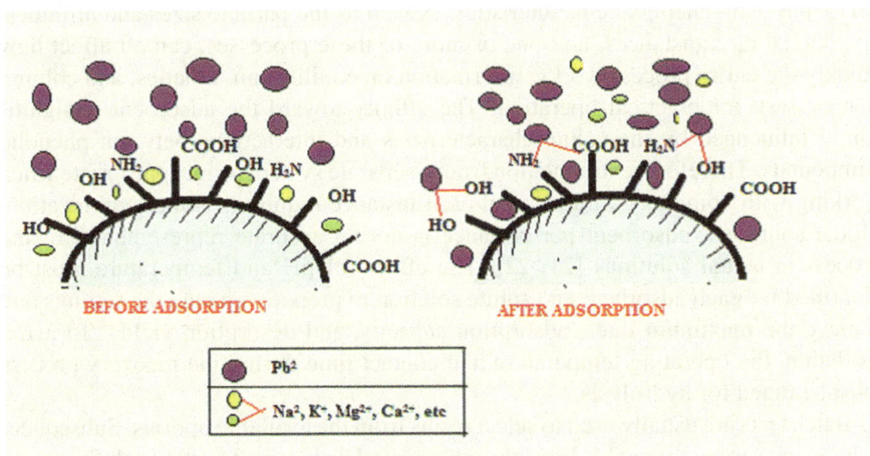

Fig. 1 Pb adsorption using an ion exchange process

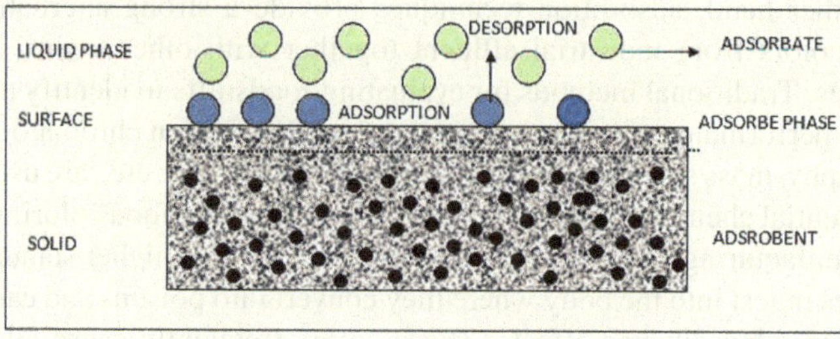

Fig. 2 Basic adsorption concepts

2 Process of Adsorption and Its Mechanism

Reversible adsorption is more effective when the goal is to recover the solute because it allows for a variety of interactions between the adsorbent and the molecule in the aqueous phase, including van der Waals, covalent bonds, electrical, H bonds, and acid–base interactions. The fundamental adsorption mechanics and technical design methods are discussed in classical publications [19, 20]. The stages of the adsorption process are solute-solid contact, porosity diffusion within the solid phase, and bulk transfer from the aqueous phase to the particle's outer surface. The solute and adsorbent's physical–chemical characteristics, as well as the particle sizes and affinities, operational circumstances, and one or more of these processes, can all affect how quickly the entire process works. Information on equilibrium, kinetics, and column is necessary for practical operation. The affinity toward the adsorbent is significantly influenced by the solute characteristics and interactions between phenolic compounds. Therefore, extrapolation from a separate system is not appropriate when working with complex solutions, and each instance requires distinct optimization. Model solutions' adsorbent performance is not an accurate representation of the process in actual solutions [21, 22]. The effects of pH and temperature must be identified for each adsorbent and solute solution to properly change the settings and achieve the maximum rates, adsorption capacity, and desorption yields. To avoid oxidation, the operating temperature and contact time during the recovery process must be tuned for hydrolysis.

Batch tests are usually used to select resins from the available options. Subsequent column investigations enable breakthrough curve development, suitable eluting agent selection, and process scale-up. Some materials benefit from phenolic compounds' selective adsorption. The recovery of thermo-sensitive molecules is limited by the irreversible nature of the process and the necessity of regeneration using chemicals or at high temperatures, although regeneration of activated carbon employing extreme fluids may be able to overcome this restriction [23, 24].

Covalent bonding produces electrostatic attraction, which results in chemical adsorption, also known as chemisorption. In chemisorption, the material being absorbed interacts chemically with the adsorbent, which had been previously subjected to certain chemical alterations to form such exterior functional groups (i.e. acidic functional groups like carboxylic anhydride, carbonyl, lactone, carboxyl, and hydroxyl), which are the most prevalent types of surface functional groups. As a result, these functional groups adsorb the material considerably more firmly, requiring much more energy to release it.

Resins are the most extensively researched adsorbents due to their chemical stability, low toxicity, selectivity, high adsorption capacity, and ease of regeneration at moderate temperatures. When extracting vanillin and syringaldehyde from lignin in alkaline solutions, for instance, one can avoid the considerable acid consumption necessary for direct acidification of the solution by using cation exchangers. Ion-exchange resins have also been suggested for the adsorption of pure phenolics and naturalistic extracts in addition to the frequently employed non-ionic resins. The enrichment and purification of phenolic chemicals from extraction and waste streams is one potential use for the synthetic resins now employed commercially in food processing to remove undesirable polyphenols responsible for astringency, bitterness, or browning [25]. By using cation exchangers, for instance, to extract syringaldehyde and vanillin from lignin in alkaline solutions, the solution is avoided being directly acidified, which would have required the use of a large amount of acid. In industrial food processing, synthetic resins are often used to remove astringent and bitter polyphenols. Browning may be suggested for the growth and purity of phenolic components from extraction and waste streams. Molecularly impressed polymers, which are cross-linked 3-D structures with specialized binding capabilities to maintain the necessary targeted molecules, have been proposed for the analysis or purification of phenolic compounds [26, 27].

Many research have employed chemical modification with alkali or acid treatment, or both, and found that when onion skin was treated with thioglycolic acid to increase ion adsorption on different agricultural waste surfaces, the onion dust did not change. Citrate is grafted onto pomelo peels in order to remove methylene blue from an aqueous solution [28]. Peels from bananas and peapods were chemically pretreated with HCl and NaOH, and the results in terms of adsorption capacity were good. Rice straw carbon (RSC) is created by adding 10% KOH solution to rice straw after it has been pretreated with NaOH, ethanol, HCL, and deionized water to create rice straw biochar. Sugarcane bagasse was treated with citric acid to increase the adsorption capacity to 13.5 mg/g.

3 Adsorption Characteristics of Physisorption and Chemisorption

Adsorption takes place on the surface of a substrate. Adsorption is the process by which a substance builds up on a surface in progressively higher concentrations of a particular molecular species. Activated charcoal is where gases like oxygen, nitrates, and hydrogen adsorb. The adsorption process requires the following two elements:

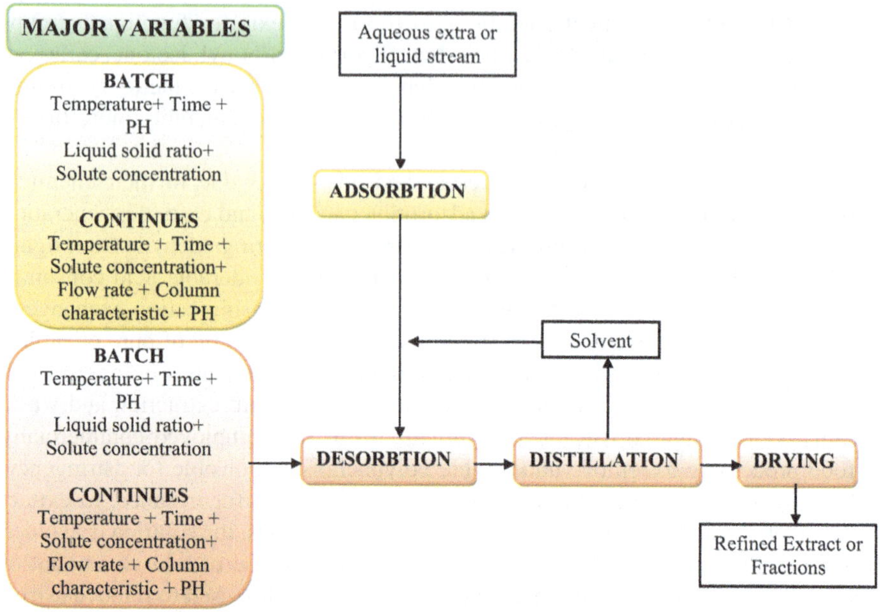

Fig. 3 Flowchart of the adsorption and desorption processes

Adsorbate: a substance that sticks to the surface of another substance, such as the gases H_2, N_2, and O_2. Adsorbent: A substance that adsorbs on the surface of an adsorbent. Several examples include alumina, silica gel, and charcoal.

Physisorption characteristics: This type of adsorption is primarily caused by physical forces, which are also the cause of weak physisorption phenomena, numerous layers in this adsorption, and physical adsorption that can occur anywhere on the adsorbent.

Chemisorption has the following characteristics: (1) This type of adsorption is caused by chemical forces; (2) It is a powerful process; (3) It almost always occurs in a single layer; (4) Chemisorption occurs at reaction centers on the adsorbent and is very selective; (5) Surface area, temperature, and the type of adsorbate affect chemisorption; and (6) The activation energy ranges from 40 to 400 kJ/mol (Fig. 3).

4 Different Types of Adsorbent Resins

Adsorbent resin, also known as polymer adsorbent, is a form of high molecular polymer that not only acts as an adsorbent but also concentrates and separates organic materials. Depending on the surface characteristics of the resin, which is manufactured from a monomer polymer with a tiny dipole moment, the adsorption resin is frequently characterized as a non-polar adsorption resin. A common example of an adsorption resin with any functional group is the styrene–divinylbenzene system; a medium polarity adsorption resin is an ester-containing adsorption resin like acrylate or methacrylate and dimethacrylate. A specific form of cross-linked copolymer known as a polar adsorption resin is one that has a nitrogen, oxygen, or sulfur polar functional group, such as an amide group, a cyano group, or a phenolic hydroxyl group. The customizable Adsorbent resins, which are macroporous polymer adsorbents, are frequently used in the decolorization of juice, pharmaceutical purification, and plant extraction. Activated carbon, which has a huge specific surface area (500–1700 m^2/g), was the first adsorbent used by humans. Water pollutants may be removed using ion exchange, chemical adsorption, and physical adsorption. It also has effective gas adsorption capabilities. Numerous studies and review articles have examined the different adsorbents' methods for extracting pollutants from contaminated water. Such adsorbents fall within the categories of natural or artificial adsorbents.

The different types of natural adsorbents include cement, chitin, ash, charcoal, clay, minerals, chitosan, zeolites, ores, and peat. The availability and cost of these natural adsorbents are benefits, and they also offer a lot of potential for modification so that they can eventually have superior adsorption capabilities.

Industrial waste, sewage sludge, home-use adsorbents, marine adsorbents, and adsorbents made of polymers are all examples of synthetic adsorbents [29]. Every adsorbent has different properties, such as pores, structural porosity, and the composition of its adsorbing surfaces. Red mud, ore minerals, peat moss, petrochemical wastes, fruit wastes, coconut shells, sugar industry wastes, fertilizer wastes, chitosan, blast furnace slag, sedimentary soil, clays, zeolites, seaweed, algae, sediment and soil, ore minerals, and a variety of other waste products are just a few examples of the waste materials that are produced.

5 Activated Carbon (AC)

Activated carbon (AC), which has a high internal surface area, is pourable, and has excellent micro-porousness, is said to be the best heavy metal adsorbent. Large surface areas, high adsorption capacities, porous sorbents, high rates, functional groups, the capacity to efficiently adsorb a variety of pollutants, superior kinetics, and high-quality treated effluent are some of the main benefits of AC [30].

The main disadvantages of AC, on the other hand, are its high cost and the need for multifunctional chemicals to increase its value as a disposal method. The phenols in industrial wastewater generated by the petrochemical, pharmaceutical, and chemical pesticide industries are also removed using AC. The main suppliers include paints, steel mills, petroleum refineries, coal gas, coke oven facilities, plywood manufacturing businesses, pharmaceuticals, and synthetic resins [31].

6 Natural Adsorbent

The Macroporous adsorbent is a type of high molecular polymer that uses the concepts of molecular screening and adsorption to concentrate and filter organic compounds. By physically adsorbing specific organic compounds from the solution, the macroporous adsorbent achieves separation and purification goals. It can be absorbed by swelling in water and organic solvents have good selectivity to organic materials and are unaffected by the presence of inorganic salts, strong ions, and low molecular molecules. When separating the target object, the molecule polarity in addition to the particle size must be taken into consideration.

Although adsorbent resins with tiny pores have more surface area, molecules must be able to enter the bead for them to be adsorbed. Chemical interactions, solubility, size, weight, and capacity are some of the elements that affect adsorption efficiency. A variety of polymer matrices, such as aromatic polystyrene or polydivinylbenzene, aromatic halogenated polystyrene/divinyl benzene, and aliphatic methacrylate, are available for adsorbent resins. Selecting the right adsorbent resin requires consideration of the distinctive physical characteristics of each polymer matrix.

7 Nanomaterial Adsorbent

An example of a particular kind of nanomaterial employed in the adsorption of contaminants from polluted water and wastes is carbon nanotubes. CNTs are made up of 1 or more cylindrical sheets with a radius under 100 nm and a length of more than 20 m. CNTs are functionalized to boost their efficacy. Metal-oxides that are extremely tiny, up to the Nanoscale, are known as nanomaterials [32]. Cadmium, iron, lead, copper, arsenic, mercury, chromium, cobalt, selenium, and other heavy metals and organic contaminants can be effectively removed from aqueous solutions using these adsorbents. Due to their high surface-to-volume ratio, large surface area, catalytic properties, and other beneficial characteristics, these adsorbents demonstrated exceptional adsorption performance against many pollutants in water and wastewater. Ferric oxides, Aluminium oxides, titanium oxides, cerium oxides, manganese oxides, magnesium oxides, etc., are the most widely utilized metal oxide nanomaterials [33] (Table 1).

Table 1 Recovery of biofunctional and techno functional molecules from food processing byproducts

S. no	Adsorbent resins	Nature and function
1	SEPLITE@LWT510	Water treatment resins for removing COD and organic substance
2	MN102	Color, taste and odor removal
3	PAD600	Caffeine removal
4	MN100, MN102, MN502, MN152	Decolorization sugar solution
5	PAD400	Natural color extraction (Taste and odour)
6	SEPLITE@LX1600	Protein recovery
7	SEPLITE@LXA81	Removal of phenolic compounds
8	SEPLITE@LXA81	Herbal extract purification

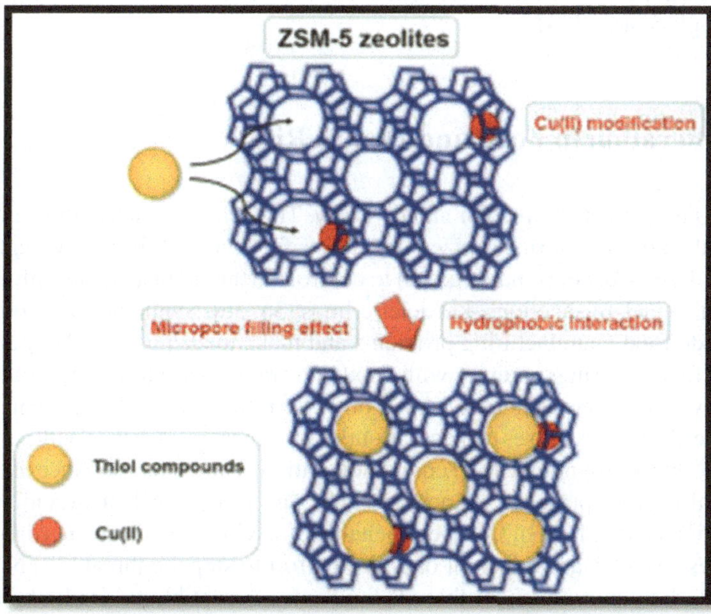

Structure of Natural Adsorbents

(i) **Recovery of Proteins and Peptides**

Ion exchangers are one method for isolating and purifying proteins and amino acids, including whey proteins, which are produced in huge amounts as cheese-making byproducts. The majority of the proteins in whey are R-lactalbumin and -lactoglobulin, with lesser amounts of immunoglobulin, bovine serum albumin, glycomacropeptides, lactoferrin, and lacto peroxidase.

(ii) **Recovery of Polyphenols from Food Processing Waste**

Plant secondary metabolites, notably polyphenols, are frequently found in high concentrations in the byproducts of the processing of plant foods. Phenolic compounds are highly sought-after by the pharmaceutical, food, and cosmetics industries due to their biofunctional characteristics, such as their anti-carcinogenic, antithrombotic, anti-inflammatory, and antioxidant qualities, as well as their techno-functional characteristics, such as their antioxidant and antibacterial potential or their preserving and regulating effects on food color and scent.

(iii) **Phenolic Extracts from Fruit**

Even though leftover peels, cells, cores, and membranes from the production of citrus juice are commonly used as cow feed, pectin can also be extracted from them. Molasses, ethanol, pulp wash, flavonoids, D-limonene, natural flavors, and scents can all be derived from citrus byproducts. Citrus fruits and the products made from them are well known for having a particularly high concentration of flavanones and flavanone glycosides [34].

8 Adulteration in Food and Health Risk

Customers need foods with color and diversity. The dealers enhance the quality and appeal of their products to maximize sales and profit. Consumer illiteracy, negligence, apathy, and a lack of coordinated action to control the threat. Inadequate enforcement of laws and a lack of sanctions that deter crimes [35, 36]. Synthetic food colours are used to hide food manufacturing problems and make inadequate food appear great. Artificial food colorings interact with food and transform into toxins in the body, where they cause cancer, mutations, and other harmful consequences. Some people experience allergic responses to these colours. Customers want dishes with colour and diversity. Causes of synthetic food color adulteration: The dealers enhance the quality and appeal of their products to increase sales and profit. Inadequate enforcement of regulations and a lack of deterrent penalties for violators, Consumer ignorance, carelessness, apathy, and a lack of organized effort to stop the threat [37] N.

The latest food crises show that all foods are susceptible to food adulteration. Since milk was tested for nitrogen content, melamine, a nitrogen-rich chemical with 67% nitrogen per mass unit, was added to milk to increase the protein content and prevent detection. Food adulteration frequently has a detrimental effect on people's health some of the issues are (i) cancer and paralysis caused by the addition of mineral oils to fats and edible oils; (ii) liver damage, harm to an unborn child's brain, allergies, and abortion. (iii) giddiness, joint discomfort, and stomach issues caused by switching from coffee to chicory powder; (iv) diarrhea and vomiting caused by zinc-containing drugs, and so on [38]. To identify adulteration, a variety of methods (chemical/biochemical, physical, and molecular) are used, depending on the sort of adulteration that needs to be found. Physical methods include studying the physical characteristics of food and performing macroscopic visual structure analysis.

Immunologic, spectroscopic, electrophoretic, and chromatographic procedures are among the methodologies utilized in chemical and biochemical research.

Some techniques, such as Raman, mass spectrometry techniques, vibrational spectroscopies, mid-infrared, near-infrared, Nuclear Magnetic Resonance Spectroscopy, mass spectrometry techniques, and mid-infrared are created to address significant challenges including food security, bioterrorism, and climate change in addition to the ongoing concern with contaminated food. Since they take a lot of time and are unsuitable for color combinations, adsorptive voltammetry, thin-layer chromatography, and spectrophotometry methods are no longer utilized to detect food dyes. Reversed-phase chromatography (RPLC), capillary electrophoresis, and ion pair RPLC are utilized instead. Aquatic items that include a food preservative like formaldehyde can be found using surface-enhanced Raman spectroscopy using Au/SiO2 as the activator substrates. While concentrated gas chromatography is used to determine sorbate and benzoate in tomato and orange beverages, sequence injection analysis is utilized to identify nitrate and nitrite in cured meat [39, 40].

They harm the kidneys and adrenals and are carcinogenic. Artificial food coloring decreases hemoglobin concentration, red cell count, and allergic reactions. They prevent nerve endings from taking up dopamine (reduced dopamine turnover). These colors have an impact on the liver and intestine, as well as restlessness, irritability, and disturbed sleep in atopic or hypertensive children more so than adults. They produce glossitis, which shows symptoms (Inflation of the tongue). Different allergic reactions might cause urticaria, dermatitis, angioedema, etc. When consumed in excessive concentrations, artificial food colors can cause ear infections, asthma, dyslexia, eczema, and other conditions [41].

9 Operating Design

In a laboratory context, certain undesirable hazardous substances, such as synthetic colors and heavy metals, can be adsorbed on various adsorbents utilizing batch mode or adsorption techniques for fixed bed columns. When utilizing adsorption to remove specific pollutant components, batch laboratory adsorption studies can be beneficial.

Batch studies: Using batch approaches, it is possible to examine how several factors, such as pH, adsorbent dosage, pollutant beginning concentration, surface shape, modification methodology, temperature, and contact time, affect the adsorption process. The information regarding the interaction between adsorbate and adsorbent can be learned from the batch equilibrium studies' determined adsorption capacity.

The pollutant's initial concentration (Co) and the final concentration (C1) are determined via the batch adsorption method. In the adsorption process, W stands for the mass of the adsorbent used, V for the volume of the solution being studied, and qo and q1 for the first and last stages, respectively, of adsorption capability (pollutant concentration on solid).

The equation contains the process's mass balance.

Under the condition of equilibrium, **C1 Ce and q1 qe,**

$$V(C_0 - C_1) = W(q_0 - q_1) \quad (1)$$

Under equilibrium conditions: C_1 C_e and q_1 q_e so the equation will be

$$V(C0 - Ce) = W(q0 - qe) \quad (2)$$

Batch Adsorption equations is $V(C0 - Ce) = W(q0 - qe)$.

10 Testing for Static Adsorption and Desorption to Screen Resins

The following steps were taken to conduct the static adsorption and desorption studies: 1 g of resin was put into a 125 mL Erlenmeyer flask. Then, each flask received 50 mL of fruit aqueous extract. To rule out any influences on the observed absorbance, a control sample was used to monitor any changes in the initial concentration values. The flasks were then shaken for 24 h at 120 rpm at room temperature (25 °C) in a shaking water bath. Adsorption was followed by filtering and washing resins in 50 mL of pure water. For ethanol solution that had been 70% and 95% acidified (with 1% formic acid), desorption was evaluated. The resins filled with adsorbate were placed in flasks and were then given 50 ml of an ethanol–water solution.

To calculate the ratios between the adsorption and desorption capabilities, the following formulas were utilized. Adsorption evaluation

Desorption Assessment,

$$D = (CdVD) \div (C0 - Ce)V0 \times 100$$

R is equal to $Cd * Vd \div C0V0 \times 100$.

where Cd is the solute concentrations in the desorption solutions (in milligrams per liter), D is the desorption ratio (in percent), Vd is the volume of the desorption solution (in milliliters), and C0, Ce and V0 are present.

11 Adsorption Kinetics

On a first batch of chosen resins, the adsorption kinetics were studied. Sorption kinetics, which governs the residence time of the sorption reaction and establishes the efficiency of sorption, describes the rate of solute adsorption. The rate of phenolic chemical absorption on the selected resins was determined using pseudo-first-order and pseudo-second-order kinetic models [42].

Pseudo-First Order

Lagregren created the pseudo-first-order equation in 1898 to determine the adsorption rate. The equation provides the adsorption rate,

The equation's linear form is as follows after integration with boundaries conditions from **t = 0 to t = t and qt = 0 to qt = qt:**

$$\log(qe - qt) = \log - qe - k/2.303 \times t$$

The slope and intercept of the log (qe–qt) vs time plots, which result in a straight line, are used to compute the values of qe and k at various starting concentrations if the kinetics data are consistent with pseudo-first-order kinetics.

Pseudo-Second Order

The equation provides the pseudo-second-order adsorption rate, where the adsorbent's capacity (in mg/g) is represented by q ((1–5) %) %.

$$t/qt = 1/k2qe^2 + 1/qe \times t$$

When the solution concentration reaches Cb, Q is the flow rate (ml/min) and M is the mass of adsorbent packed across the column (g).

12 Adsorption Isotherm

The retention (or release) or movement of a material from aqueous porosity media or aquatic environments to a solid layer at a constant temperature and pH is often represented by a broad curve known as an adsorption isotherm. Adsorption equilibrium (the ratio of the amount of adsorbed to the residual of the solution) is achieved when the phase-containing adsorbent has been in connection with the adsorbent for enough time and its adsorbate concentrations in the bulk solution are dynamically balanced with the interface concentration [43–45].

At room temperature (25 °C), 30 °C, and 35 °C, the equilibrium isotherms of adsorption and the effects of anthocyanins on three different resins were examined. Two of the most popular models for characterizing adsorption isotherms are the Langmuir and Freundlich isotherms. According to the Langmuir model, the adsorbed molecules do not interact with one another and have energetically equivalent sorption sites. The Freundlich model predicts that molecules will adsorb to surfaces with heterogeneous sorption regions and a range of sorption energies. A monomolecular's adsorption activity in both a single-molecular layer and a multimolecular layer can be explained by this model. Correlation coefficients for each model on three different resins at well-chosen temperatures. Due to greater correlation coefficients, the Langmuir model was thought to be a better model for describing adsorption equilibrium for all three resins.

13 Recovery of Adulterants from the Food Sources

An extremely valuable enzyme called papain has been employed in some sectors, including the food business and the pharmaceutical sector. Papain is used to alleviate swelling, discomfort, and inflammation following surgery or injury. Using reversed-phase expanded bed adsorption chromatography, it has been made possible to directly recover papain enzyme from unclarified Carica papaya juice. At linear flow velocity, bed expansion of two, and feedstock viscosity, the dynamic binding capability for the RP-EBAC at 10% breakthrough adsorbent was attained. The investigation of papain extraction by RPLC from unclarified papaya juice used the AmberliteTM XAD7HP equipment. The purification of papain in RP-EBAC was increased using a two-step elution process, which resulted in great papain purity and a large purifying factor of 7.04. This work demonstrates the enormous potential of purifying papain utilizing RP-EBAC and an optimization stage during the extraction of juice from unclarified Carica papaya [46].

14 Applications of the DESI for Examining Chemical Pollutants in Food

Involatile chemical substances can be promptly evaluated and their structure revealed using desorption mass spectrometry (MS), a technique that has been utilized for years. Successful applications include laser desorption ionization (LDI), secondary ion mass spectrometry (SIMS), fast atom bombardment (FAB), desorption chemical ionization (DCI), matrix-assisted or surface-enhanced laser desorption/ionization (MALDI or SELDI), and field desorption (FD). All of these approaches have the limitation of probing and ionizing the material in a vacuum, which first modifies the environment. Even though MALDI can be used at atmospheric pressure, the presence of the matrix considerably alters the sample, making it difficult to detect compounds with low molecular masses [47].

We propose DESI as a quick, qualitative or semi-quantitative screening tool that may be used on the spot before collecting samples and transporting them to the food-control laboratory. By adding DESI chemicals to the electrospray solvent, it is simple to modify the selectivity and sensitivity of the electrospray process. Examples of reactions using betaine aldehyde, boric acid, phenylboronic acid, hydroxylamine, and acylium ions that have improved the identification of cholesterol, cis-diols, phosphonate esters, anabolic steroids, and cyclic acetals have been published. The atmospheric pressure electrospray ionization method (MS), which has a chemical-ionization (DAPCI) and photoionization (DAPPI) versions have been created as adaptations to the DESI principle. Sudan dyes may not be added as colorants. **Chen et al.,** published a report on the quick DESI-MS2 analysis of Sudan I, II, III, and IV in sausages and chili powder, Fried eggs, and tomato sauce. According to the English summary of this Chinese study, the LODs, or limits of detection, for these dietary matrices ranged from 0.01 to 1.0 pg/mm2. In an aerosol, the powdered fake sweetener was found. Applications like these could support preserving food sector workplace hygiene requirements [48, 49].

The DESI-MS2 analysis of a naturally produced diterpene glycoside sweetener found in Stevia plants and dietary products was recently disclosed by Jackson et al. Rubusoside, steviolbioside, stevioside, rebaudioside A–F, and glucoside were found in raw leaves, and accurate-mass data from DESILTQ–Orbitrap analysis confirmed this discovery. The identical diterpene glycosides were still present in a sample of a marketed supplement, despite the presence of many fructose oligomers. Hartmanova et al. used Nano-DESI to recognize tainted wine. Red wine samples' major anthocyanin patterns as revealed by DESI contrasted favorably with the results of LC-MS2. More importantly, DESI analysis of the anthocyanin profiles swiftly detected illegal wine coloring or the combination of different wine varietals. The anthocyanins in grape slices and the wine stains on the cotton fabric were also given early information [50].

15 Probe-Mass Spectrometry for the Study of Atmospheric Solids

Fussell et al. assessed the use of ASAP in food analysis. The detection of hazardous colors in spices and the detection of pesticides in cereals were the two main areas of inquiry. Additionally, according to Fussell et al., papaya was found to contain the coumarins found in cinnamon as well as the EU-approved food additives bixin and norbixin. While ASAP-MS generates robust qualitative results, the method struggles

Table 2 Applications for desorption electrospray Ionization (DESI) mass spectrometry in the detection of chemical adulterants in Food

Chemical analytes	Food sample matrix forms
Adulteration	
Melamine, cyanuric acid	Milk
Anthocyanins	Wine, slice of wine grapes, wine stain on cotton
Food forensics	
Steroids	Seized hormone cocktail
Steroid esters	Bovine hair
Tylosin	Residue from medicine mixture
Budesonide	Dust particle
Food additives	
Sudan dyes	Sausages, chilli powder, fried eggs, tomato sauce
Natural sweeter	Dietary supplement, leaf
Artificial sweetner	Powder like aerosol particle

to generate enough quantitative results, making it potentially insufficient for detecting food adulteration [51] (Table 2).

16 Paper Spray Mass Spectrometry (PS-MS)

According to some, PS-MS was the first ambient mass spectrometry method developed. In order for paper spray ionization to work, a triangle of paper that has been slightly wet with solution is exposed to a high voltage. Charged particles are produced when the high voltage is delivered, which is typical of an ESI process. A lot of work has gone into using PS-MS to analyze food, according to the literature. Zhang et al. provided a summary of some of the previous research, which included the discovery of melamine in milk powder and infant formula, Sudan colors in chili powder, plasticizers in sports drinks, and salbutamol, clenbuterol, ractopamine, and terbutaline in pork and beef [52].

17 Regeneration and Reuse

The general development of adulterants on the surface of the adsorbent over time causes the efficiency of the chemical to decrease throughout the adsorption process. The depleted surface ions must also be replaced in order for the process to be lucrative. Numerous eluents, including NaOH, H2SO4, HCL, NA2CO3, etc., have been utilized in studies to remove pollutants from adsorbent surfaces [53].

18 Conclusion

Adsorption is a method that has the potential to get removal of adulterants. The batch mood is beneficial for assessing an adsorbent's adsorption capacity even though it cannot be used while formulating adsorption column experiments. a method for removing both inorganic contaminants like synthetic colors and dyes and organic pollutants like phenols, methylene blue, methyl orange, and benzene from polluted water or food. The most popular adsorbents are examined in this overview along with research on their effectiveness, price, and regeneration. These include bio-sorbents, natural adsorbents, industrial byproducts, activated carbon, nanomaterials, and agricultural wastes. This study also covers the adsorption process, which makes use of kinetics and isotherms. The two isotherm models that are most frequently used to explain how organic and inorganic chemicals adsorb are Langmuir and Freundlich. Pseudo-first-order and pseudo-second-order kinetics theories are frequently used to describe the adsorption process. Adsorption mechanism is determined by the type of functional groups on the adsorbent's surface, and chemical modification procedures can increase adsorption capacity.

References

1. Galanakis, C.M.: Recovery of high added-value components from food wastes: conventional, emerging technologies and commercialized applications. Trends Food Sci. Technol. **26**(2), 68–87 (2012). https://doi.org/10.1016/j.tifs.2012.03.003.(1)
2. Bérot, S., Compoint, J.P., Larré, C., Malabat, C., Guéguen, J.: Large-scale purification of Rapeseed proteins (Brassica napus L.), J of Chromatography B **818**(1), 35–42 (2005). https://doi.org/10.1016/j.jchromb.2004.08.001
3. Wanasundara, J.P.: Proteins of Brassicaceae oilseeds and their potential as a plant Protein source. Crit. Rev. Food Sci. Nutr. **51**(7), 635–677 (2011). https://doi.org/10.1080/10408391003749942
4. Fetzer, A., Herfellner, T., Stäbler, A. et al.: Influence of process conditions during aqueous protein extraction upon yield from pre-pressed and cold-pressed rapeseed press cake. Ind. Crop Prod **112**, 236–246 (2018). https://doi.org/10.1016/j.indcrop.2017.12.011
5. Ali, A.: A review article on edible pigments properties and sources as natural biocolorants in foodstuff an food industry. World J. Dairy Food Sci. **6**(1), 71–78 (2011)
6. McCann, D., Barrett, A., Cooper, A., Crumpler, D., Dalen, L., Grimshaw, K., Kitchin, E., Lok, K., Porteous, L., Prince, E., et al.: Food additives and hyperactive behavior in 3-year-old and 8/9-year-old children in the community: a randomized, double-blinded, placebo-controlled trial. Lancet **370**(9598), 1560–1567 (2007). https://doi.org/10.1016/S0140-6736(07)61306-3.PMid:17825405
7. Sezer kirala, S., Tekin, A.: Reducing polycyclic aromatic hydrocarbons (PHAs) in olive pomace oil using short-path molecular distillation. Food Additives Contaminants: part A **37**(3), 401–407 (2019). https://doi.org/10.1080/19440049.2019.1704444
8. Eksi-Kocak, H., Mentes-Yilmaz, O., Boyaci, I.H.: Detection of green pea adulteration in pistachio nut granules by using Raman hyperspectral imaging. Eur. Food Res. Technol. **242**(2), 271–277 (2016)
9. Scordino, M., Mauro, A., Passerini, A., Maccarone, E.: Adsorption of flavonoids on resins: Cyanidin 3-glucoside. J. Agric. Food Chem **52**(7), 1965–1972 (2004)

10. Geankoplis, C.J.: Transport Processes and Unit Operations, 2nd edn. Allyn and Bacon, Boston (1993)
11. Toma-Barbera, F.A., Bla´zquez, M.A., Garcı-Viguera, C., Ferreres, F. et al.: A comparative study of different Amberlite XAD resins in flavonoid analysis. Phytochem. Anal. **3**, 178–181 (1992)
12. Bilgili, A.S.: Adsorption of 4-chlorophenol from aqueous solutions by XAD-4 resin: Isotherm, kinetic, and thermodynamic analysis. J. Hazard. Mater. **137**(1), 157–164 (2006)
13. Ho, Y.S.: Review of second-order models for adsorption systems. J. Hazard. Mater. **136**(3), 681–689 (2006)
14. Chen, Z., Wu, T., Xiang, C., Xu, X., et al.: Rapid identification of rainbow trout adulteration in Atlantic salmon by Raman spectroscopy combined with machine learning. Molecules **24**(15), 2851–2864 (2019). https://doi.org/10.3390/molecules24152851
15. Liu, C.Y., Zhao, L.L, Sun, Z., Cheng, N.X.F. et al.: Determination of three flavor enhancers using HPLCECD and its application in detecting adulteration of honey. Analyt. Methods **10**(7), 743–8 (2018). https://doi.org/10.1039/C7AY02248D
16. Mi, J.P., Xu, Y.J., Zhu, P.C., Wu, D.R., Zhang, W.Y.: Application and research progress of high-performance liquid chromatography-mass spectrometry in the determination of the abuse of food additives and illegal additives. Genomics Appl. Biol. **34**(7), 1579–1586 (2015). https://doi.org/10.13417/j.gab.034.001579
17. El-Halwany, M.M.: Study of adsorption isotherms and kinetic models for Methylene Blue adsorption on activated carbon developed from Egyptian rice hull (Part II). Desalination **250**(1), 208–213 (2010)
18. Ofomaja, A.E., Ho, Y.S.: Effect of temperatures and pH on methyl violet biosorption. Bio resoul Technol **99**(13), 5411–5417 (2008)
19. Suzuki, M.: Adsorption Engineering. Elsevier, Tokyo, Japan (1990)
20. Toth, J.: Adsorption: Theory, Modeling, and Analysis. Surfactant Science Series; 107. Marcel Dekker, NY (2002)
21. Kammerer, J., Kammerer, D.R., Jensen, U., Carle, R.: Interaction of apple polyphenols in a multi-compound system upon adsorption onto a food-grade resin. J. Food Eng. **96**, 544–554 (2010)
22. Kühn, S., Wollseifen, H.R., Galena, R., Schulze-Kaysers, N., Kunz, B.: Adsorption of flavonols from onion (Allium cepa L.) processing residues on a macroporous acrylic resin. Food Research International 7th World Congress on Polyphenols Applications 65A, pp. 103–108 (2014)
23. Kammerer, D.R., Carle, R., Stanley, R.A., Saleh, Z.S: Pilot-scale resin adsorption as a means to recover and fractionate apple polyphenols. J. Agric. Food Chem. **58**(11), 6787–96 (2010)
24. Frascari, D., Bacca, A.E.M., Zama, F., Bertin, L., Fava, F., Pinelli, D.: Olive mill wastewater valorization through phenolic compounds adsorption in a continuous flow column. Chem. Eng. J. **283**, 293–303 (2016)
25. Ren, Y., Cui, C., Wang, P.: Pomelo peel modified with citrate as a sustainable adsorbent for removal of methylene blue from aqueous solution. Molecules **23**, 1342 (2018). https://doi.org/10.3390/molecules23061342
26. Mota, M.I.F., Pinto, P.C.R., Loureiro, J.M., Rodrigues, A.E.: Recovery of vanillin and syringaldehyde from lignin oxidation: a review of separation and purification processes. Sep. Purif. Rev. **45**(3), 227–259 (2016)
27. Larrán, P.P., Reinoso, B.D., Moure, A., Alonso, J.L., Domínguez, H.: Adsorption technologies to recover and concentrate food polyphenols. COFS (2018).https://doi.org/10.1016/j.cofs.2017.10.005
28. Olasehinde, E.F., Adegunloye, A.V., Adebayo, M.A., Oshodi, A.A.: Cadmium (II) adsorption from aqueous solutions using onion skins. Water Conserv. Sci. Eng. (2019). https://doi.org/10.1007/s41101-019-00077-2
29. Raval, N.P., Shah, P.U., Shah, N.K.: Adsorptive removal of nickel (II) ions from the aqueous environment: a review. J. Environ. Manage. **179**, 1–20 (2016). https://doi.org/10.1016/j.jenvman.2016.04.045

30. Ioannidou, O., Zabaniotou, A.: Agricultural residues as precursors for activated carbon production-a review. Renew. Sustain. Energy Rev. **11**(9), 1966–2005 (2007). https://doi.org/10.1016/j.rser.2006.03.013
31. Valentín-Reyes, J., García-Reyes, R.B., García-González, A., Soto-Regalado, E., Cerino-Córdova, F.: Adsorption mechanisms of hexavalent chromium from aqueous solutions on modified activated carbons. J. Environ. Manage. **236**, 815–822 (2019). https://doi.org/10.1016/j.jenvman.2019.02.014
32. Pacheco, S., Tapia, J., Medina, M., Rodriguez, R.: Cadmium ions adsorption in simulated wastewater using structured alumina-silica nanoparticles. J. Non Cryst. Solids **352**, 5475–5481 (2006). https://doi.org/10.1016/j.jnoncrysol.2006.09.007
33. Jegadeesan, G., Al-Abed, S.R., Sundaram, V., Choi, H., et al.: Arsenic sorption on TiO_2 nanoparticles: Size and crystallinity effects. Water Res. **44**(3), 965–973 (2010). https://doi.org/10.1016/j.watres.2009.10.047
34. Schieber, A., Stintzing, F.C.: Carle R By-products of plant food processing as a source of functional compounds—recent developments. Trends Food Sci. Technol. **12**(11), 401–413 (2001)
35. Vianai, K.W.C., Oliveira, I.S.N., Rochai, J.C.C. et al.: Study of adsorption isotherms and physicochemical characterization of natural food colorants. Food Sci. Technol. Campinas **39**(1). 74–80 (2019)
36. Zatar, N.A.: Simultaneous determination of seven synthetic water-soluble food colorants by Ion-pair Reversed-phase High-performance Liquid chromatography. J. Food Technol. **5**(3), 220–224 (2007)
37. Bachalla, N.: Identification of synthetic food colors adulteration by paper chromatography and spectrophotometric methods. IAIM **3**(6), 182–191 (2016)
38. Vyralakshmi, G., Jayasheela, G.: Food adulteration and contamination-a catastrophe. IOSR J. Environ. Sci. Toxicology Food Technol. (IOSR-JESTFT) **11**(7), 62–70 (2017)
39. El Sheikha, A.F.: DNAFoil: a novel technology for the rapid detection of food adulteration. Trends Food Sci. Technol. **86**, 544–552 (2019)
40. Ellis, D.I., Brewster, V.L., Dunn, W.B., Allwood, J.W., Golovanov, A.P., Goodacre, R.: Fingerprinting food: current technologies for the detection of food adulteration and contamination. Chem. Soc. Rev. **41**(17), 5706–5727 (2012)
41. Jonnalagadda Padmaja, R., Rao, P., Bhat, R.V., Nadamuni, N.: Type, extent and use of colors in ready to eat (RTE) foods prepared in the nonindustrial sector a case study from Hyderabad. Int. J. Food Sci. Technol. **39**(2), 125–131 (2004)
42. Onyancha, D., Mathura, W., Ngila, J.C., Ongoma, P., Chacha, J.: Studies of chromium removal from tannery wastewaters by algae biosorbents, Spirogyra condensata, and Rhizoclonium hieroglyphical. J. Hazard. Mater. **158**(2–3), 605–614 (2008). https://doi.org/10.1016/j.jhazmat.2008.02.043
43. Limousin, G., Gaudet, J.P., Charlet, L., Szenknect, S., Barthès, V., Krimissa, M.: Sorption isotherms: a review on physical bases, modeling, and measurement. Appl. Geochem. **22**(2), 249–275 (2007). https://doi.org/10.1016/j.apgeochem.2006.09.010
44. Allen, S.J., Mckay, G., Porter, J.F.: Adsorption isotherm models for basic dye adsorption by peat in single and binary component systems. J. Colloid Interface Sci. **280**(2), 322–333 (2004). https://doi.org/10.1016/j.jcis.2004.08.078
45. Pompeu, D.R., Moura, F.G., Silva, E.M.: Equilibria, kinetics, and mechanisms for the adsorption of four classes of phenolic compounds onto synthetic resins. Separat. Sci. Technol. **45**(5), 700–709 (2010). https://doi.org/10.1080/01496390903562274
46. Anspach, F.B., Curbelo, D., Hartmann, R., Garke, G., Deckwer, W.D.: Expanded-bed chromatography in primary protein purification. J. Chromatogr. A **865**, 129–144 (1999)
47. Cotte-Rodriguez, I., Takats, Z., Talaty, N., Chen, H., Cooks, R.G.: Desorption electrospray ionization of explosives on surfaces: sensitivity and selectivity enhancement by reactive desorption electrospray ionization. Anal. Chem. **77**(21), 6755–6764 (2005)
48. Chen, H., Zhang, X., Luo, M.: Fenxi Huaxue **34**, 464 (2006)

49. Dong, J., Rezenom, Y.H., Murray, K.K.: Implementation of DART and DESI ionization on a fieldable mass spectrometer. J. Amer. Soc. Mass Spectrometry **19**, 1419–1424 (2008)
50. Hartmanova, L., Ranc, V., Papouskova, B., Bednar, P., Havlicek, V., Lemr, K., et al.: Fast profiling of anthocyanins in wine by desorption Nano-electrospray ionization mass spectrometry. J. Chromatogr. **1217**(25), 4223–4228 (2010)
51. https://www.food.gov.uk/science/asdditives/enumberlist (2014)
52. Zhang, Z., Cooks, R.G., Ouyang, Z.: Paper spray: a simple and efficient means of analysis of different contaminants in foodstuffs. Analyst **137**(11), 2556–2558 (2012)
53. Elmolla, E.S., Hamdy, W., Kassem, A., Abdel, H.A.: Comparison of different rice straw based adsorbents for chromium removal from aqueous solutions. Desalin. Water Treat. **57**(15), 6991–6999 (2016). https://doi.org/10.1080/19443994.2015.1015175

Adsorption and Desorption of Micropollutants

K. Hemkumar, P. Ananthi, M. Kanagalakshmi, and Anitha Pius

Abstract Micropollutant (MP) contamination of water resources has sparked extensive environmental research. These MPs are frequently found in sewage, groundwater, wastewater, drinking water and surface water. This chapter gives the brief overview of the isotherms of MPs from aqueous solution from various adsorbents such as agricultural solid wastes, activated carbons, clays, and metal–organic-framework. Several isotherm models including Langmuir, Temkin, Freundlich, Dubinin- Radushkevich, Sips, Redlich-Peterson and Toth isotherm were discussed. This chapter demonstrates that the equilibrium graph fits the Langmuir isotherm in most cases and has found successful application in many monolayer adsorption sorption processes. Most of the literatures available perform batch process but this review provides the platform for developing continuous flow systems with large-scale applications at the small-scale also. The majority of the reported studies are done in batch mode, but this review provides a foundation for designing continuous flow systems with industrial applications at the commercial level as well.

Keywords Micropollutant · Antibiotics · UV treatment · Pharmaceutical waste · Freundlich · Desorption

1 Introduction

Environmental pollution has become the severe hazard to human health and water resources over the last decade and it has become necessary to eliminate the harmful contaminants from water matrices. Organic pollutants, also classified as "emerging contaminants," includes medications, hormones, personal care items, surfactants, illicit narcotics and pesticides [1]. Pharmaceutical chemicals are often prescribed and provided to the humans and domestic livestock in the modern period and their subsequent excretion results in the drug accumulation in the surrounding water resources

K. Hemkumar · P. Ananthi · M. Kanagalakshmi · A. Pius (✉)
Department of Chemistry, The Gandhigram Rural Institute – Deemed to be University, Tamil Nadu, Gandhigram, Dindigul 624 302, India
e-mail: dranithapius@gmail.com

J. Tharini and S. Thomas (eds.), *Carbon Nanomaterials and their Composites as Adsorbents*, Carbon Nanostructures, https://doi.org/10.1007/978-3-031-48719-4_10

[2]. Many methods for treating water pollution were caused by the pharmaceuticals and personal care products (PPCPs) including advanced oxidation process (AOP), photodegradation, enzymatic systems, membrane filtration, and adsorption. Most methods have drawbacks or challenges, such as high energy consumption and toxic intermediates etc. Because of its ease of operation and lack of secondary pollution, adsorption method that directly removes PPCPs from polluted water was currently regarded as the one of the most feasible and effective methods [3]. Carbon adsorbents, clay, minerals, polymeric resins and other adsorbents such as metals and their oxides, molecularly imprinted polymers, mesoporous material and gels are the most commonly used adsorbents for the removal of pharmaceutical compounds from aqueous solution. Despite the foregoing, it is worthwhile to investigate modified adsorbents for the removal of pharmaceutical compounds from water than conventional systems [4]. Metal–organic-frameworks (MOFs) are formed by strong coordination bonds between inorganic clusters and organic ligands. Over the last two decades, the diversity of organic and metal units has resulted in over 20,000 different types of MOFs with specific surface areas ranging from 100 to 10,000 $m^2\ g^{-1}$. MOFs have a distinct advantage over conventional porous materials due to its pore size, high specific surface area and its properties could be controlled through synthesis and rational design.

Excellent chemical properties and pore structure of MOF's make them suitable for the usage as the catalysis, adsorption/separation, sensing, energy storage and biological applications. MOFs with good water stability would enhance the stability through post-synthetic modification strategies and in-situ synthesis. The use of MOFs in the water phase adsorption and separation is being promoted significantly by these ongoing studies. As a result, these MOFs have been used to separate and adsorb contaminants including inorganic and organic compounds. Aside from those mesoporous MOFs, some microporous MOFs with small vent such as UiO-66, ZIF-67 and ZIF-8 have been extensively studied in terms of surface adsorption; tunable chemical properties. The presence of active sites on the MOF allows the complete interaction with guest molecules and hence enhancing the adsorption activity. This chapter concentrates on advancements in the application of MOF-based antibacterial adsorption materials. The superior quality of MOFs provides necessary conditions for preparing MOFs as an adsorbing materials with excellent activity and knowledge for the removal of PPCPs [5].

2 Common Source of MPs (Antibiotics)

Antibiotics are primary components of present-day medicines that are used to treat a various diseases caused by the various pathogens. They are widely used in human and animal medicines, as well as in aquaculture to improve the growth and development. Due to their inappropriate behavior, overusage and discharge in the environment have the negative impact on the environment [6].

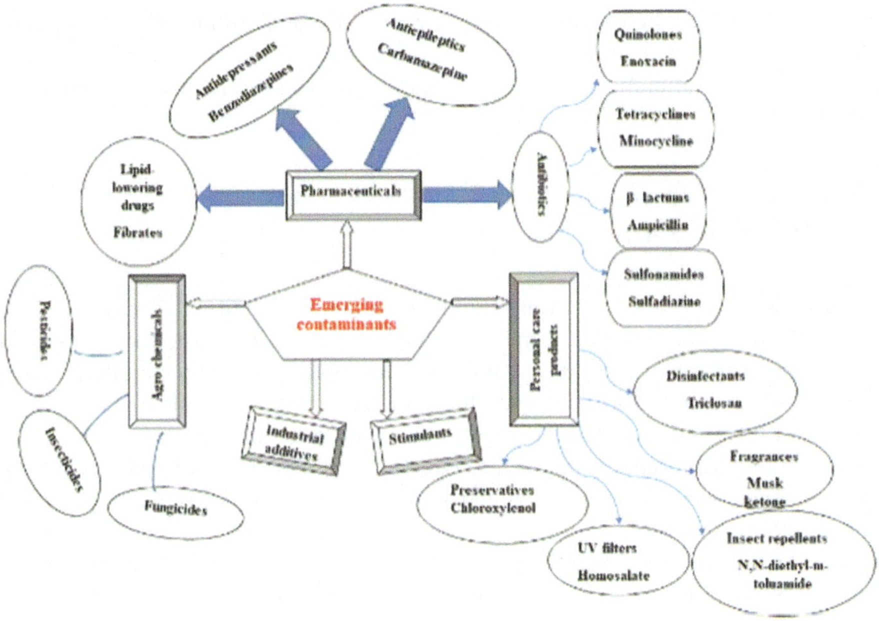

Fig. 1 Emerging contaminates

2.1 *Hospitals*

Hospital wastewaters are a complex mixture of many antibiotics that cause problems to the environment which are 10–20 times more toxic than typical municipal waste. The majority of hospital wastes are antibiotics which are usually un-metabolized medicines from patients, mainly through feces and urine. The amount of antibiotics released from the hospital sewage wastewater varies according to the antibiotic cluster. For example, it is approximately 67% for azithromycin, 49% for fluoroquinolones, and 94% for clarithromycin [7]. Fluoroquinolones are present in high concentrations in hospital wastewater because these antibiotics are commonly used as the first line of treatment in hospitals (Fig. 1).

2.2 *Pharmaceutical Waste*

Pharmaceutical companies have been identified as a potential source of antibiotics in the environment. Common antibiotics such as oxytetracycline (600 mg L^{-1}) and ciprofloxacin (10 mg L^{-1}) in wastewater effluents from the pharmaceutical industry could be several mg L^{-1} higher than permissible values [8]. Large amounts of sulfonamides, quinolones, and macrolides were found in water bodies with the majority of antibiotics being cefotaxime, azithromycin, ofloxacin, sulfamethoxazole, clarithromycin and trimethoprim present in 1.3, 2.8, 1.2, 1.1, 2.4 and 0.9 g L^{-1} respectively. β lactam antibiotic such as Cephalexin was found in higher concentrations which is around 4.2 g L^{-1} in river water [9].

2.3 Domestic Waste

Human's excretory product has the significant amount of antibiotics in the form of host molecules and unsettled forms, which eventually reach wastewater treatment plants. Conventional wastewater treatment plants, which primarily stimulate sludge progression, does not remove these antibiotic molecules. As the result, higher concentrations of host molecule and their by-products are generally detected in the water bodies. In India, antibiotics such as norfloxacin, levofloxacin and ciprofloxacin were found in higher concentrations ranging from 44.5 to 311.0 g L^{-1}. In case of the United States, sulfamethoxazole, ciprofloxacin and cephalexin were found in the concentration ranging from 0.2 to 13.8 g L^{-1} [10]. Sulfamethoxazole and trimethoprim antibiotics are commonly found in household wastewater.

2.4 Animal Production

Cattle farms generate significant amounts of unabsorbed antibiotics. China is the world's largest consumer of veterinary antibiotics. Beef cattle farms uses erythromycin and sulfathiazole, while swine farms uses sulfadimethoxine, lincomycin, sulfathiazole and sulfamethazine [11].

2.5 Aquaculture

Marine species cultivation is the fast-growing calorie rich sources, accounting for the half of global fish consumption. Due to exhaustive farming practices and the fast transmission of diseases has resulted in the high antibiotic levels in feedstuffs. The majority of antibiotics are found in fish feedstuffs which are released into the water bodies as urine, faecal matter and sewage feedstuffs. Several classes of antibiotics are seen in water streams and accumulate in the sewage water. For example, the highest observed antibiotics content in water samples from shrimp ponds in Vietnam were found to be 2.02 and 5.53 mg L^{-1} for trimethoprim and sulfamethoxazole, respectively, whereas the maximum content of oxytetracycline were found to be 17.9 mg L^{1} in the freshwater bodies [12].

3 Different Treatment Technologies

The efficient and cost-effective PPCPs removal from wastewater is an important and difficult job for restoring environment. Many removal techniques are known in the literatures depend on the several factors namely amount of PPCP, amount of effluent and the cost. Wastewater treatment plants (WWT) are made up of a series of chemical, biological and physical unit activities and processes that are classed as primary, secondary, and tertiary. Secondary treatment methods are generally biological and helps in decreasing the biochemical oxygen demand or organic content of influent via trickling filters, activated sludge, etc. Tertiary treatment often employs adsorption and advanced oxidation processes (AOPs) (Fig. 2).

3.1 Pharmaceutical Waste Treatment by Activated Carbons

Many scientists have made efforts to find alternate carbon sources for producing activated carbons. Activated carbons (AC) are typically produced through the pyrolysis of biomass in an inert atmosphere. Freundlich, Langmuir, and Dubinin- Raduskevitch models have been used to simulate adsorption isotherms. The Langmuir model better represent the adsorption of sulfamethoxazole and metronidazole which shows

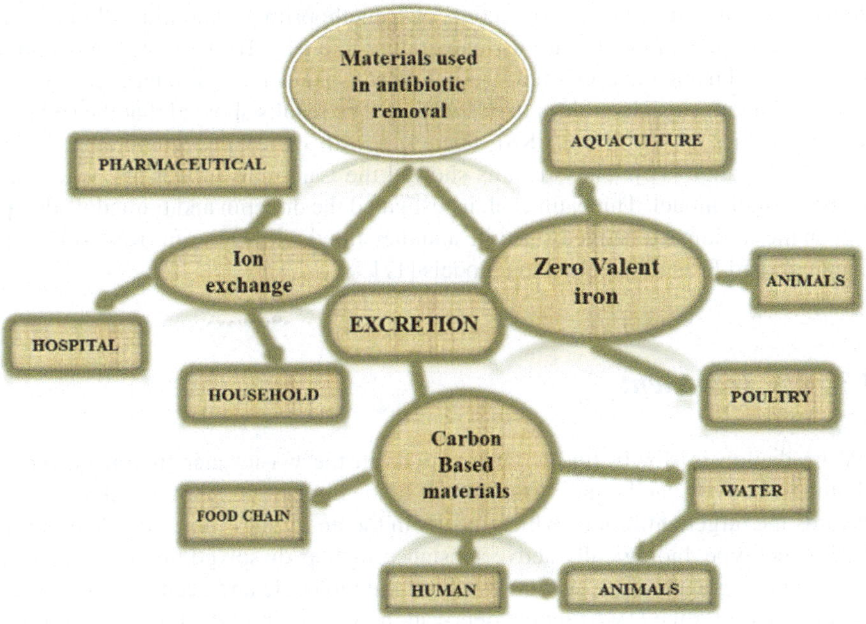

Fig. 2 Material used in antibiotic removal

the maximum adsorption capacities of 184.66 and 145.89 mg g^{-1}, respectively. The adsorption of naproxen (NPX), diclofenac (DCF), ketoprofen (KPN) and ibuprofen (IBF) on AC made from waste olive cakes was reported by Baccar et al. [13]. The results revealed that the Langmuir isotherm fits well with adsorption of four pharmaceuticals compounds namely Ibuprofen, ketoprofen, naproxen, diclofenac and shows maximum adsorption capacities of 12.6, 24.7, 39.5, 56.2 mg g^{-1}, respectively. Ferreira et al. investigated the adsorption of paracetamol using AC from Babassu and Dende Coconut Mesocarp [14]. The single layer adsorption were found to be 71.39 and 70.62 mg g^{-1} at the AC derived from babassu coconut mesocarp and dende coconut mesocarp respectively.

3.2 Pharmaceutical Products Treatment by Clay Minerals

Clays are abundant and thus inexpensive three dimensional materials. Diatomite, Illite, montmorillonite, serpentine, saponite, kaolinite, bentonite, pyrophyllite, sepiolite, vermiculite and fuller's earth are some well-known clay minerals. Many researchers have demonstrated that clays have a relatively good ability to remove pharmaceutical products. Bekci et al. used montmorillonite KSF to remove trimethoprim under various conditions such as pH, temperature and ionic strength [15]. To determine the characteristic parameters of each model, the adsorption data could be fitted with Freundlich, Langmuir, and Dubinin-Radushkevich equation models. Bekci et al. investigated the adsorption of trimethoprim by montmorillonite clay using a batch technique at various temperature and pH. Trimethoprim adsorption was described using the Langmuir, Dubinin-Radushkevich and Freundlich isotherm models to attain capacity values of adsorption. The results showed that the (relative capacity values of adsorption) KSF decreases with increasing temperature in the 298–318 K range [16]. These results showed the Langmuir model fits better than the Freundlich model. Thiebault et al. investigated the doxepin and tramadol adsorption on the sodium exchanged smectite and they fitted with Dubinin- Radushkevich, Langmuir and Freundlich equation models [17].

3.3 UV Treatment

UV irradiation is slowly finding applications in the wastewater treatment, where photolytic activity can be indirect or direct. In the direct type, photons are projected towards the target antibiotics which results in the antibiotics molecule cleavage. In the indirect type, biologically active substance such as dissolved organic matter act as an oxidizing agent. The quantum yield of the molecule and accumulation of UV light usually alter the UV-triggered deprivation of an antibiotic. During the removal of anitibiotics during photolyis, UV dosage, organic matter, molecular chemical assembly and reaction period plays the major role [18].

3.4 *Photochemical and Chlorination Technique*

There are only limited literature available for antibiotic removal from the water bodies using chlorinating agents. Chlorination is an effective water detoxifying technique used around the world before water bodies get released from various pathogens. Among the different chlorinating agents, hypochlorite has the highest oxidation capacity of 1.47 V, followed by chlorine gas at 1.35 V and chlorine dioxide 0.96 V. The main disadvantages of using Cl based detoxifiers includes the production of detoxifier fragments [19].

3.5 *Biological Treatment*

Biological treatment is the process of converting composite organic matter (antibiotics) into various metabolic compounds via bio-transformation or mineralization using a contaminated or uncombined microorganism medium. Flavin-based monooxygenase and flavin reductase was demonstrated by the SADA and SADC genes respectively, results in the formation of p-aminophenol. The resultant p-aminophenol was converted into hydroxyquinol prior to mineralization [20]. Several techniques can be used to generate robust algal species for antibiotic elimination. The abundance of algae in the environment allows for the development of algal strains for antibiotic removal.

4 Agricultural Solid Wastes

Waste from agricultural has no economic value and frequently causes issues while disposing. Raw agricultural solid wastes such as leaves and seeds, as well as from forest waste residues, have been used as the adsorbent due to the low cost and unique physicochemical properties. Araujo et al. discussed the removal of Diclofenac using seed husk of Moringa Oleifera [21]. As a adsorbent material, coffee and rice husk wastes are used by Paredes-Laverde et al. showed the removal of norfloxacin a widely used antibiotic drug [22]. Langmuir, Freundlich and Redlich-Peterson isotherms were used to analyze the equilibrium adsorption data. A monolayer-type adsorption model was suggested which follows Langmuir and Redlich-Peterson isotherms. N'diaye and Kankou prepared a low-cost adsorbent using Balanites aegyptiaca seeds for adsorption of caffeine from an aqueous solution [23]. The purpose of batch sorption experiments was to determine caffeine adsorption isotherms on the Balanites aegyptiaca seeds. Four isotherm models Sips, Freundlich, Redlich-Peterson, and Langmuir were examined for nonlinear modelling of adsorption. N'diaye and Kankou investigated the use of a Zizyphus mauritiana seed as the adsorbent for caffeine removal from an aqueous solution. The Langmuir, Toth isotherms, Freundlich, Sips, Temkin and

Redlich–Peterson were used to calculate and analyze equilibrium isotherms using a nonlinear method [24]. The Langmuir isotherm model was found to be the best fit for the experimental data and the single layer adsorption maximum capacity was found to be 2.29 mg g^{-1}. Batch equilibrium method was carried out by N'diaye et al. to investigate the paracetamol adsorption on the shells of groundnut. The Langmuir, Freundlich, Temkin, Sips, Redlich–Peterson and Toth equations were used to fit the experimental data [25].

5 Adsorption—A Versatile Treatment Option

Carbonaceous materials such as carbon nanotubes(CNT), AC and graphene oxide (GO); mesoporous clay, MOF and bio-sorbents like agricultural soil, sludge have all been explored for PPCP removal [26]. Hasan et al. (2013) investigated various adsorbents for the removal of PPCP; however, the maximum removal efficiency was obtained using MIL-101 functionalized ethylene diamine, while using acidic amino methane sulfonic acid shows poor performance [27].

6 Adsorption of Micropollutants by MOFs

MOF have high surface area and porosity, highly stable and easy to synthesis with the tunable properties. The MOFs' distinct properties make them ideal material for the variety of applications, including hydrogen storage, purification and filtration, catalysis, optical devices and sensor [28]. For adsorbing antibiotics from the aqueous solutions, various kinds of adsorbent using MOF-based materials, which includes bare MOFs and their composite materials, have been reported. Microporous MOFs act as the perfect materials for the antibiotic removal due to the large surface area, good number of functional groups on the outer surface and small size [29].

Quinolones such as enoxacin, ciprofloxacin, norfloxacin and ofloxacin are globally used in the treatment of bacterial infection caused in animal and human. It has been used for decades, ever since the first-generation quinolone was synthesised synthetically in the 1960s. However, prolonged use and production of these drugs has resulted in the water pollution. The use of β-lactam and penicillin antibiotics in the clinical settings has increased significantly which includes ceftriaxone, ampicillin (AMP), cefradine (RAD), cloxacillin (CLX), and amoxicillin (AMX) [30]. β-lactam antibiotics, particularly penicillin and amoxicillin account for the approximately 60–75% of global antibiotic usage due to their high bactericidal activity, low toxicity, and broad application.

Tetracyclines, which include oxytetracycline (OTC), tetracycline (TC), and chlortetracycline (CTC) have 4 hexatomic rings and several functional groups, particularly the carbonyl and hydroxyl groups [31]. This kind of medicines is used to treat bacterial infections in animals and humans by blocking most Gram-negative and

Gram-positive bacteria. These antibiotics have a low absorption efficiency and a sluggish metabolism, and they are discharged in the vast quantities into the environment via the excretory system.

Furantoin (NFT) and furacillin (NZF) were primarily used to cure infections in intestinal infections, urinary tract, wound on skin and as food additives to prevent infection on intestinal region in poultry. Similarly, some health organizations have classed nitrofurans and their by-products compounds as a non-edible one, and they are currently utilized in the animals only. Because of their lower pKa values, NZF and NFT have a more controlled electrostatic interaction than β-lactams and tetracyclines [32].

Nitroimidazoles are a class of medications that have a specific nitroimidazole ring structure, such as tinidazole (TNZ), metronidazole (MNZ), dimetridazole (DMZ) and menidazole (MZ). Nitroimidazoles could be utilized as anti-viral, anti-tumor, anti-tuberculosis and anti-parasitic sensitizers. Most of the microporous MOFs were also used in the trapping of these pollutants due to their less molecular sizes. Furthermore, the lower pKa values 3.0 favor the use of MOFs which is positively charged.

7 Desorption of MPs from MOFs

MOFs have demonstrated exceptional adsorptive removal applicability for the variety of antibiotics. Adsorption technology still confronts hurdles, particularly with regard to recycling. After dozens of cycles, incomplete desorption would eventually induce inadequate adsorption performance, resulting in the secondary waste and high cost. As the result, contaminant desorption from adsorbent as well as adsorbent regeneration is a critical and basic unit. Unlike adsorption, desorption involves mass transfer process in which pollutants migrate from the liquid phase of adsorbent's surface or channel. Desorption efficiency strongly dependent on the intrinsic qualities of the adsorbent, adsorbate, and, most critically, the eluent. Various eluents have been used to obtain effective antibacterial desorption from MOF-based adsorbents. Organic solvents such as acetone and alcohols, aqueous solutions like acid and alkaline solutions, salts, water and hybrid mixtures of organic solvent and water are among the eluents chosen. Based on the following rules, the majority of regeneration percentages can reach >84%: First and foremost, adsorbate concentration differences in the eluent and adsorbent are required; secondly most antibiotics have varying solubility in water and organic solvents [33].

As a result, the polarity of the chosen eluent should be compatible with the antimicrobial agent. In this regard, acetone, methanol, and ethanol perform the desorption of nitroimidazoles, sulfonamides, and tetracyclines. For example, ammonia, HCl and NaOH were used to limit the desorption of tetracyclines, quinolones and sulphonamides.

8 Conclusion

This chapter discusses the adsorptive removal of the antibiotics using various adsorbents. Antibiotics are the emerging pollutants that are toxic even at the trace levels and it is persistent nature in the environment which posses the major treats to the ecosystem. Hence it is essential to remove the pollutants using simple, low cost and appropriate methods. The current chapter outlined the environmental fates and behaviours of antibiotics, as well as the challenges. We have discussed the current antibiotic removal technology as well as the advances in that field. A review of various technologies revealed that adsorption using MOFs and their combination have shown the effective technologies for the removal of antibiotics. Thus, MOF-based materials shows the outstanding performance for antibiotics removal. We hope that this chapter will be useful to researchers and will encourage the use of high-efficiency MOF-based adsorbents for the commercial treatment of pharmaceutical wastewater.

References

1. Balakrishna, K., Rath, A., Praveenkumarreddy, Y., Guruge, K.S., Subedi, B.: A review of the occurrence of pharmaceuticals and personal care products in Indian water bodies. Ecotoxicol. Environ. Saf. **137**, 113–120 (2017). https://doi.org/10.1016/j.ecoenv.2016.11.014
2. Wang, J., Gardinali, P.R.: Uptake and depuration of pharmaceuticals in reclaimed water by mosquito fish (Gambusia holbrooki): a worst-case, multiple-exposure scenario. Environ. Toxicol. Chem. **32**, 1752–1758 (2013). https://doi.org/10.1002/etc.2238
3. Watkinson, A.J., Murby, E.J., Kolpin, D.W., Costanzo, S.D.: The occurrence of antibiotics in an urban watershed: from wastewater to drinking water. Sci. Total. Environ. **407**, 2711–2723 (2009). https://doi.org/10.1016/j.scitotenv.2008.11.059
4. Khandelwal, G., Maria Joseph Raj, N.P., Vivekananthan, V., Kim, S.J.: Biodegradable metal-organic framework MIL-88A for triboelectric nanogenerator. iScience. **24**, 102064 (2021). https://doi.org/10.1016/j.isci.2021.102064
5. Carraro, E., Bonetta, S., Bertino, C., Lorenzi, E., Bonetta, S., Gilli, G.: Hospital effluents management: chemical, physical, microbiological risks and legislation in different countries. J. Environ. Manage. **168**, 185–199 (2016). https://doi.org/10.1016/j.jenvman.2015.11.021
6. Malakootian, M., Nasiri, A., Amiri Gharaghani, M.: Photocatalytic degradation of ciprofloxacin antibiotic by TiO_2 nanoparticles immobilized on a glass plate. Chem. Eng. Commun. **207**, 56–72 (2020). https://doi.org/10.1080/00986445.2019.1573168
7. Perini, J.A.L., Tonetti, A.L., Vidal, C., Montagner, C.C., Nogueira, R.F.P.: Simultaneous degradation of ciprofloxacin, amoxicillin, sulfathiazole and sulfamethazine, and disinfection of hospital effluent after biological treatment via photo-Fenton process under ultraviolet germicidal irradiation. Appl. Catal. B Environ. **224**, 761–771 (2018). https://doi.org/10.1016/j.apcatb.2017.11.021
8. Cardoso, O., Porcher, J.M., Sanchez, W.: Factory-discharged pharmaceuticals could be a relevant source of aquatic environment contamination: review of evidence and need for knowledge. Chemosphere **115**, 20–30 (2014). https://doi.org/10.1016/j.chemosphere.2014.02.004
9. Fick, J., Söderström, H., Lindberg, R.H., Phan, C., Tysklind, M., Larsson, D.G.J.: Contamination of surface, ground, and drinking water from pharmaceutical production. Environ. Toxicol. Chem. **28**, 2522–2527 (2009). https://doi.org/10.1897/09-073.1

10. Huang, Y., Cheng, M., Li, W., Wu, L., Chen, Y., Luo, Y., Christie, P., Zhang, H.: Simultaneous extraction of four classes of antibiotics in soil, manure and sewage sludge and analysis by liquid chromatography-tandem mass spectrometry with the isotope-labelled internal standard method. Anal. Methods **5**, 3721–3731 (2013). https://doi.org/10.1039/c3ay40220g
11. Chen, J., Liu, Y.S., Zhang, J.N., Yang, Y.Q., Hu, L.X., Yang, Y.Y., Zhao, J.L., Chen, F.R., Ying, G.G.: Removal of antibiotics from piggery wastewater by biological aerated filter system: treatment efficiency and biodegradation kinetics. Bioresour. Technol. **238**, 70–77 (2017). https://doi.org/10.1016/j.biortech.2017.04.023
12. Rico, A., Satapornvanit, K., Haque, M.M., Min, J., Nguyen, P.T., Telfer, T.C., van den Brink, P.J.: Use of chemicals and biological products in Asian aquaculture and their potential environmental risks: a critical review. Rev. Aquac. **4**, 75–93 (2012). https://doi.org/10.1111/j.1753-5131.2012.01062.x
13. Grover, D.P., Zhou, J.L., Frickers, P.E., Readman, J.W.: Improved removal of estrogenic and pharmaceutical compounds in sewage effluent by full scale granular activated carbon: impact on receiving river water. J. Hazard. Mater. **185**, 1005–1011 (2011). https://doi.org/10.1016/j.jhazmat.2010.10.005
14. Zhang, X., Guo, W., Ngo, H.H., Wen, H., Li, N., Wu, W.: Performance evaluation of powdered activated carbon for removing 28 types of antibiotics from water. J. Environ. Manage. **172**, 193–200 (2016). https://doi.org/10.1016/j.jenvman.2016.02.038
15. Wang, H., Ding, J., Ding, L., Ren, N.: Analysis of sulfonamides in soil, sediment, and sludge based on dynamic microwave-assisted micellar extraction. Environ. Sci. Pollut. Res. **23**, 12954–12965 (2016). https://doi.org/10.1007/s11356-016-6383-0
16. Song, Y., Sackey, E.A., Wang, H., Wang, H.: Adsorption of oxytetracycline on kaolinite. PLoS ONE **14**, 1–13 (2019). https://doi.org/10.1371/journal.pone.0225335
17. N'diaye, A.D., Kankou, M.S.: Valorization of balanites aegyptiaca seeds from mauritania: Modeling of adsorption isotherms of caffeine from aqueous solution. J. Environ. Treat. Tech. **7**, 450–455 (2019)
18. Fatta-Kassinos, D., Vasquez, M.I., Kümmerer, K.: Transformation products of pharmaceuticals in surface waters and wastewater formed during photolysis and advanced oxidation processes—Degradation, elucidation of byproducts and assessment of their biological potency. Chemosphere **85**, 693–709 (2011). https://doi.org/10.1016/j.chemosphere.2011.06.082
19. Homem, V., Santos, L.: Degradation and removal methods of antibiotics from aqueous matrices—A review. J. Environ. Manage. **92**, 2304–2347 (2011). https://doi.org/10.1016/j.jenvman.2011.05.023
20. Ricken, B., Corvini, P.F.X., Cichocka, D., Parisi, M., Lenz, M., Wyss, D., Martínez-Lavanchy, P.M., Müller, J.A., Shahgaldian, P., Tulli, L.G., Kohler, H.P.E., Kolvenbacha, B.A.: Ipso-hydroxylation and subsequent fragmentation: a novel microbial strategy to eliminate sulfonamide antibiotics. Appl. Environ. Microbiol. **79**, 5550–5558 (2013). https://doi.org/10.1128/AEM.00911-13
21. Rafatullah, M., Sulaiman, O., Hashim, R., Ahmad, A.: Adsorption of methylene blue on low-cost adsorbents: a review. J. Hazard. Mater. **177**, 70–80 (2010). https://doi.org/10.1016/j.jhazmat.2009.12.047
22. Paredes-Laverde, M., Silva-Agredo, J., Torres-Palma, R.A.: Removal of norfloxacin in deionized, municipal water and urine using rice (Oryza sativa) and coffee (Coffea arabica) husk wastes as natural adsorbents. J. Environ. Manage. **213**, 98–108 (2018). https://doi.org/10.1016/j.jenvman.2018.02.047
23. Demba N'diaye, A., Bollahi, M.A., Sid', M., Kankou, A.: Sorption of paracetamol from aqueous solution using groundnut shell as a low cost sorbent. J. Mater. Environ. Sci. **10**, 553 (2019)
24. Demba, A.N., Sid, M., Kankou, A.: Modeling of adsorption isotherms of pharmaceutical products onto various adsorbents: a Short Review. J. Mater. Environ. Sci. **11**, 1264–1276 (2020)
25. Baccar, R., Sarrà, M., Bouzid, J., Feki, M., Blánquez, P.: Removal of pharmaceutical compounds by activated carbon prepared from agricultural by-product. Chem. Eng. J. **211–212**, 310–317 (2012). https://doi.org/10.1016/j.cej.2012.09.099

26. Dhamodharan, K., Aneesh, T., Pius, A.: Reductive dissolution studies of CeO <inf>2</inf> and simulated mixed oxide fuel of (Ce, U)O <inf>2</inf> in nitric acid medium. J. Radioanal. Nucl. Chem. 319 (2019). https://doi.org/10.1007/s10967-018-6372-5
27. Jiang, J.Q., Ashekuzzaman, S.M.: Development of novel inorganic adsorbent for water treatment. Curr. Opin. Chem. Eng. **1**, 191–199 (2012). https://doi.org/10.1016/j.coche.2012.03.008
28. Mason, J.A., Veenstra, M., Long, J.R.: Evaluating metal-organic frameworks for natural gas storage. Chem. Sci. **5**, 32–51 (2014). https://doi.org/10.1039/c3sc52633j
29. Suksaengrat, P., Amornkitbamrung, V., Srepusharawoot, P.: Enhancements of hydrogen adsorption energy in M-MOF-525 (M= Ti, V, Zr and Hf): A DFT study. Chinese J. Phys. **64**, 326–332 (2020). https://doi.org/10.1016/j.cjph.2020.01.011
30. Li, H., Cao, X., Zhang, C., Yu, Q., Zhao, Z., Niu, X., Sun, X., Liu, Y., Ma, L., Li, Z.: Enhanced adsorptive removal of anionic and cationic dyes from single or mixed dye solutions using MOF PCN-222. RSC Adv. **7**, 16273–16281 (2017). https://doi.org/10.1039/c7ra01647f
31. Mirsoleimani-Azizi, S.M., Setoodeh, P., Zeinali, S., Rahimpour, M.R.: Tetracycline antibiotic removal from aqueous solutions by MOF-5: Adsorption isotherm, kinetic and thermodynamic studies. J. Environ. Chem. Eng. **6**, 6118–6130 (2018). https://doi.org/10.1016/j.jece.2018.09.017
32. Wang, B., Lv, X.L., Feng, D., Xie, L.H., Zhang, J., Li, M., Xie, Y., Li, J.R., Zhou, H.C.: Highly stable Zr(IV)-based metal-organic frameworks for the detection and removal of antibiotics and organic explosives in water. J. Am. Chem. Soc. **138**, 6204–6216 (2016). https://doi.org/10.1021/jacs.6b01663
33. Zhao, R., Ma, T., Zhao, S., Rong, H., Tian, Y., Zhu, G.: Uniform and stable immobilization of metal-organic frameworks into chitosan matrix for enhanced tetracycline removal from water. Chem. Eng. J. **382**, 122893 (2020). https://doi.org/10.1016/j.cej.2019.122893

Recent Trends in the Adsorption and Desorption of Heavy Metals

P. Ananthi, K. Hemkumar, S. Subasini, and Anitha Pius

Abstract The biggest issue in the globe is the environmental pollution, mainly from heavy metals and minerals from industrial effluents. Their concentration has reached risky levels due to the widespread anthropogenic activities such as industrial operations, mining, agricultural processes, and industrial waste materials. Several adsorbents have been utilised to remove heavy metals in recent years due to the rising pollution levels. Industrial effluent contains heavy metals such as Ni, Cr, Pb, Zn, Ar, Cr, Se, and U and other organic pollutants. Adsorbents have been used in the variety of successful heavy metal removal procedures because it is highly straightforward, affordable, efficient, and adaptable. It has emerged as the method of choice for removing hazardous pollutants from wastewater. Here, we discuss about how to remove heavy metals from wastewater using freshly prepared nanoparticles. The good investigation would be useful for figuring out how to conduct adsorption–desorption cycles for the removal of heavy metal from the water sources.

Keywords Adsorption · Desorption · Heavy metal · Water · Industrial effluents · Cr

1 Introduction

Environmental pollutants and their toxicity are major issues all over the world. New pollutants develop over time, causing serious health and scientific problems. Water contamination is one of the most important environmental issues that affect all living things. In the last few decades, numerous scientists and researchers throughout the world have been focusing on removing various contaminants from water and wastewater. Due to their mobility in the aqueous ecosystem, toxicity to higher life forms and their non-biodegradable nature, heavy metals are considered to be the most significant inorganic contaminant in the aquatic environment. Heavy metals are still the most

P. Ananthi · K. Hemkumar · S. Subasini · A. Pius (✉)
Department of Chemistry, The Gandhigram Rural Institute—Deemed to be University, Gandhigram, Dindigul 624 302, Tamil Nadu, India
e-mail: dranithapius@gmail.com

J. Tharini and S. Thomas (eds.), *Carbon Nanomaterials and their Composites as Adsorbents*, Carbon Nanostructures, https://doi.org/10.1007/978-3-031-48719-4_11

common inorganic contaminants, and their presence in sludge, fertilizer, pesticides, municipal trash, mine residues, and smelting industries could cause huge damage to the environment [1]. The exposure of heavy metals As, Pb, Cd, Cr, Zn, Hg, Cu, Cd and cause cancer, bone damage and tissue damage. Hence it is must to remove the heavy from the aquatic ecosystem. Many heavy metal removal methods have been used namely mica precipitation, lime coagulation, ion exchange, reverse osmosis and solvent extraction. When the metal concentration in the effluent is minimal, these so-called traditional procedures are ineffective as they are also non-selective [2]. Adsorption is a well-known method used for the effective removal of heavy metal ions [3]. Chemical and biological sorbents could be used in the adsorption to remove and recover heavy metals. All living things on the Earth, including trees, plants, algae, fish, and plankton in the ocean and freshwater are considered as biosorbents. The variety of adsorbents, such as red mud, zeolite, clay, fly ash and activated carbon could be used in the chemisorption approach. The adsorbents should satisfy the following requirements,

(i) It should be inexpensive and reusable;
(ii) accurate and reliably capture and release revelations;
(iii) Efficient and easy desorption of metal ions from the sorbents [4].

The high affinity of the adsorption process to the surface of adsorbents made desorption difficult even though the adsorption of contaminants from aqueous solutions is easy. Desorption is generally used to rejuvenate adsorbents and to remove reversibly adsorbed species molecules from them [5]. One of the important qualities of an efficient adsorbent for practical applications is rejuvenation. Additionally, the expense of creating the adsorbent increases the significance of rejuvenation. Adsorbent rejuvenation generally results in the recovery of adsorbate molecules, the reusing of adsorbents in the adsorption process and reducing secondary waste [6]. The adsorbents will be used in a continuous sorption–desorption cycle for total metal removal and recovery. Recycling involves repeating the adsorption and desorption processes, which has substantial advantages for both the economy and the environment [7]. Reviewing the adsorption–desorption properties of several heavy metals from their respective adsorbents is the primary goal of the current chapter.

2 Heavy Metal Toxicity

An element with an atomic mass between 63.5 and 200.6 with a specific gravity of at least 5.0 is referred to be heavy metal. The majority of heavy metals are hazardous to the environment and human health. In industrial wastewater, heavy metals such as lead, chromium, mercury, uranium, selenium, zinc, arsenic, cadmium, silver, gold, and nickel are present. The main dangers associated with exposure to heavy metals for human health includes lead, cadmium, mercury, and arsenic. International organisations like the World Health Organization constantly assess the impact of these metals on human health after significant research on them (WHO).The central nervous system, heart, lungs, kidneys, liver, endocrine glands, and bones suffer severe damage from the transient heavy metal intoxications. Chronic exposure to heavy metals causes several disorders that impact the same systems and may increase the risk of acquiring some cancers [8] (Table 1).

Table 1 Heavy metal ions' sources and toxic effects

S. No	Heavy metal	Major source	Toxic effect	References
1	Chromium	Steel fabrication, dyeing, paints and textile, pigments	Lung tumours, epigastric discomfort, nausea, vomiting, severe diarrhoea, teratogenicity, and vomiting	Gadd [9]
2	Zinc	Refineries, mining, plumping, brass manufacturing	Gastrointestinal distress causes short term "metal-fume fever"	Manohar et al. [10]
3	Copper	Printing operations, plating, painting, and copper polishing	Diarrhoea, acute toxicity, neurotoxicity and dizziness	Biswas et al. [11]
4	Nickel	Non-ferrous metal, enamelling porcelain, making paint, and electroplating	Lung cancer, chronic bronchitis, and a decrease in lung function	Robertson [12]
5	Cadmium	Welding, refining, fertilizer, plastic, pesticide, mining	Hypertension, weight loss, bone marrow, gastrointestinal problem, kidney damage, bronchitis, lung disorder, cancer, Itai–Itai disease	Singh et al. [13]
6	Lead	Mining, burning of coal, painting, use of pigments, and electroplating	Mental retardation in children, liver, kidney, brain, and gastrointestinal damage, anaemia and appetite loss	Chen and Hao [14]; Papandreou et al. [15]
7	Mercury	Batteries, mining, the paper industry, and the paint industry	skin corrosiveness, protoplasm poisoning, damage to the neurological system, eyes, and muscles, dermatitis, kidney damage	Sumesh et al. [16]
8	Arsenic	Pesticides, smelting, rock sedimentation, mining	Hepatomegaly, Bronchitis, dermatitis, hemolysis, bone marrow depression	Granados-Correa and Serrano-Gómez [17]

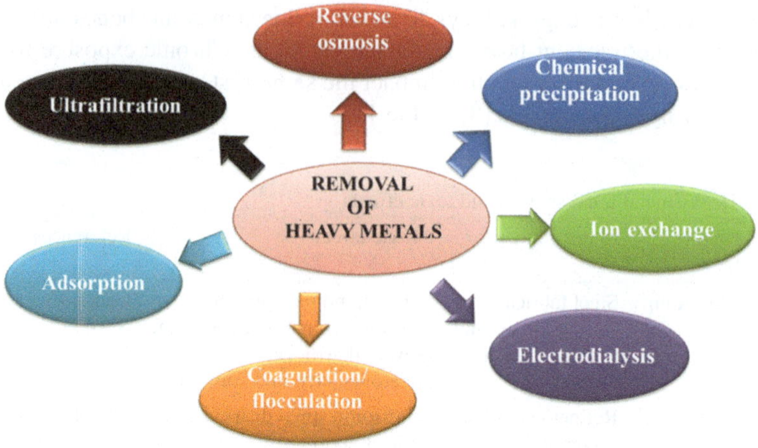

Fig. 1 Several typical methods of heavy metal removal

3 Methods of Heavy Metal Removal

3.1 *Chemical Precipitation*

This method procedure begins with the addition of chemical reagents and ends with the separation of the precipitated solids from the cleaned water. By adding coagulants such as salts, alum, iron, lime, and other metals, organic polymers can be precipitated. Using a combined air flotation and hydroxide precipitation approach, Gopalratnam et al. (1988) reported that Cu, Zn and Pand up to 96.2% of the oil were removed from industrial wastewaters [18] (Fig. 1).

3.2 *Ion Exchange*

Over the last few decades, the physical/chemical method of ion exchange, which removes impurities from water, has become more widely used method. Cations are replaced with metal ions in wastewater during the ion exchange process. In the ion exchange procedure, synthetic resin and natural zeolites are employed. Zeolites are frequently employed because of their low cost and ability to remove the metal selectively. Strong base anion exchange resins can be used to extract As(V) with ease. The pH of the solution, ion concentration, alkalinity, and type of resin all have a role in the ion exchange process' ability to remove As (V). Its restrictions include the assumption that the pH of the solution affects this method and the fact that it cannot be applied on a big scale [19].

3.3 Electrodialysis

Ions are moved through a semi-permeable membrane by an electric voltage during the membrane electrodialysis (ED) process. Since the membranes are cation- or anion-selective, only positive ions or only negative ions will effectively pass through. Positively charged ions can pass through negatively charged polyelectrolytes known as cation-selective membranes whereas negatively charged ions are rejected.

3.4 Flocculation/Coagulation

Both flocculation and coagulation play crucial roles in the treatment of wastewater and drinking water. When a coagulant or chemical is added to water, a coagulation chemical reaction takes place. The coagulant encourages water's gelatinous particles to group together forming microscopic clusters called "flocs". Then, these flocs attract any suspended materials. During flocculation, water is mixed slowly and softly to allow the flocs to grow and reach a size where they may easily settle out [20].

3.5 Ultrafiltration

Ultrafiltration is a separation technique that makes use of membranes having pores that range in size from 0.1 to 0.001 microns. Ultrafiltration is often used to eliminate high-molecular-weight compounds, inorganic polymeric molecules, organic and colloidal materials and other types of molecules. Particles, colloids, and macromolecules are preserved but low molecular weight substances and water pass through the membrane during this pressure-driven purification process. The primary mechanism of removal is size exclusion, but membrane or particle surface chemistry and electrical charge can also be important.

3.6 Reverse Osmosis

Reverse osmosis uses membranes that resemble cellophane to separate dirty water from the pure water. In the RO process, the concentrated side of the membrane is compressed to drive clean water into the diluted side while rejected water is used to wash away impurities from the concentrated side. Some applications for RO processes includes the treatment of radioactive wastewater, municipal wastewater, contaminated groundwater, organic wastewater, wastewater from electroplating and metal finishing, pulp and paper, mining and petrochemical, textile, food processing industries, radioactive wastewater, and organic wastewater [20].

3.7 Adsorption

During the adsorption process, a gas or liquid solution forms an atomic or molecular film on the surface of a solid adsorbent (adsorbate). Adsorption is frequently used in commercial operations such as water filtration and waste water treatment. Due to its simplicity and affordability, adsorption is currently one of the most effective treatment method for the wastewater [21]. This is the common technique for eliminating metal ions from the range of industrial effluents. The most commonly used adsorbent are activated carbon, which appear in the form of small pellets or a powder. This is highly porous, amorphous solid made up of microscopic crystallites with a graphite lattice.

4 Biology in the Heavy Metal Adsorption Process

Bioaccumulation and biosorption are the two primary methods for eliminating metals from the environment. Biosorption is the term used to describe any biological interaction that results in the adsorption of xenobiotic substances from the environment. Biological substances in biosorption can either be genuine living beings or anything derived from them. Bioaccumulation is the metabolically active process that uses living organisms to remove the harmful substances. Biosorption is a metabolically inactive process requires the interaction of sorbent and sorbate. Bioaccumulation involves intracellular and extracellular mechanisms and is primarily influenced by physical, chemical, and biological variables [22].

4.1 Biosorption

The biosorbent consists of the solid phase and the sorbate which help in eliminating the heavy metals using biological components. Essentially, the sorbate is still disseminated in the liquid phase. For the long period of time, plant leftovers and agricultural waste are used to make biosorbents. However, it has been demonstrated that microorganisms act as the superior biosorbent compared to plant and agricultural waste for heavy metals biosorption. Many microorganisms, including bacteria, algae, fungi, yeast, and cyanobacteria have been used in these experiments. However, due to their diversity, ease of cultivation, high yield, the existence of the greatest number of cell wall components, and functional groups, fungi are the most frequently used adsorbents [23].

4.2 *Bioaccumulation*

The process of bioaccumulation sometimes referred to as active biosorption, combines the wide range of techniques, metal ion binding to the intracellular components as well as methylation. Regarding kinetics and activation energy, bioaccumulation and biosorption are the distinct processes. Physical parameters and biological traits both affect the bioaccumulation. The factors that influence bioaccumulation includes nutrition, temperature, pH and the presence or absence of light. Biomass is the non-recyclable process because metal ions are deposited inside the microbial cell during bioaccumulation, which increases the possibility of creating new waste. Biosorption and bioaccumulation have activation energies of around 21 and 63 kilo Joule/mol, respectively. Since dead biomass absorbs heavy metals, biosorption is typically thought of as a passive process, whereas fundamentally, bioaccumulation involves live microorganisms actively securing metal ions [24].

5 Types of Nanoparticles for Removing Heavy Metals

Based on the types of nanomaterial, nanoadsorbent could be classified as metal—oxide nanoparticles, carbon based nanoparticles, silica based nanoparticles, zerovalent particles and nanocomposite (Fig. 2).

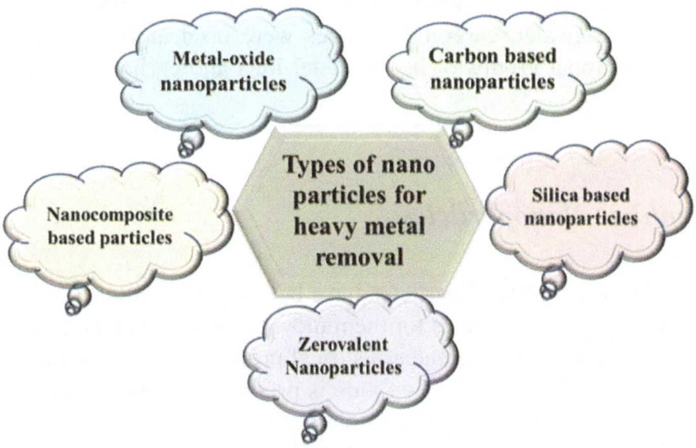

Fig. 2 Types of nanoparticles for heavy metal removal

5.1 Carbon-Based Nanoparticles

Carbon-based nanomaterials such as fullerenes, carbon nanotubes, activated carbon, and graphene and their derivatives have been widely used in the recent years for the exclusion of heavy metals [25]. Carbon adsorbents have a unique ability for the effective removal of heavy metals due to their abundance of active, flexible surface functional groups, which are important for the surface chemistry of carbon particles and the accumulation of heavy metals.

5.2 Carbon Nanotubes

Carbon nanotubes, the most common form of carbon, can be considered structural forms of carbon compounds. They are mostly cylindrical in shape and range in size from 100 to 1000 nm in length and 1–3 nm in width. Multiwalled carbon nanotubes are made up of several graphene sheets as opposed to single-walled carbon nanotubes, which are made up of a single graphene sheet that has the shape of a pipe [26]. They can be divided into two groups: single-walled and multi-walled carbon nanotubes. Due to the high active surface area to volume ratio and precise measurements of the aperture size, carbon nanotubes are effective in removing heavy metals. In contrast to granular activated carbon and conventional powder have inherent drawbacks such as a smaller active surface area and a higher adsorption activation energy and good adsorption capacity. For the highly complicated process of removing heavy metal ions from the wastewater, carbon nanotubes were used as the adsorbent. In this process, the functional groups of heavy metal ions and carbon nanotubes interact electrostatically [29].

5.3 Graphene Nanoparticles

Graphene and graphene-based particles have been more popular in recent years as an alternative to carbon nanotubes for the removal of heavy metals from wastewater due to their different features, greater electrical properties, mechanical strength and thermal conductivity [27]. Graphene oxide is produced when graphene reacts with different oxygen functional groups, like epoxy and carbonyl groups. Graphene oxide and activated carbon could be used as the base materials which effectively bind with the metal ions. The heavy metal removal by graphene depend on the electrostatic interactions and the precipitation of surface metal hydroxide [25].

5.4 Silica-Based Nanoparticles

Silica-based nanoparticles are the excellent candidates for the application because of their unique properties, such as variable surface characteristics, well-defined pore size and large surface area and selective adsorption. Heavy metal removal was greatly aided by the silica-based nanoparticles because of their better surface qualities and lack of toxicity [28, 29]. The 3-D web configuration of silica (SiO_2) permeable has a greater surface area which enables the metal ions to enter the interior area with a high response percentage. It also offers superior water stability and mechanical stability. These nanoparticles could be modified with amide or sulfhydryl groups to provide nanocomposite strength. The NH_2-SNHS silica nano hollow sphere act as the effective adsorbent for the removal of Cd^{2+}, $Ni^{2+}$and Pb^{2+} metal ions. The heavy metal such as Cd^{2+}, $Ni^{2+}$and Pb^{2+} could be removed at the neutral pH in the nanomolar concentration. It has been proposed that silica-based nanoparticles' capacity to chelate can be enhanced by the addition of an amino group. The heavy metal namely Cu^{2+}, Cd^{2+}, P^{b2+}andCd^{2+} metal ions could be removed from the water samples by the nano polyaniline and nanocomposite of nano silica. This is due to the polyaniline and silica nanocomposites' hydroxyl functional groups form complex with heavy metal ions (Sil-Phy-NPANI) [30, 31].

5.5 Metal Oxide-Based Nanoparticles

Based on metal oxide nanoparticles, which are thought to be promising moieties, have amazing properties like increased elimination capacity and a preference towards heavy metals. Among this nanoscale, oxides are the oxides of manganese, zirconium, titanium, cerium, zinc, iron, and magnesium.

5.6 Iron Oxide-Based Nanoparticles

Due to their strong paramagnetism, economic viability, straightforward manufacturing process, enhanced surface-to-volume ratio, and low noxiousness, iron oxide nanoparticles offer a huge promise for the removal of many pollutants. Iron-based nanoparticles have been prepared in the variety of shapes, including nano-ovals, nanorings, and nanobelts, for heavy metal exclusion. Metal adsorption on iron oxide nanomaterials removed the variety of organic and inorganic contaminants from various water sources, which resulted in the buildup of nanoparticles that altered their magnetic characteristics [32, 33]. Different chelating agents and functional groups, such as amine, hydroxyl, carboxyl, and sulfhydryl groups, have been added to magnetite Fe_3O_4 nanoparticles to decrease metal buildup in the aqueous solution.

Goethite (–FeOOH), a critical form of iron oxide-based nanoparticles, is recognised as an efficient adsorbent for removing the heavy metals from the aqueous medium. When compared to non-nanoscale –FeOOH, FeOOH nanoparticles made from various ferrous and ferric salts showed outstanding removal effectiveness for uranium, especially in the pH range between 5.5 and 7.5.

5.7 Zinc Oxide-Based Nanoparticles

Zinc oxide (ZnO) nanoparticles are used as adsorbents due to their substantial surface area, low cost, and good exclusion capability. The average size of the casein-capped ZnO nanoparticles was found to be 10 nm and they fitted with the Langmuir model. Using the batch approach, ZnO nanoparticles removed Zn^{2+}, Cd^{2+}, and Hg^{2+} with respective adsorption capacities of 357, 387, and 714 mg g^{-1} [37]. The average size of the casein-capped ZnO nanoparticles was found to be 10 nm and they fitted with the Langmuir model.

5.8 Aluminium Oxide, Mn Oxide and TiO_2-Based Nanoparticles

Additionally, it has been found that manganese oxide (MnO) nanoparticles, hydrous manganese oxide and manganese dioxide at nanoscales could remove the heavy metals from waste water [34, 35] due to the sizeable surface area of MnO and M–O $^+$ and M–O units present on its surfaces. Hydrous manganese oxide is the good adsorbent for the removal of heavy metals because of its larger outer surface, adsorption permeable nature, and robust metal ions' electrostatic interaction. Alumina (Al_2O_3) exists in four different crystalline forms, with -Al_2O_3 being the most widely utilised due to its incredibly high stability. Their intriguing characteristic includes excellent thermal conductivity, resistance to corrosion and wear, electrical insulation and compressive strength [39].

TiO_2 nanoparticles are a potential asset for the removal of heavy metals from wastewater due to the higher chemical stability, simple manufacturing process, low cost, photocatalytic nature, and lack of abrasiveness. TiO_2 nanoparticles was prepared from the Jatropha curcas L. leaf extract which eliminate 76.48 per cent of Cr^{6+}at a pH of 5 and a dose of 1 mg mL^{-1} from sewage by photocatalytic activity [36].

5.9 Zero-Valent Metal-Based Nanoparticles

Hazardous contaminants can only be removed from wastewater with the help of nanoparticles of zero-valent metal. These days, zero valent zinc nanoparticles have a remarkable ability to remove dioxins, while Ag nanomaterials are employed to carry out the antibacterial activity in wastewater purification. Zero-valent iron is created by combining iron nanoparticles with 0 valency Fe(0) with a ferric oxide layer, with Fe serving as the reducing potential (0). The ferric oxide coating, on the other hand, offers binding sites for metal ion electrostatic communication. It has generated considerable interest as the novel adsorbent for the removal of several heavy metals, such as Hg^{2+}, Cr^{6+}, Cu^{2+}, Ni^{2+}, Cd^{2+}, and many others. This is primarily brought by its higher reduction potential in the presence of numerous active sites [37].

5.10 Ag and Au Nanoparticles

Numerous investigations have shown the relationship between Ag nanoparticles and Hg^{2+} metal. Due to Ag's reduced reduction capacity, it has little contact between Hg^{2+} and bulk Ag. Ag nanoparticles still perform better by forming compounds like AuHg, $AuHg_3$, and Au_3Hg. Au nanoparticles demonstrate a higher affinity for Hg with an adsorption capacity of 4.065 g g^{-1}. Citrate-coated Au nanoparticles were successful in removing Hg^{2+} ions, while citrate ions act as the reducing agent to convert Hg^{2+} to Hg^0[16].

6 Adsorption–Desorption Investigations Were Conducted Using the Following Experimental Setup

In adsorption–desorption investigations, heavy metals are typically removed from the adsorbents using batch and column methods. At the laboratory scale, it was found that column experiments was perfect for desorption studies. A liquid phase is pushed through a column that is filled with a specific substance, in the case of desorption, heavy metal-loaded adsorbents, to conduct the column experiment. The eluent solution is then injected through the column after achieving equilibrium between the adsorbent and neutral liquid. Columns are made of stainless steel or acrylic glass. Studies on the adsorption and desorption of heavy metals frequently employ glass columns [38].

Placing the column where percolation occurs is the most common technique. The pump helps the liquid move from top to bottom in this configuration. Using sawdust as a biosorbent, cadmium ions were adsorbed and desorbed using this technique. The fixed-vertical column was used to push the heavy metals solution. After the sorbents had been saturated with metal ions, the eluent was poured through the column, and the effluent was then gathered. This process was performed for each of the adsorption–desorption cycles. Eluent can also move upward through a column where a pump helps the liquid move from the bottom to the top of the column. However, occasionally, the eluent is propagated across distinct columns that are connected in series to one another. This kind of arrangement has been successful in completing adsorption–desorption experiments simultaneously. The adsorption and regeneration capacities of ion exchange resin were evaluated using a column experiment. A column filled with ion exchange resin was fed with aqueous solutions of heavy metal ions in a downflow mode to achieve adsorption. The eluent flow direction was reversed to conduct the desorption experiment after the adsorption technique was finished, running both experiments simultaneously in the same setup. For continuous experimental operations with high effluent flow rates, adsorption–desorption in a column mode is advantageous [39]. When doing research in batch mode, the setup is created by mixing the desorbing eluent with the heavy metal-loaded adsorbent. The adsorbents and eluents are then separated after the setup has been shaken for a predefined period in an orbital shaker. The recovered eluent's heavy metal ion concentration is then calculated. To carry out the nitric acid desorption activity for removing cadmium ions from the sesame waste biosorbent, the biosorbent and adsorbent solution were combined, and the mixture was then kept in a shaking condition for 30 min. Although column mode facilitates continuous heavy metal adsorption and desorption, the batch mode has several advantages over column mode, which are listed below:

Expensive machinery is needed to build column experimental set–up.

- It is challenging to keep the eluent flow rate through the column constant.
- It is challenging to keep the temperature uniform across the column.
- It is challenging to maintain a constant temperature throughout the column.
- A variation in particle size in the column's bed may affect the stability of the result.
- Compared to the batch experimental mode, the column experimental technique is more complex and labour-intensive.

7 Conclusion

In the last several decades, rapid industrialization has led to an increase in the usage of heavy metal ions and creating major environmental issues on the global scale. There have been various attempts to discover the suitable corrective solution because of the substance's toxic and bio-accumulative characteristics. Heavy metal ions have been removed using the variety of traditional techniques but these techniques have been ineffective for many reasons. For the elimination of heavy metal ions, even at low concentrations, the adsorption process is one of the most beneficial and efficient method. It is an alternative to the traditional methods for removing harmful heavy metals from industrial effluents, and it has several benefits, including low cost, high efficiency, minimal chemical/biological sludge, and renewal of adsorbent. The danger and influence of nanomaterials on the ecosystem, however, cannot be disregarded. When taking into account their cost, synthesis, reusability, and separation, there is still a long way to go before adopting nanomaterials in the practical treatment of heavy metals.

It's essential to choose adsorbents that are affordable, efficient, and recyclable. Adsorbents have high capacity for solubilization, inexpensive, easy to separate from aqueous solutions and can be reused. Similarly, selecting an appropriate eluent is the crucial factor in the desorption study. The eluent should be metal-selective, economically viable, and have a high desorption rate. Collaboration and technology transfer among professionals in the relevant field will lead to successful adsorption studies.

References

1. Halim, M., Conte, P., Piccolo, A.: Potential availability of heavy metals to phytoextraction from contaminated soils induced by exogenous humic substances. Chemosphere **52**, 265–275 (2003). https://doi.org/10.1016/S0045-6535(03)00185-1
2. Wang, S., McGuirk, C.M., d'Aquino, A., Mason, J.A., Mirkin, C.A.: Metal-organic framework nanoparticles. Adv. Mater. **30**, 1–14 (2018). https://doi.org/10.1002/adma.201800202
3. Sabela, M.I., Kunene, K., Kanchi, S., Xhakaza, N.M., Bathinapatla, A., Mdluli, P., Sharma, D., Bisetty, K.: Removal of copper (II) from wastewater using green vegetable waste derived activated carbon: an approach to equilibrium and kinetic study. Arab. J. Chem. **12**, 4331–4339 (2019). https://doi.org/10.1016/j.arabjc.2016.06.001
4. Muraleedharan, T., Iyengar, L., Venkobachar, C.: Biosorption: an attractive alternative for metal removal and recovery. Curr. Sci. **61**, 379–385 (1991)
5. Alinnor, I.J.: Adsorption of heavy metal ions from aqueous solution by fly ash. Fuel **86**, 853–857 (2007). https://doi.org/10.1016/j.fuel.2006.08.019
6. Duran, A., Soylak, M., Tuncel, S.A.: Poly(vinyl pyridine-poly ethylene glycol methacrylate-ethylene glycol dimethacrylate) beads for heavy metal removal. J. Hazard. Mater. **155**, 114–120 (2008). https://doi.org/10.1016/j.jhazmat.2007.11.037
7. Kara, A., Uzun, L., Beşirli, N., Denizli, A.: Poly(ethylene glycol dimethacrylate-n-vinyl imidazole) beads for heavy metal removal. J. Hazard. Mater. **106**, 93–99 (2004). https://doi.org/10.1016/j.jhazmat.2003.08.016
8. Babel, S., Kurniawan, T.A.: Low-cost adsorbents for heavy metals uptake from contaminated water: a review. J. Hazard. Mater. **97**, 219–243 (2003). https://doi.org/10.1016/S0304-3894(02)00263-7
9. Gadd, G.M.: Metals, minerals and microbes: geomicrobiology and bioremediation. Microbiology **156**, 609–643 (2010). https://doi.org/10.1099/mic.0.037143-0
10. Manohar, D.M., Anoop Krishnan, K., Anirudhan, T.S.: Removal of mercury(II) from aqueous solutions and chlor-alkali industry wastewater using 2-mercaptobenzimidazole-clay. Water Res. **36**, 1609–1619 (2002). https://doi.org/10.1016/S0043-1354(01)00362-1
11. Biswas, C., Mukherjee, P., Drew, M.G.B., Gómez-García, C.J., Clemente-Juan, J.M., Ghosh, A.: Anion-directed synthesis of metal-organic frameworks based on 2-picolinate Cu(II) complexes: a ferromagnetic alternating chain and two unprecedented ferromagnetic fish backbone chains. Inorg. Chem. **46**, 10771–10780 (2007). https://doi.org/10.1021/ic701440x
12. Robertson, F.N.: Arsenic in ground-water under oxidizing conditions, South-west United States. Environ. Geochem. Health **11**, 171–185 (1989). https://doi.org/10.1007/BF01758668
13. Singh, K.K., Singh, A.K., Hasan, S.H.: Low cost bio-sorbent "wheat bran" for the removal of cadmium from wastewater: kinetic and equilibrium studies. Bioresour. Technol. **97**, 994–1001 (2006). https://doi.org/10.1016/j.biortech.2005.04.043

14. Chen, J.M., Hao, O.J.: Microbial chromium (VI) reduction. Crit. Rev. Environ. Sci. Technol. **28**, 219–251 (1998). https://doi.org/10.1080/10643389891254214
15. Papandreou, A., Stournaras, C.J., Panias, D.: Copper and cadmium adsorption on pellets made from fired coal fly ash. J. Hazard. Mater. **148**, 538–547 (2007). https://doi.org/10.1016/j.jhazmat.2007.03.020
16. Sumesh, E., Bootharaju, M.S., Anshup, Pradeep, T.: A practical silver nanoparticle-based adsorbent for the removal of Hg2+ from water. J. Hazard. Mater. **189**, 450–457 (2011). https://doi.org/10.1016/j.jhazmat.2011.02.061
17. Granados-Correa, F., Serrano-Gómez, J.: CrO42- ions adsorption by Fe-modified pozzolane. Sep. Sci. Technol. **44**, 924–936 (2009). https://doi.org/10.1080/01496390802691240
18. Peters, R.W., Bennett, G.F.: The simultaneous removal of oil and heavy metals from industrial wastewaters using hydroxide or sulfide precipitation coupled with air flotation. Hazard Waste Hazard. Mater. **6**, 327–345 (1989). https://doi.org/10.1089/hwm.1989.6.327
19. Kononova, O.N., Kholmogorov, A.G., Kachin, S.V., Mytykh, O.V., Kononov, Y.S., Kalyakina, O.P., Pashkov, G.L.: Ion exchange recovery of nickel from manganese nitrate solutions. Hydrometallurgy **54**, 107–115 (2000). https://doi.org/10.1016/S0304-386X(99)00052-3
20. Slater, C.S., Ahlert, R.C., Uchrin, C.G.: Applications of reverse osmosis to complex industrial wastewater treatment. Desalination **48**, 171–187 (1983). https://doi.org/10.1016/0011-9164(83)80015-0
21. Yadanaparthi, S.K.R., Graybill, D., von Wandruszka, R.: Adsorbents for the removal of arsenic, cadmium, and lead from contaminated waters. J. Hazard. Mater. **171**, 1–15 (2009). https://doi.org/10.1016/j.jhazmat.2009.05.103
22. Gadd, G.M.: Biosorption: Critical review of scientific rationale, environmental importance and significance for pollution treatment. J. Chem. Technol. Biotechnol. **84**, 13–28 (2009). https://doi.org/10.1002/jctb.1999
23. Dhankhar, R., Hooda, A.: Fungal biosorption-an alternative to meet the challenges of heavy metal pollution in aqueous solutions. Environ. Technol. **32**, 467–491 (2011). https://doi.org/10.1080/09593330.2011.572922
24. Chatterjee, A., Abraham, J.: Efficient management of e-wastes. Int. J. Environ. Sci. Technol. **14**, 211–222 (2017). https://doi.org/10.1007/s13762-016-1072-6
25. Wang, H., Yuan, X., Wu, Y., Huang, H., Zeng, G., Liu, Y., Wang, X., Lin, N., Qi, Y.: Adsorption characteristics and behaviors of graphene oxide for Zn(II) removal from aqueous solution. Appl. Surf. Sci. **279**, 432–440 (2013). https://doi.org/10.1016/j.apsusc.2013.04.133
26. Yu, G., Lu, Y., Guo, J., Patel, M., Bafana, A., Wang, X., Qiu, B., Jeffryes, C., Wei, S., Guo, Z., Wujcik, E.K.: Carbon nanotubes, graphene, and their derivatives for heavy metal removal. Adv. Compos. Hybrid Mater. **1**, 56–78 (2018). https://doi.org/10.1007/s42114-017-0004-3
27. Santhosh, C., Velmurugan, V., Jacob, G., Jeong, S.K., Grace, A.N., Bhatnagar, A.: Role of nanomaterials in water treatment applications: A review. Chem. Eng. J. **306**, 1116–1137 (2016). https://doi.org/10.1016/j.cej.2016.08.053
28. Zhao, G., Li, J., Ren, X., Chen, C., Wang, X.: Few-layered graphene oxide nanosheets as superior sorbents for heavy metal ion pollution management. Environ. Sci. Technol. **45**, 10454–10462 (2011). https://doi.org/10.1021/es203439v
29. Ahmed, M., Elektorowicz, M., Hasan, S.W.: GO, SiO_2, and SnO_2 nanomaterials as highly efficient adsorbents for Zn2+ from industrial wastewater—a second stage treatment to electrically enhanced membrane bioreactor. J. Water Process Eng. **31**, 100815 (2019). https://doi.org/10.1016/j.jwpe.2019.100815
30. Jawed, A., Saxena, V., Pandey, L.M.: Engineered nanomaterials and their surface functionalization for the removal of heavy metals: a review. J. Water Process Eng. **33**, 101009 (2020). https://doi.org/10.1016/j.jwpe.2019.101009
31. He, Y., Luo, L., Liang, S., Long, M., Xu, H.: Synthesis of mesoporous silica-calcium phosphate hybrid nanoparticles and their potential as efficient adsorbent for cadmium ions removal from aqueous solution. J. Colloid Interface Sci. **525**, 126–135 (2018). https://doi.org/10.1016/j.jcis.2018.04.037

32. Mahmoud, M.E., Fekry, N.A., El-Latif, M.M.A.: Nanocomposites of nanosilica-immobilized-nanopolyaniline and crosslinked nanopolyaniline for removal of heavy metals. Chem. Eng. J. **304**, 679–691 (2016). https://doi.org/10.1016/j.cej.2016.06.110
33. Ghiloufi, I., Khezami, L., El Mir, L.: Preparation and characterization of nanoporous resin for heavy metal removal from aqueous solution. J. Water Supply Res. Technol. Aqua. **64**, 316–325 (2015). https://doi.org/10.2166/aqua.2014.086
34. Pham, T.D., Do, T.T., Ha, V.L., Doan, T.H.Y., Nguyen, T.A.H., Mai, T.D., Kobayashi, M., Adachi, Y.: Adsorptive removal of ammonium ion from aqueous solution using surfactant-modified alumina. Environ. Chem. **14**, 327–337 (2017). https://doi.org/10.1071/EN17102
35. Saravanan, A., Kumar, P.S., Vo, D.V.N., Yaashikaa, P.R., Karishma, S., Jeevanantham, S., Gayathri, B., Bharathi, V.D.: Photocatalysis for removal of environmental pollutants and fuel production: a review. Environ. Chem. Lett. **19**, 441–463 (2021). https://doi.org/10.1007/s10311-020-01077-8
36. Goutam, S.P., Saxena, G., Singh, V., Yadav, A.K., Bharagava, R.N., Thapa, K.B.: Green synthesis of TiO_2 nanoparticles using leaf extract of *Jatropha curcas L.* for photocatalytic degradation of tannery wastewater. Chem. Eng. J. **336**, 386–396 (2018). https://doi.org/10.1016/j.cej.2017.12.029
37. Fu, F., Dionysiou, D.D., Liu, H.: The use of zero-valent iron for groundwater remediation and wastewater treatment: a review. J. Hazard. Mater. **267**, 194–205 (2014). https://doi.org/10.1016/j.jhazmat.2013.12.062
38. Banzhaf, S., Hebig, K.H.: Use of column experiments to investigate the fate of organic micropollutants—a review. Hydrol. Earth Syst. Sci. **20**, 3719–3737 (2016). https://doi.org/10.5194/hess-20-3719-2016
39. Vijayaraghavan, K., Jegan, J., Palanivelu, K., Velan, M.: Batch and column removal of copper from aqueous solution using a brown marine alga Turbinaria ornata. Chem. Eng. J. **106**, 177–184 (2005). https://doi.org/10.1016/j.cej.2004.12.039

Carbon Nanomaterials for Adsorption and Desorption of Pesticides

Astha Tripathi, Seema Lal, and Pratibha Kumari

Abstract Pesticides represent a major class of water contaminants that are released in huge amounts to water bodies through agricultural activities. The accumulation of pesticides in soil and water causes severe health problems including endocrine disruption, damage of the kidney, liver, and other body organs, reproductive problems, and cancer. It is crucial to remove pesticides from the environment to protect biodiversity and reduce their hazardous effect on human beings. Adsorption is the most simple, practical, economical, feasible, and efficient method to remove pesticides from the environment. Carbon-based nanomaterials are gaining an increasing upsurge of interest as effective adsorbent materials due to their high surface area, biocompatibility, thermal stability, and high adsorption efficiency. The application of various carbon-based nanomaterials such as graphene, carbon nanotubes, chitosan, and cellulose has been demonstrated in this chapter for the development of efficient adsorbent material/membrane for the removal of all types of pesticides present in water. The regeneration of these materials can be performed through the desorption of pesticides and several recycling runs have been reported using these materials under mild conditions.

Keywords Pesticides · Adsorption · Carbon materials · Graphene · Chitosan · Cellulose · Carbon nanotubes

1 Introduction

The drastic increase in the world population has pressurized the agriculture sector to increase production so that the food demand can be fulfilled. Agriculture production is improved by utilizing pesticides in fields. The excessive use of pesticides in agriculture has made water bodies highly contaminated. The agrochemicals industryis also releasing a huge amount of wastewater containing pesticides. It is essential to

A. Tripathi · S. Lal · P. Kumari (✉)
Bioorganic Material Research Laboratory, Deshbandhu College, University of Delhi, Kalkaji, New Delhi, India
e-mail: pkumarichemistry@gmail.com

J. Tharini and S. Thomas (eds.), *Carbon Nanomaterials and their Composites as Adsorbents*, Carbon Nanostructures, https://doi.org/10.1007/978-3-031-48719-4_12

eliminate pesticides from water to protect people and the environment from potential harm. Pesticides can cause various health problems, from minor skin and eye irritation to more serious conditions like cancer, endocrine disruption, and reproductive issues. Pesticides can also contaminate soil and water, leading to a decrease in biodiversity. Removing pesticides from the water helps to reduce the risks of these negative effects to people and the environment. Although many strategies have been developed for the remediation of pesticides including photocatalysis, redox degradation, adsorption, and electrochemical reactions, there is an urgent need for more facile technology to solve the problem of water pollution.

Adsorption is a process by which pesticides can be effectively removed from water bodies. In this process, the pesticide molecules are attracted to and bind to the surface of the adsorbent material. Nanomaterials are promising adsorbent materials due to their high surface area and unique physical and chemical properties. Therefore, nano-adsorbents are being developed with specific functionalization to induce specificity in adsorption. Adsorption is an effective process and nano-adsorbents can be used to remove pesticides from the water bodies. Zeolites and mesoporous silica are the most widely used adsorbent materials for water treatment.

Carbon nanomaterials have been widely used in adsorption applications due to their unique physicochemical properties, and high tunable surface area. Various pollutants such as heavy metals, organic compounds, pesticides, and radioactive materials have been removed using carbon-based nanomaterials. Carbon nanomaterials are also being explored for their application in targeted drug delivery. The adsorption and desorption using carbon nanomaterials is a promising area of research that could lead to more effective and sustainable methods for environmental remediation. Graphene derivatives, carbon nanotubes, cellulose, chitosan, and graphitic carbon nitride are the most commonly used carbon-based nanomaterials which have been explored in the adsorptive removal of pesticides as discussed in the following sections.

2 Graphene

Graphene, a wonder material made of a single layer of carbon atoms is a 2D-graphite sheet in its 2D-honeycomb lattice. The sp^2-bonded C-atoms are arranged in a regular hexagonal pattern as in a graphite sheet [1]. The common method of isolation of graphene sheets is simply cleaving the 3D graphite with sticky tape. The honeycomb 2D-lattice (one atom thick) of graphene with the network of pi-electrons can be considered as graphitic allotropes of carbon. Graphene can be modified to graphene oxide and reduced graphene by various chemical methods (Fig. 1) and it can be transformed into different carbon allotropes (Fig. 2).

Graphene offers pivotal properties (Fig. 3) such as high capacitance, outstanding carrier mobility, fast electron transfer rate, high surface area (2630 m^2g^{-1}), high surface-to-volume ratio, greater mechanical strength, and superior electrical

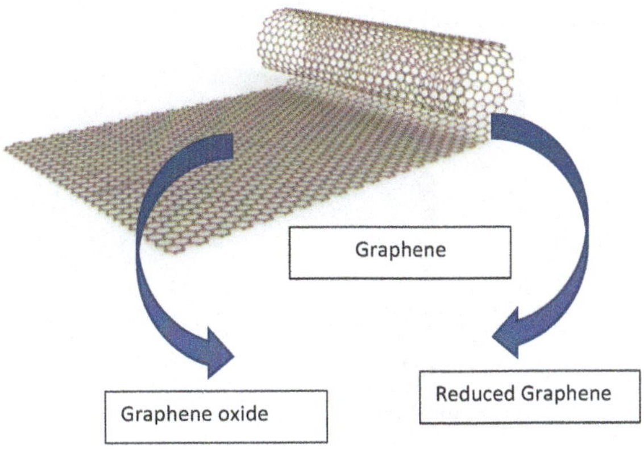

Fig. 1 Structure of graphene

conductivity [2]. Graphene provides infinite possibilities for functionalization and modification of its carbon backbone which makes it a versatile adsorbent.

2.1 *Graphene-Based Materials for Pesticide Removal*

Graphene exhibits a high adsorption tendency and is thereby used as an excellent material for the removal of various pesticides from water. In a recent study, a distinct core/shell structured composite of graphene oxide, magnetite, and diatomite was synthesized by dry coating technique and was used for the adsorption of organochlorine pesticides [3]. The surface area, pore volume, and pore diameter of synthesized composites were 23.4 m^2/g, 0.0026 cm^3/g, and 4.5 nm, respectively. The pH-independent 97% adsorption with 7.78 mg/g maximum adsorption efficiency was observed for organochlorine pesticides within 2 h. These magnetic GO composites offered seven regeneration cycles with 90% removal efficiency.

Kumari et al. [4] reported the synthesis and characterization of azo-functionalized calix[4]pyrrole decorated graphene oxide as an adsorbent for the removal of flonicamide insecticide. The batch studies as a function of the concentration of flonicamide solution, contact time, adsorbent dosage, temperature, and pH were evaluated. Maximum adsorption efficiency of 11.43 mg/g with 93.38% adsorption capacity within 40 min was attributed to the synergistic effect capable of offering more adsorption sites. The adsorption data were found to follow Langmuir adsorption isotherm with pseudo-second-order kinetics. In another study by the same group of researchers, ionic-liquid modified calix[4]arene magnetite nanoparticles have been reported as nano adsorbent for nitrogenous pesticides, metribuzin (MET), and dicloran (DIC) from aqueous media. The maximum adsorption with the physisorption phenomenon was found to be 5.88 and 1.94 mg/g for MET and DIC respectively. The adsorption data were reported to follow Freundlich isotherm and pseudo-second-order kinetics

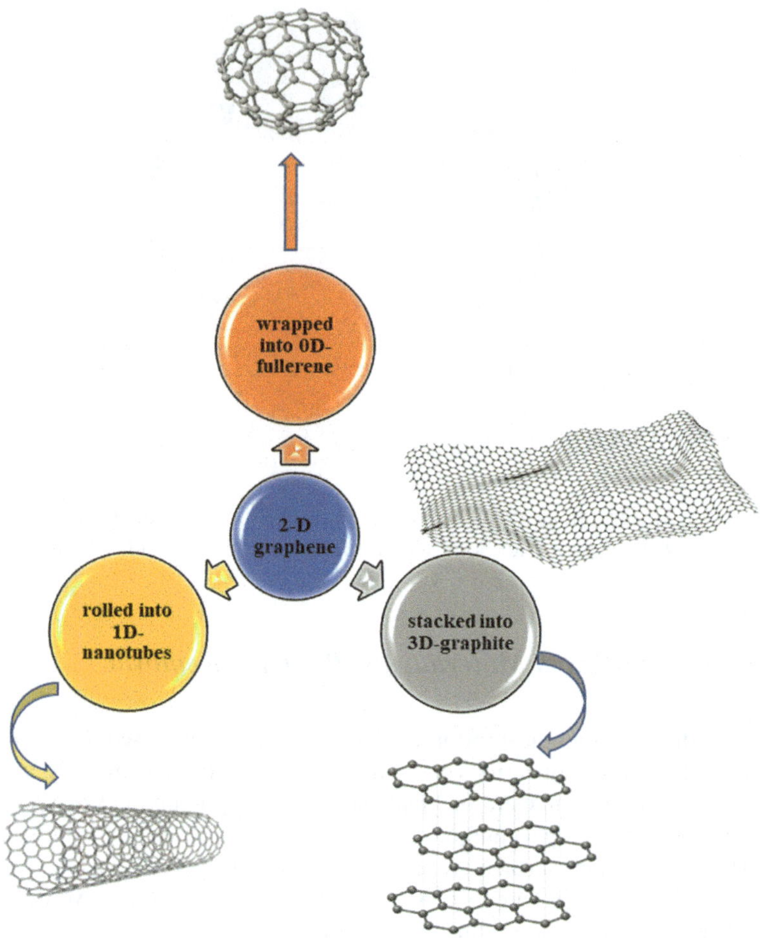

Fig. 2 Graphene transformation to various carbon allotropes

Fig. 3 Unique properties and applications of 2D-graphene

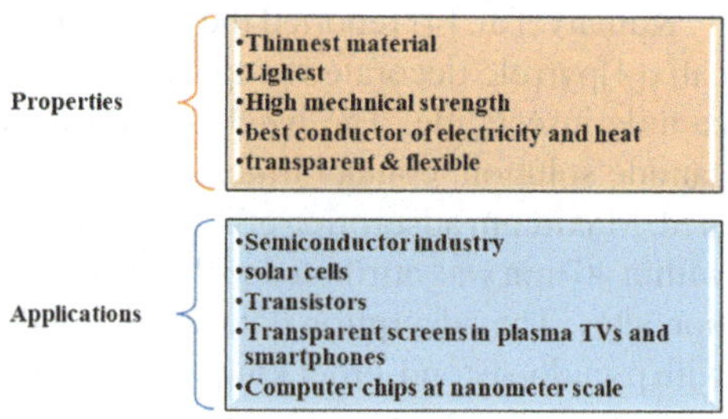

with three regeneration cycles [5]. The modified Fischer esterification protocol was also reported for the synthesis of silica monolith anchored graphene oxide composite as an efficient adsorbent for carbofuran and imidacloprid pesticides from wastewater [6]. The synthesized adsorbent/composite material was found to be highly efficient with a maximum adsorption capacity of 342.46 mg g^{-1} for imidacloprid and 37.15 mg g^{-1} for carbofuran. The adsorption data followed the pseudo-second-order kinetic model and the Freundlich adsorption isotherm. The application of various graphene oxide-modified nanomaterials for the adsorption of pesticides is summarized in Table 1.

3 Carbon Nanotubes (CNT)

Carbon nanotubes (CNTs) are rolled-up tubes of graphite hexagonal sheets (2D). The CNTs can be capped at ends by carbon pentagons. Such enclosed CNTs may be described as cylindrical fullerenes. However, the open carbon CNTs are also known.

There are two types of CNTs: Single-walled carbon nanotubes (SWCNTs) and multi-walled carbon nanotubes (MWCNTs) (Fig. 4). A SWCNT consist of one cylindrical tube (diameter of 0.4-5 nm) and an MWCNT consist consists of several concentric tubes (outer diameter in the range of 2–25 nm, and inner diameter of 1–3 nm). The interlayer distance in MWCNT is about 0.35 nm. The energy stored in the dangling bonds of a 2-D graphene sheet is reduced by forming the CNTs through the rolling up of the graphene sheet. The CNTs may show thermal and electrical conductivities like graphite. However, the electrical conductivity of a CNT depends on both the chirality and diameter of the tube. Thus, depending on the condition, the CNTs may be metallic, semi metallic (Eg = 0), semiconducting (Finite Eg value), or insulating (Eg very large).

The unique, well-defined, and adjustable nano structural properties of CNTs offer a wide potential for water remediation by adsorption of various toxic contaminants present in the water. These hollow nanotubes provide easy transfer of water molecules and make them appropriate for the separation of various pollutants including pesticides [2, 17, 18].

3.1 Carbon Nanotubes for Pesticide Removal

There are many reports on the application of carbon nanotubes in the adsorptive removal of many water contaminants. In a study [19], researchers reported the synthesis, characterization, and application of magnetic multiwalled carbon nanotubes with zeolitic imidazolate frameworks-67 for the adsorption of pesticide butachlor from water bodies. In this study, response surface morphology was used for the optimization of experimental parameters. The maximum adsorption of 86% was observed for the butachlor within 95 min at pH 4.5 with an initial concentration

Table 1 Graphene-based materials as adsorbent for pesticides removal

Graphene-based materials	Pesticide removed	Adsorption capacity (mg/g)	Adsorption isotherm followed and the kinetic model	Regeneration cycles	Remarks	Refs.
Magnetic graphene oxide and carboxymethyl cellulose	Chlorpyrifos	108.3	Langmuir and Pseudo second order	4	Response surface methodology (RSM) and central composite design (CCD) were used	Dolatabadi et al. [7]
MIL 101(Cr) MOF decorated with graphene oxide nano-layers	2, 4-dichloro phenoxy acetic acid	476.9	Langmuir and Pseudo second order	4	The one-pot hydrothermal method was used for the synthesis	Khaloo et al. [8]
Graphene and hexagonal boron nitride	Diazinon, parathion, oxacillin and ciprofloxacin	2680	Langmuir and Pseudo second order	5	*L*-ascorbic acid as a reducing agent	Chi et al. [9]
Graphene oxide (GO), reduced graphene oxide (rGO), and graphene nanoplatelets (GNP)	Atrazine	rGO (1084) GO (1011.9) GNP (1006)	The Langmuir, Freundlich, Temkin, Dubinin– Radushkevich and Sips models Elovich kinetic model	–	The adsorption process was driven synergistically by H-bonding and π-π conjugation interactions	Azizzadeh et al. [10]
Graphene oxide (GO) and reduced graphene oxide (rGO)	Methomyl (Met), Acetamiprid (Ace), Azoxystrobin (Azo)	For Met, Ace and Azo with GO are 106.22, 285.96,and 2896.84 mg/g and with rGO are 96.86, 357.65, and 2818.04 mg/g respectively	Sips model, pseudo-second-order, and Elovich models		Face-centered composite design (FCCD) through response surface methodology (RSM), Synergestic adsorption mechanism	Shi et al. [11]

(continued)

Table 1 (continued)

Graphene-based materials	Pesticide removed	Adsorption capacity (mg/g)	Adsorption isotherm followed and the kinetic model	Regeneration cycles	Remarks	Refs.
Graphene oxide grafted polymer	Sulfonylurea herbicides		Pseudo-second-order kinetics equation and Freundlich model		n-π, π-π, hydrogen bonding, hydrophobic and electrostatic interaction	Li et al. [12]
Nanocomposite of Fe_3O_4 nanoparticles, graphene oxide, and *N*-methyl-D-glucamine functionalized calix[4]arene	Hexaconazole and chlorpyrifos	78.74 and 93.46 mg g^{-1} were obtained for chlorpyrifos and hexaconazole, respectively	Langmuir and Pseudo second order		Thermodynamic and free energy data indicated the physisorption mechanism for the adsorption process	Nodeh et al. [13]
Activated carbon derived from sieve-like cellulose/graphene oxide composites	Organophosphorus pesticides	152.5 mg g^{-1} for chlorpyrifos	Langmuir	8	Adsorption mechanism depends on the electron-donating abilities of the S and P atoms of pesticides	Suo et al. [14]
Graphene-based tetraethoxysilane-methyltrimethoxysilane sol–gel hybrid magnetic nanocomposite	Organophosphorus pesticides (phosphamidon, dimethoate, chlorpyrifos, diazinon)	54.4–76.3 mg g^{-1}				Rashidi Nodeh et al. [15]
Graphene-based silica-coated magnetic nanoparticles	Lindane, chlorpyrifos, hexaconazole, and azaconazole	13.04–18.69 mg g^{-1}				Nodeh et al. [16]

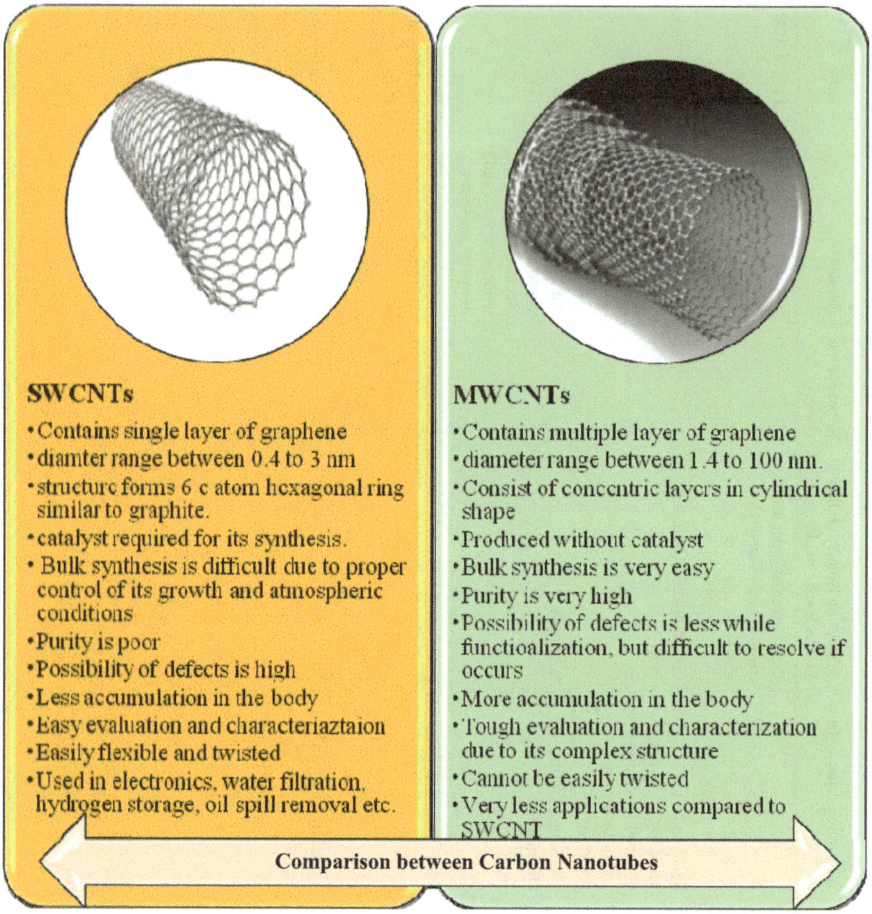

Fig. 4 Comparison between single-walled carbon nanotubes (SWCNTs) and multiple walled carbon nanotubes (MWCNTs)

of 5.75 ppm. The adsorption data were well-fitted by the Temkin adsorption model and followed pseudo-second-order adsorption kinetics. Lee and Chang [20] reported a quantum chemical approach for the adsorption of organophosphorus pesticides chlorpyrifos and fenitrothion on four materials based on boron nitride nanotubes (namely, pristine BNNT, C-doped BNNT, tetrapeptide/BNNT, and tetrapeptide/C-doped BNNT). It has been found that the decoration of tetrapeptide on BNNT has enhanced the sensing of both the pesticides with inferior adsorption than carbon-doped BNNT. The C-doping method attributed to the reduction of bandgap and formation of hydrogen bonds with enhanced non-covalent interactions. These materials offer applications not only as sensors but as good adsorbents for the purification of water by selectivity toward organophosphorus pesticides.

Lung et al. [21] reported the devrinol and triadimefon pesticides adsorption using iron and manganese oxides modified and carboxylic functionalized carbon nanotubes. The batch studies were performed considering the effect of pH, initial concentration of pesticides, amount of CNT doses, temperature, and contact time. The temperature-dependent maximum adsorption capacity was found in the range of 14.925 and 20.492 mg/g for devrinol and between 12.723 and 14.706 mg/g for triadimefon. The experimental data were best fitted to the Langmuir isotherm model and follow the pseudo-second-order kinetic model. The reusability of this nanocomposite was determined via five adsorption–desorption cycles and the degree of removal of pesticides decreased up to 79.4% for devrinol and 60.7% for triadimefon pesticide.

In a recent study, researchers reported the loading of two metal–organic frameworks (ZIF-8 or UiO66-NH2) on CNT aerogels by in situ nucleation process as MOFs@MPCA aerogels. The synthesized products were evaluated for the adsorption of chipton and alachlor herbicides. The maximum adsorption capacity reported for Chipton using UiO66-NH2@MPCA was 227.3 mg/g with 5 recycle processes without much decrease in the adsorption process [22]. In another study, malathion pesticide adsorption in the presence of dissolved organic matter (DOM) in water using 2-phenylethylamine-modified MWCNT is reported [23]. The modified MWCNT showed enhanced adsorption efficiency (98.34%) in 50 min. than unmodified MWCNTs (92.14%) in 30 min. The presence of DOM in aqueous media significantly reduced the adsorption capacity by up to 52.74% without affecting other physical properties. The adsorption data were found to be chemisorption supported by Langmuir adsorption isotherm and followed a pseudo-second-order kinetic model. The reusability efficacy of modified MWCNT decreased to 3.22% in two cycles in the presence of DOM.

In another study [24], MWCNTs were efficiently used for 100% adsorption of diazinon, one of the most commonly used organophosphorus pesticides from water bodies. The specific surface area (SBET) and the inside diameter of the multi-walled carbon nanotubes were 370 m^2 /g and 3–5 nm, respectively. The adsorption parameters included an initial diazinon pesticide concentration of (0.3 mg/L) and MWCNT as nanoadsorbent with a concentration of 0.1 g/L at pH 4 and 7 within a contact time of 15 min. The highest efficiency of diazinon pesticide removal was observed by multi-walled nanotubes at pH = 4 (100%) and the lowest removal efficiency (1%) at pH = 7. The high removal efficiency of diazinon in acidic pH is attributed to the protonation of the hydroxyl group on the adsorbent and the protonation of nitrogen atoms on the diazinon pyrimidine groups.

In a similar study [25], MWCNT modified by chitosan using the crosslinking method was explored for the adsorption of diazinon from water media. It has been found that the synthesized adsorbate has 222.86 mg/g adsorption capacity within 60 min at pH 5.5. The adsorption data were well fitted in the Sips isotherm model and followed pseudo-second-order adsorption kinetics. The synthesized adsorbent can be successfully regenerated for up to four cycles with a 7.82 mg/g adsorption capacity.

Malathion, the first contact broad-spectrum organophosphorus pesticide with a rapid and relatively constant effect is the most used pesticide for the control of sucking and chewing insects. Repeated or prolonged contact with this pesticide can lead to acute effects. Reports are available for the adsorptive removal of malathion (57%) using MWCNT [26]. The optimization of experimental parameters was done based on three-factor surface modeling (RSM) approach and achieved as high as 100% adsorption of malathion pesticide with an initial concentration of malathion and MWCNT of 6 mg/L and 0.5 g/L respectively in a contact time of 30 min.

Magnetic CNT consisting of metal–organic framework ZIF-8 and magnetic multi-walled carbon nanotubes with high surface area and large pore volume has also been reported for the removal of organophosphorus pesticides from water [27]. In a study by Barbosa et al., iron and aluminum oxide incorporated and nitrogen-doped magnetic CNT prepared by the chemical vapor deposition method was utilized for pesticide removal [28]. In another study [29], CNT was compared with pulverized activated carbon and granular activated carbon for the adsorption of three pesticides including atrazine, simazine, and diuron from water treatment plants in their commercial and analytical forms. It has been observed that pristine CNT was highly efficient for the adsorption of atrazine and simazine as functionalization doesn't favor enhanced adsorption.

4 Chitosan

Chitosan, poly [β-(1 → 4)-2-amino-2-deoxy-D-glucopyranose], is the second most abundant and widely used natural polymer obtained from the deacetylation of chitin [30]. Chitosan exhibits a unique set of properties, making it an excellent candidate for the development of water treatment systems. Among them, their high biodegradability, low toxicity, non-allergenic, and natural availability are the most relevant. However, this biopolymer has some limitations such as low acid stability, poor mechanical properties, low thermal stability, resistance to mass transfer, low porosity, and surface areas [31–34]. Chitosan can exhibit molecular weights between 5 and 1000 kg/mol. Depending on the level of acetylation, chitosan is a semicrystalline polymer that exists in many allomorphs in the solid state [35].

A significant portion of the units in chitosan has hydroxy and amino functional groups [36, 37]. These groups are accountable for chitosan's reactivity as an exceptional natural and potent adsorbent. The greater reactivity of these groups enables chitosan applications in a variety of domains, including wastewater treatment, ion exchange, and functional matrixes [38, 39]. The amino group on chitosan is particularly an advantageous functional group that can be modified chemically to obtain chitosan derivatives with desired adsorption characteristics. Several researchers worked largely on the modification of chelating functionalities, mainly by increasing amino groups [40–42]. The applications of some chitosan-based materials in the adsorptive removal of pesticides are presented in Table 2.

Table 2 Carbon-based materials for pesticide removal through adsorption

Adsorbent	Pesticide removed	Adsorbent capacity (q_e) mg/g	Remarks and isotherms	Refs.
Chitosan-sodium alginate	Malathion	52.08	–	Sabbagh et al. [43]
Chitosan-glycidyl methacrylate	2,4-Dichlorophenoxyacetic acid (2,4-D)	–	–	Silvestro et al. [44]
Chitosan-nanoclay hydrogel beads	Paraquat	–	Non-spontaneous, physisorption takes place	Baigorria and Fraceto [45]
Activated carbon	Bentazon	169.4	At 20–40 °C, the adsorption was exothermic, spontaneous, and practical. The adsorption data are fit by the pseudo-second-order model, providing evidence for the chemisorption mechanism	Omri et al. [46]
Activated carbon	Carbofuran	164.0	Adsorption was efficient and exothermic	Salman and Hameed [47]
Activated carbon	2,4-D	33.5 to 171.32	Exothermic, viable adsorption was possible. It was a physisorption process and a diffusion kind of adsorption	Salman and Njoku [48]
Activated carbon	MCPA (4-chloro-2-methylphenoxyacetic acid)	11.51 to 63.38	Exothermic and feasible adsorption	Gimeno et al. [49]
Cellulose/graphene oxide	Chlorpyrifos	152.50	The ability of the S and P atoms to donate electrons was crucial to the adsorption mechanism	Suo et al. [14]
Cellulose/graphene	Ametryn	6.45 to 9.58	Endothermic, spontaneous adsorption takes place	Zhang et al. [50]
Chitosan	Glyphosate	35.08	The $–NH_3$ group of chitosan and pesticide anion were found to interact electrostatically	Rissouli et al. [51]

(continued)

Table 2 (continued)

Adsorbent	Pesticide removed	Adsorbent capacity (q_e) mg/g	Remarks and isotherms	Refs.
Chitosan	Oxadiazon	5.02	Oxadiazon exhibited a potent chemisorption to the adsorbent chitosan	Arvand et al. [52]
Coconut shell-based activated carbon (NP-5)	Chloroorganic compound HCH	650	Exothermic adsorption took place. Since HCH was adsorbed in a monolayer, there was little to no competition for adsorption surface sites with water molecules or pesticide molecules	Ignatowicz [53]
Molecularly imprinted polymer	1-naphthyl methyl carbamate	50.0	Strong hydrogen bonds between the pesticide's amide group and carbocyclic groups occur at acidic pH levels	So et al. [54]
Crosslinked modified chitosan	Pantachlorophenol	36.63	Exothermic, feasible, and spontaneous adsorption occurred Values of ΔH° suggested that physical adsorption would be possible	Shankar et al. [55]
Magnetic-nanomodified activated carbon	Thiamethoxam	42.44	–	de Freitas et al. [56]
Chitosan-activated carbon	Oxadizon	–	–	Arvand et al. [52]
Walnut shell-modified activated carbon	Diazinon	4.95 to 156.25	The amount $1/n < 1$ demonstrated that diazinon adsorption was physisorption based on the Freundlich isotherm	Bayat et al. [57]
$MnFe_2O_4$@cellulose-activated carbon	Glyphosate	167.2	Chemisorption played a major role in the adsorption process. The π- electrons from the adsorbent and the d electrons from the adsorbate combine to form new chemical bonds	Chen et al. [58]

4.1 *Chitosan Modified with Metal or Metal Oxide Nanoparticles*

Metal and metal oxide nanoparticles, due to their small size and high density at their corner or edge surface sites exhibit distinctive physical and chemical properties [59]. The use of metals and metal oxides in composites of chitosan- has received increased attention in recent years as an alternative adsorbent in the treatment of water.

Chitosan and chitosan functionalized with silver nanoparticles (AgNPs) were synthesized by Moustafa et al. as adsorbents for the removal of imidacloprid. It was observed that at a slightly acidic pH, chitosan and AgNP@chitosan membranes successfully removed 40 and 85% of imidacloprid, respectively. Moreover, the amount of imidacloprid removed was proportional to its original concentration, showing that the AgNP@chitosan membrane can effectively remove imidacloprid even at higher pesticide concentrations. Dehaghi et al. [59] examined the effectiveness of composite chitosan/ZnOnanoparticles (CS/ZnONP) as an adsorbent in batch studies. The removal capacities of nanocomposite beads were investigated. The good adsorption performance of the CS/ZnONP beads was demonstrated. The removal effectiveness by CS/ZnONPs beads increased from 49 to 99% for permethrin solution (0.1 mg L^{-1}) when chitosan beads were replaced by this nanocomposite. In this study, desorption was done in the presence of 0.1 M of sodium hydroxide solution. The permethrin adsorption and regeneration investigations showed that the CS/ZnONP beads could be recycled successfully, with 56% adsorption efficiency after three cycles in a column adsorption method. Additionally, CS-ZnONP beads provide a new biocompatible and environmentally benign technique for pesticide removal and can be employed in water treatment.

Shankar et al. [55] studied the role of chitosan and modified chitosan as an adsorbent for the adsorption of pentachlorophenol, a fungicide used in the wood protection sector that is responsible for soil and groundwater contamination, which can cause leukemia and peripheral neuropathy [60]. Their group synthesized chitosan (CS), chitosan crosslinked with 2-hydroxy-1-naphthaldehyde (CSH), and functionalized chitosan grafted with $CuCl_2$ (CSHC) and studied their efficiency for pesticide adsorption. The high adsorption capacity of CSH (39.1 mg/g), CSHC (35.4 mg/g), and CS (24.4 mg/g) was observed. The desorption studies were conducted in the presence of NaOH, NaCl, H_2SO_4, and H_3PO_4. It was observed that NaOH works better than other desorbing agents to regenerate the worn-out chitosan composites. Jaiswal et al. [61] examined the performance of synthesized novel copper-coated chitosan for the removal of malathion from agricultural runoff. At an optimum pH of 2,the maximum adsorption capacity of malathion was observed to be 322.6 ± 3.5 mg g^{-1}.

4.2 Chitosan Modified with Organoclay

Due to its substantial surface area and high surface energy, organoclay has a great adsorption capacity. Their efficiency can be improved by modification with carbon-based nanomaterials. Because of the intrinsic water molecule cover layer in clay including bentonite, and montmorillonite, these materials become hydrophilic and thereby put limitations to their application for the adsorption of organic contaminants. It has been suggested that the surface modifications of clay with chitosan or organic cations, such as cationic surfactants, can affect the surface's hydrophilicity. As a result, organically modified organoclays are garnering much more interest as adsorbents to remove organic contaminants from the environment.

For the adsorptive removal of clopyralid, Celis et al. [62] created a bio-nanocomposite based on montmorillonite and chitosan. It was found that the physical and chemical properties of chitosan get enhanced after incorporating the clay and could be employed as an efficient adsorbent. Depending on the cationic exchange capacity (CEC) of the clay, chitosan chains formed mono- or bilayer structures within the clay mineral interlayer. The adsorption of chitosan on montmorillonite results in structures with strong adsorption properties for anions because the $-NH^{3+}$ groups operate as anionic exchange sites.

Hnana et al. [63] studied the adsorption of glyphosate by developing an adsorbent with Moroccan Ghassoul (stevensite) doped with cationic surfactant CTAB which gets impregnated into its inter lamellar spaces by cation exchange reaction and by chitosan direct interaction with clay sheet. The adsorption capacity of Stv/CTAB (128.49 mg/g) was lesser than that of Stv/CS (159.10 mg/g).

4.3 Chitosan Decorated with Metal–Organic Framework

Huang et al. [64] synthesized different nanocomposites by varying the concentration of ionic liquids and UiO-66 with chitosan and used it as an adsorbent for the removal of 2,4-D. Ionic liquid provides stability to chitosan and UiO-66 brings the availability of oxygen groups which enhances the adsorption capacity of the adsorbent. The other factor which affects the adsorption capacity is the electrostatic interaction between the positive charge on the surface of the adsorbent and the 2,4-D anion. The high adsorption efficiency of 336 mg/g was obtained using ILCS/U-10 adsorbent.

Motaghi et al. [65] developed a nanocomposite of magnetic chitosan doped with activated carbon and UiO-66 (MCS/AC@UiO-66) for the ultrasound-assisted adsorption of imidacloprid. Three different forms of eluents, including nitric acid, ethanol, and deionized water, were used to regenerate MCS/AC@UiO-66. The adsorbent was first gathered using an external magnet, and nitric acid was then used to regenerate the adsorbent. The removal efficiency of the adsorbent was 88.56 percent, 93.45 percent, and 88.45 percent for Co (II), malachite green, and imidacloprid, respectively, even after five consecutive cycles, demonstrating the high stability and efficiency of the bio-nanocomposite.

The removal of acephate (AC), omthosate (OM), and methyl parathion (MP) was investigated by Mostafa et al. [66] using chitosan/zeolite-A (CS/ZA) as an adsorbent. Higher stability, surface reactivity, and a high adsorption capacity were provided by zeolite when used as a carrier with chitosan. For Acephate (AC), omthosate (OM), and methyl parathion (MP), adsorption capacities of 650.7 mg/g, 506.5 mg/g, and 560.8 mg/g, respectively, were observed during the adsorption experiments using fixed bed columns. The recyclability of CS/ZAcomposite as an adsorbent for AC, OM, and MP pesticides was examined. For the desorption, the CS/ZA nanocomposite was washed with diluted NaOH solution (0.02 M) at a constant temperature of 50° C for 120 min. The regenerated CS/ZA particles were utilized in five adsorption runs for the three pesticides.

4.4 Chitosan-Based Other Materials for Pesticide Adsorption

Using chitosan and a chitosan/alginate membrane, glyphosate herbicide was removed from water in a study by Carneiro et al. [67]. They found that chitosan functions as a better adsorbent than chitosan-alginate with an adsorption capacity of 10.88 μg/g and 8.70 μg respectively, which follows pseudo-second-order kinetics. The electrostatic interaction between the herbicides and chitosan in this adsorption is the basis for its operation. Dinu et al. [68] synthesized chitosan cryogels (CSGA) for pesticide adsorption. CSGA sponges were prepared using a facile and cost-effective approach that consisted of a three-step cryogenic process: (i) freezing at −18 °C, (ii) storage in a frozen state for a certain period, and (iii) thawing at room temperature. The batch adsorption experiments were performed under varied conditions, where the effects of several parameters, such as pH, contact time, and initial pollutant concentration, were evaluated to identify the appropriate adsorption conditions for maximum pesticide removal. The CSGA-based cryogel sponges exhibited a maximum adsorption capacity of 160.82 mg/g for Fastac 10EC pesticide and good recyclability at room temperature.

Sabbagh et al. [69] synthesized chitosan alginate nanocomposites by using the microemulsion method. Alginate with chitosan can be produced as a biopolymer through electrostatic forces between alginate carboxylate polyanions and chitosan amine polycations. To effectively remove pollutants from effluents, the produced biopolymer is used as an adsorbent. The highest rate of clearance for malathion is 82.35% by using this material. Attallah et al. [70] designed a chitosan/gelatin (CS/Gel) polymeric composite and employed it as an adsorbent for pesticide adsorptive removal from an aqueous solution. As model pesticides, atrazine and fenitrothion were used. The CS/Gel composite had excellent results in terms of reusability, and the reduction in adsorption capacity after three cycles was negligible. The developed adsorption approach using CS/Gel composite as adsorbent was evaluated against synthetic pesticides wastewater samples created by fortifying distilled water with varying quantities of atrazine and fenitrothion as a trial for large-scale wastewater treatment. After 120 min, a removal efficiency range of 85.65 to 96.45%of the model pesticides was observed. Chitosan gel beads and porous crab shell powder were also employed for the removal of seventeen organochlorine pesticides in water [71].

5 Cellulose

Cellulose, a natural and abundant biopolymer, has been studied as a potential adsorbent for the removal of pesticides from contaminated soil and water resources. The high surface area, hydrophilicity, and biocompatibility of cellulose make it a promising candidate for use in pesticide removal [72–74]. The adsorption of various pesticide formulations by cellulose-based materials has been examined in numerous research studies (Table 2).

5.1 Cellulose-Based Materials for Pesticide Removal

Kodali et al. [75] developed an entirely novel cellulose-based composite and evaluated its effectiveness in removing the 2,4-dichloro phenoxy acetic acid (2,4-D) from water. The outcomes demonstrated the strong 2,4-D adsorption ability of the cellulose-based composite, with a maximum adsorption capacity of 428.18 mg/g. When using the adsorbent to scale up the adsorption process, recycling or reusability of the adsorbent is crucial. The possibility of using diluted formic acid as a desorption reagent was investigated. After three cycles, attempts at continued recurrent regeneration resulted in a decline, with regeneration efficiencies of 99 percent, 97 percent, and 96 percent for the first, second, and third cycles, respectively. AgNPs were embedded in nanofabricated cellulose, which was then employed as an adsorbent to remove paraquat, cartap, and cypermethrin [76].

Tüysüz et al. [77] examined the usage of poly(2-hydroxyethyl methacrylate-glycidyl methacrylate (HEMA-GMA) polymeric cryogels modified with cellulose nanofibers as an adsorbent to remove atrazine. The findings demonstrated that the synthesized adsorbent performed exceptionally well in eliminating target micropollutants from an alkaline medium through electrostatic interactions based on the interaction of -OH groups of the cellulose linked to the polymeric material with partial positive areas on atrazine. The maximum adsorption capacity of 95.76 mg/g was obtained using this adsorbent. Dolatabadi et al. [7] reported the use of magnetic graphene oxide and carboxy methyl cellulose composites (MGOC) for the adsorption of chlorpyrifos. The maximum adsorption capacity of 108.3 mg/g was observed. A solution of NaOH (50 mM) was used to conduct the desorption experiments. Five adsorption and desorption cycles were performed which demonstrated that after five consecutive cycles, the removal efficiency percentage reduced from 100 to 86.1%. The 50 mM NaOH solution was unable to release the physiochemically adsorbed chlorpyrifos throughout each adsorption cycle. As a result, the sorption capacity for the following cycle was reduced.

Surface-initiated atom transfer radical polymerization method was employed by Lin et al. [78] to develop photo-responsive cellulose-based engraved smart imprinted adsorbent materials (P-Cell-MI), which was used for the removal of 2,4-dichloro phenoxy acetic acid. The produced materials outperformed non-imprinted cellulose materials in terms of absorption capacity (11.039 mg/g). The desorption study was done after treating the adsorbent with UV light. Moreover, P-Cell-MIP demonstrated remarkable stability, improved reusability after 8 cycles, and UV light recovery. Xu et al. [79] designed a bi-functionalized Zr-amino-based adsorbent for the elimination of glyphosate. They showed in their investigation that higher adsorption efficiency was partially due to the protonated amino group's electrostatic attraction and mainly due to the ligation between the zirconium and P-O group. The adsorption capacity of 107.38 mg/g was observed in studies.

According to Takács et al. [80] an adsorbent produced by incorporating GMA (glycidyl methacrylate) onto cellulose was employed for the removal of phenol and pesticides such as 2,4-dichloro phenoxy acetic acid. Phenol was adsorbed more quickly and reached saturation in 5–6 h. The adsorption capacity was relatively low (4–8 mg/g). The addition of β -CD greatly increased the pesticide's ability to adsorb as reported by Desmet et al. [81].

The mechanisms of pesticide adsorption by cellulose involve several factors, including the surface charge of the adsorbent, the pH of the solution, and the chemical structure of the pesticide. Kumar et al. found that the adsorption process of chlorpyrifos by cellulose nanofibers was mainly driven by hydrogen bonding and electrostatic interactions between the negatively charged cellulose nanofibers and the positively charged chlorpyrifos molecules. By grafting polyvinyl amine onto cellulose nanocrystals, the amino group was introduced for the elimination of chlorpyrifos [82]. The maximum adsorption capacity for chlorpyrifos was 98.116 mg/g, which suggested that the nanocomposite of nanocellulose with polyvinyl amine grafting had a much higher adsorption capacity than other adsorbents.

To increase the adsorption capacity of pure cadmium sulfide (CdS) nanoparticles, Komal et al. [83] investigated a novel and simple design method employing silanized cellulose nanofibers (SCNF) generated from sugarcane bagasse as a template. For all contaminants, the CdS@10%SCNF nanocomposite showed good adsorption efficiency. Even after six adsorptive runs, there was no discernible change in the adsorption performance. This result confirms the remarkable stability of the CdS@10% SCNF nanocomposite.

A positively charged $MnFe_2O_4$@cellulose activated carbon (CAC) was also synthesized for the removal of negatively charged pesticides [58]. $MnFe_2O_4$ @CAC had a substantially greater maximum glyphosate adsorption capacity (167.2 mg/g) than both CAC (61.44 mg/g) and $MnFe_2O_4$ nanoparticles (93.48 mg/g).The cellulose-graphene oxide (CCE/G) nanocomposite, which resembles a sieve was modified with KOH to develop alkaline cellulose-graphene oxide(ACCE/G) for the adsorption of chlorpyrifos [14]. A maximum adsorption capacity of 152.5 mg/g for

ACCE/G was noted. It was observed that the electron-donating properties of the sulfur and phosphorus groups were responsible for the adsorption mechanisms.

The composite Cu-BTC@cotton was employed for the removal of Ethion [84]. The maximum sorption capacity of Cu-BTC@cotton composite reached 182 mg/g and the removal percent of ethion exceeded 97%. Abdelhameed et al. employed cellulose acetate doped with Cu-MOF as an adsorbent to remove the herbicide dimethoate from wastewater [85]. When 40% Copper-MOF with cellulose acetate (Cu-BTC@CA) membrane was utilized in place of the cellulose acetate membrane, the adsorption capabilities significantly increased from 207.8 mg/g to 321.9 mg/g. By dissolving the dimethoate in acetonitrile and then drying it on air at room temperature, the dimethoate was separated from the membrane. For a total of five cycles, the recycled membrane was used for dimethoate adsorption. The adsorption capacity of dimethoate decreased from 321.9 mg/g to 249.4 mg/g after 5 iterations. As a result, the Cu-BTC@CA membrane demonstrated effective reusability in the adsorption of dimethoate, reflecting its broad applicability in the removal of pesticides.

Using modified carbonized bacterial cellulose, Zhu et al. [86] created different adsorbents. They found that the hydrophobic interaction between the organophosphate pesticides and the carbonized bacterial cellulose treated with hydrazine hydrate resulted in a 13-fold improvement in adsorption effectiveness. Cellulose-based materials have shown promising results as effective adsorbents for the removal of pesticides from contaminated soils and water resources. Further research is required to investigate cellulose-based adsorbents' potential applications and scalability for large-scale pesticide removal.

6 Graphitic-Carbon Nitride

Graphitic-carbon nitride(G-C_3N_4) is one of the strongest and most stable carbon nitride compounds.G-C_3N_4 is a compound that mostly consists of graphite-like planes of sp^2 hybridized C and N atoms. G-C_3N_4 is more appealing for photocatalysis [87, 88], energy storage [89], solar cells [90], and sensors [91] due to the incorporation of nitrogen atoms (such as s-triazine and tri-s-triazine). Due to its narrow band gap, which is excellent for the adsorption of visible light, it has frequently been used as a photocatalyst and efficiently broken-down harmful compounds when exposed to sunlight. Active sites for the adsorption of hazardous chemicals from effluent are provided by the functional groups on the surface, such as –NH_2, >N–N<, = N–, –NH–, and =C–N < [92]. The adsorption of organic and inorganic contaminants from water is improved by physical contacts (such as electrostatic interactions, Π-Π conjugated interactions, and hydrophobic interactions), as well as chemical interactions (such as complex formation or acid–base reactions).

Watcharenwong et al. [93] studied the ability of g-C_3N_4 made from melamine squandering to remove paraquat from an aqueous solution. In their findings, it was found that alkaline media are preferable for paraquat adsorption. For efficient diquat elimination, Liang et al. [94] functionalized the g-C_3N_4 with –OH and –NH_2 groups.

The g-CN-0.3 sample had an excellent capacity for DQ herbicide adsorption, with saturation adsorption capacities of 159.7 mg/g at pH = 7 and room temperature.

7 Desorption of Pesticides

The carbon-based materials exhibit good adsorption capacity for water contaminants including pesticides. However, the desorption of pesticides using carbon nanomaterials is an emerging field of research that holds great promise for environmental remediation. The regeneration and reusability of adsorbents depend on the desorption process. The desorption of pesticides from carbon nanomaterials is usually achieved by using different chemical and physical methods. The physical methods involve thermal treatment, microwave irradiation, or electromagnetic energy treatment which help to break the bonds between the pesticides and carbon nanomaterials. In chemical methods, typically solvents or surfactants are used to overcome the bonding interaction of pesticides with carbon nanomaterials. Some of the desorption systems are described in the above sections along with the adsorption of pesticides and some related materials' desorption solvents are mentioned in Table 3.

8 Conclusion

Pesticides are essential for the high production of the agriculture sector; however, their extensive use leads to the accumulation of these toxic substances in water bodies. Among several technologies and materials, the combination of nanotechnologies and carbon-based nanomaterials, such as carbon nanotubes, graphene, and

Table 3 Desorption of pesticides

Adsorbent	Pesticide	Desorbent system	Recovery (%)	Cycle	References
Chitosan-zinc oxide nanoparticles	Permethrin	NaOH	56	3	Dehaghi et al. [36]
Molecularly imprinted polymer	1-naphthyl methyl carbamate	Acetonitrile	93	8	So et al. [54]
Carbon nanotube filter coating for microextraction	Organochlorine insecticides	Thermal desorption	45 to 116		Lü et al. [95]
Walnut shells	Carbofuran and chlorpyrifos	Methanol	99	6	Memon et al. [96]
Activated carbon	Bentazon	Ethanol	73.8 to 78.3	3	Omri et al. [46]

its derivatives, is regarded as the most successful approach for environmental remediation. The advantages of using carbon nanomaterials for pesticide adsorption and desorption are their high surface area, and porosity, which may be altered to increase their adsorption/desorption capacity for certain pollutants, and strong and selective adsorption capability. Additionally, carbon nanomaterials are environmentally friendly and can be easily synthesized from renewable resources.

However, there are still challenges that need to be addressed for the usage of carbon nanomaterials for the adsorption and desorption of pesticides. Some of the challenges include their high production costs, potential toxicity to the environment and human health, and the possibility that environmental factors such as pH, temperature, and the presence of other environmental contaminants may affect how effective they are at adsorbing pesticides. Carbon-based nanomaterials can be unstable and challenging to reuse as they can degrade or lose their capacity to adsorb over time. The release of carbon nanomaterials into the environment could also harm human health.

Carbon-based nanomaterials offer tremendous possibilities for the future because of their strong adsorption capacity, substantial surface area, and reactivity. Pesticide-degrading enzymes can also be employed to functionalize carbon nanomaterials to boost their effectiveness and stability. Additionally, the investigation into the potential risks and benefits of using carbon-based nanomaterials for pesticide adsorption and desorption is still in its early phases. As a result, further investigation is necessary to fully understand the potential risks and advantages of employing carbon-based nanoparticles to remove pesticides.

Acknowledgements The authors are thankful to the Life Science Research Board, DRDO, Government of India for the research funding (No. LSRB-388/FSH&ABB/2021). The authors acknowledge the Principal, Deshbandhu College, University of Delhi for providing a research facility in the college.

References

1. Fallah, Z., Zare, E.N., Ghomi, M., Ahmadijokani, F., Amini, M., Tajbakhsh, M., Arjmand, M., Sharma, G., Ali, H., Ahmad, A., Makvandi, P., Lichtfouse, E., Sillanpää, M., Varma, R.S.: Toxicity and remediation of pharmaceuticals and pesticides using metal oxides and carbon nanomaterials. Chemosphere **275**, 130055 (2021)
2. Kesamseetty, V.R., Singampalli, R., BabuTadiboyina, A., Narasipuram, V.K.P., Prasad, K.V., Hanumanthrayappa, M., Kadiyala, C.B.N.: Role of carbon materials in the removal of organic pollutants: an abridged review. Biointerface Res. Appl. Chem. **12**, 1974–1997 (2022)
3. Sanad, M.M.S., Gaber, S., El-Aswer, E., Farahat, M.: Graphene-magnetite functionalized diatomite for efficient removal of organochlorine pesticides from aquatic environment. J. Environ. Manage. **330**, 117145 (2023)
4. Raveena, Alka, Gandhi, N., Kumari, P.: Efficacious removal of flonicamid insecticide from Water by GO@functionalizedCalix[4]pyrrole: synergistic effect in adsorption. ChemistrySelect **8**, e202203431 (2023)
5. Chatterjee, K., Alka, Kumar, S., Sharma, R.K., Kumari, P.: Effective removal of nitrogenous pesticides from water using functionalized Calix[4]arene-decorated magnetite nanoparticles. ChemistrySelect **8**, e202203426 (2023)

6. Khan, M., Muhammad, M., AlOthman, Z.A., Cheong, W.J., Ali, F.: Synthesis of monolith silica anchored graphene oxide composite with enhanced adsorption capacities for carbofuran and imidacloprid. Sci. Rep. **12**, 21027 (2022)
7. Dolatabadi, M., Naidu, H., Ahmadzadeh, S.: Adsorption characteristics in the removal of chlorpyrifos from groundwater using magnetic graphene oxide and carboxy methyl cellulose composite. Sep. Purif. Technol. **300**, 121919 (2022)
8. Khaloo, S.S., Bagheri, A., Gholamnia, R., Saeedi, R.: Graphene oxide/MIL 101(Cr) (GO/MOF) nano-composite for adsorptive removal of 2,4-dichlorophenoxyacetic acid (2,4 D) from aqueous media: synthesis, characterization, kinetic and isotherm studies. Water Sci. Technol. **86**, 1496–1509 (2022)
9. Chi, S.C., Lee, C.L., Chang, C.M.: Adsorption of pesticides, antibiotics and microcystin-LR by graphene and hexagonal boron nitride nano-systems: a semiempirical PM7 and theoretical HSAB study. Crystals (Basel) **12**, 1068 (2022)
10. Azizzadeh, A., Amooey, A.A., Ghasemi, S.: Ternary metal-organic framework/multi-walled carbon nanotube/iron oxide nanocomposite for removal of butachlor pesticide. Environ. Sci. Eur. **34**, 1–16 (2022)
11. Shi, X., Zhang, L., Cai, J.: A facile construction of supramolecular complex from polyaniline and cellulose in aqueous system. Macromolecules **44**, 4565–4568 (2011)
12. Li, C., Zhang, N., Chen, J., Ji, J., Liu, X., Wang, J., Zhu, J., Ma, Y.: Temperature and pH sensitive composite for rapid and effective removal of sulfonylurea herbicides in aqueous solution. Environ. Pollut. **255**, 113150 (2019)
13. Nodeh, H.R., Kamboh, M.A., Wan Ibrahim, W.A., Jume, B.H., Sereshti, H., Sanagi, M.M.: Equilibrium, kinetic and thermodynamic study of pesticides removal from water using novel glucamine-calix[4]arene functionalized magnetic graphene oxide. Environ. Sci. Process Impacts **21**, 714–726 (2019)
14. Suo, F., Xie, G., Zhang, J., Li, J., Li, C., Liu, X., Zhang, Y., Ma, Y., Ji, M.: A carbonised sieve-like corn straw cellulose–graphene oxide composite for organophosphorus pesticide removal. RSC Adv. **8**, 7735–7743 (2018)
15. Rashidi Nodeh, H., Wan Ibrahim, W.A., Kamboh, M.A., Sanagi, M.M.: New magnetic graphene-based inorganic–organic sol-gel hybrid nanocomposite for simultaneous analysis of polar and non-polar organophosphorus pesticides from water samples using solid-phase extraction. Chemosphere **166**, 21–30 (2016)
16. Nodeh, H.R., Wan Ibrahim, W.A., Kamboh, M.A., Sanagi, M.M.: Dispersive graphene-based silica coated magnetic nanoparticles as a new adsorbent for pre concentration of chlorinated pesticides from environmental water. RSC Adv. **5**, 76424–76434 (2015)
17. Ahmaruzzaman, Md.: Nano-materials: novel and promising adsorbents for water treatment. Asian J. Water Environ. Pollut. **16**, 43–53 (2019)
18. Agasti, N., Gautam, V., Priyanka, Manju, Pandey, N., Genwa, M., Meena, P.L., Tandon, S., Samantaray, R.: Carbon nanotube based magnetic composites for decontamination of organic chemical pollutants in water: a review. Appl. Surf. Sci. **10**, 100270 (2022)
19. Ho, S.: Low-cost adsorbents for the removal of phenol/phenolics, pesticides, and dyes from wastewater systems: a review. Water **14**, 3203 (2022)
20. Lee, C.L., Chang, C.M.: Quantum chemical approach to the adsorption of chlorpyrifos and fenitrothion on the carbon-doped boron nitride nanotube decorated with tetrapeptide. Crystals **12**, 1285 (2022)
21. Lung, I., Soran, M.L., Stegarescu, A., Opriş, O.: Devrinol and triadimefon removal from aqueous solutions using CNT-COOH/MnO_2/Fe_3O_4 nanocomposite. J. Iran. Chem. Soc. **19**, 2031–2039 (2022)
22. Liang, W., Wang, B., Cheng, J., Xiao, D., Xie, Z., Zhao, J.: 3D, eco-friendly metal-organic frameworks@carbon nanotube aerogels composite materials for removal of pesticides in water. J. Hazard. Mater. **401**, 123718 (2021)
23. Massad, Y., Hanbali, G., Jodeh, S., Hamed, O., Bzour, M., Dagdag, O., Samhan, S.: The efficiency of removal of organophosphorus malathion pesticide using functionalized multi-walled carbon nanotube: impact of Dissolved Organic Matter (DOM). Sep. Sci. Technol. **57**, 1–12 (2022)

24. Dehghani, M.H., Kamalian, S., Shayeghi, M., Yousefi, M., Heidarinejad, Z., Agarwal, S., Gupta, V.K.: High-performance removal of diazinon pesticide from water using multi-walled carbon nanotubes. Microchem. J. **145**, 486–491 (2019)
25. Firozjaee, T., Mehrdadi, N., Baghdadi, M., Nabi, R.: The removal of diazinon from aqueous solution by chitosan/carbon nanotube adsorbent. Desalin. Water Treat. **79**, 291–300 (2017)
26. Dehghani, M.H., Niasar, Z.S., Mehrnia, M.R., Shayeghi, M., Al-Ghouti, M.A., Heibati, B., McKay, G., Yetilmezsoy, K.: Optimizing the removal of organophosphorus pesticide Malathion from water using multi-walled carbon nanotubes. Chem. Eng. J. **310**, 22–32 (2017)
27. Liu, G., Li, L., Huang, X., Zheng, S., Xu, X., Liu, Z., Zhang, Y., Wang, J., Lin, H., Xu, D.: Adsorption and removal of organophosphorus pesticides from environmental water and soil samples by using magnetic multi-walled carbon nanotubes@organic framework ZIF-8. J. Mater. Sci. **53**, 10772–10783 (2018)
28. Barbosa, F.H.F., Menezes, H.C., de Carvalho Teixeira, A.P., Serp, P., Antipoff, V., de Lourdes Cardeal, Z.: Versatile magnetic carbon nanotubes for sampling and pre concentration of pesticides in environmental water. Talanta **167**, 538–543 (2017)
29. Pimenta, A.J.A., Fukumoto, F.A.A., Madeira, T.B., Alvarez Mendez, M.O., Nixdorf, S.L., Cava, C.E., Kuroda, E.K.: Adsorbent selection for pesticides removal from drinking water. Environ. Technol. **43**, 1672–1683 (2022)
30. Moustafa, M., Abu-Saied, M.A., Taha, T., Elnouby, M., El-shafeey, M., Alshehri, A.G., Alamri, S., Shati, A., Alrumman, S., Alghamdii, H., Al-Khatani, M.: Chitosan functionalized AgNPs for efficient removal of Imidacloprid pesticide through a pressure-free design. Int. J. Biol. Macromol. **168**, 116–123 (2021)
31. Badawy, M.E.I., El-Nouby, M.A.M., Marei, A.E.S.M.: Development of a solid-phase extraction (SPE) cartridge based on chitosan-metal oxide nanoparticles (Ch-MO NPs) for extraction of pesticides from water and determination by HPLC. Int. J. Anal. Chem. **2018**, 3640691 (2018)
32. Desbrières, J., Guibal, E.: Chitosan for wastewater treatment. Polym. Int. **67**, 7–14 (2018)
33. Thakur, V.K., Voicu, S.I.: Recent advances in cellulose and chitosan based membranes for water purification: a concise review. Carbohyd. Polym. **146**, 148–165 (2016)
34. Abdeen, Z., Mohammad, S.G.: Study of the adsorption efficiency of an eco-friendly carbohydrate polymer for contaminated aqueous solution by organophosphorus pesticide. Open J. Org. Polym. Mater. (2013), 41392 (2014)
35. Ahmad, N., Sultana, S., Khan, M.Z., Sabir, S.: Chitosan based nanocomposites as efficient adsorbents for water treatment. Modern Age Waste Water Probl. Solutions Using Appl. Nanotechnol. 69–83 (2020)
36. Dehaghi, M.S., Rahmanifar, B., Moradi, A.M., Azar, P.A.: Removal of permethrin pesticide from water by chitosan-zinc oxide nanoparticles composite as an adsorbent. J. Saudi Chem. Soc. **18**, 348–355 (2014)
37. Saifuddin, N., Nian, C.Y., Zhan, L.W., Ning, K.X.: Chitosan-silver nanoparticles composite as point-of-use drinking water filtration system for household to remove pesticides in water. Asian J. Biochem. **6**, 142–159 (2011)
38. Abukhadra, M.R., El-Sherbeeny, A.M., El-Meligy, M.A., Luqman, M.: Enhanced decontamination of pefloxacin and chlorpyrifos as organic pollutants using chitosan/diatomite composite as a multifunctional adsorbent; equilibrium studies. J. Solgel Sci. Technol. **99**, 652–663 (2021)
39. Kashyap, P.L., Xiang, X., Heiden, P.: Chitosan nanoparticle based delivery systems for sustainable agriculture. Int. J. Biol. Macromol. **77**, 36–51 (2015)
40. Nangia, S., Warkar, S., Katyal, D.: A review on environmental applications of chitosan biopolymeric hydrogel based composites. J. Macromol. Sci. A. **55**, 747–763 (2018)
41. Chakraborty, P., Mustafa, V., Abraham, J.: Synthesis and characterization of Chitosan nanoparticles and their application in removal of wastewater contaminants. Nat. Environ. Pollut. Technol. **17**, 469–478 (2018)
42. Rissouli, L., Benicha, M., Chafik, T., Chabbi, M.: Decontamination of water polluted with pesticide using biopolymers: adsorption of glyphosate by chitin and chitosan. J. Mater. Environ. Sci. **8**, 4591–4596 (2017)

43. Sabbagh, N., Tahvildari, K., Sharif, A.A.M.: Highly efficient and rapid removal of malathion using crosslinked chitosan-alginate nanocomposites and optimization of parameters by Box-Behnken design: isotherms and kinetic studies. J. Polym. Environ. **32**, 2595–2611 (2023)
44. Silvestro, I., Fernández-García, M., Ciarlantini, C., Francolini, I., Girelli, A., Piozzi, A.: Molecularly imprinted polymers based on chitosan for 2,4-dichlorophenoxyacetic acid removal. Int. J. Mol. Sci. **23**, 13192 (2022)
45. Baigorria, E., Fraceto, L.F.: Low-cost biosorbent hybrid hydrogels for paraquat remediation of water. J. Water Process. Eng. **49**, 103088 (2022)
46. Omri, A., Wali, A., Benzina, M.: Adsorption of bentazon on activated carbon prepared from Lawsoniainermis wood: equilibrium, kinetic and thermodynamic studies. Arab. J. Chem. **9**, S1729–S1739 (2016)
47. Salman, J.M., Hameed, B.H.: Removal of insecticide carbofuran from aqueous solutions by banana stalks activated carbon. J. Hazard. Mater. **176**, 814–819 (2010)
48. Salman, J.M., Njoku, V.O., Hameed, B.H.: Adsorption of pesticides from aqueous solution onto banana stalk activated carbon. Chem. Eng. J. **174**, 41–48 (2011)
49. Gimeno, O., Plucinski, P., Kolaczkowski, S.T., Rivas, F.J., Alvarez, P.M.: Removal of the herbicide MCPA by commercial activated carbons: equilibrium, kinetics, and reversibility. Ind. Eng. Chem. Res. **42**, 1076–1086 (2003)
50. Zhang, C., Zhang, R.Z., Ma, Y.Q., Guan, W.B., Wu, X.L., Liu, X., Li, H., Du, Y.L., Pan, C.P.: Preparation of cellulose/graphene composite and its applications for triazine pesticides adsorption from water. ACS Sustain. Chem. Eng. **3**, 396–405 (2015)
51. Rissouli, L., Benicha, M., Chafik, T., Chabbi, M.: Decontamination of water polluted with pesticide using biopolymers: adsorption of glyphosate by chitin and chitosan. J. Mater. Environ. Sci. **8**, 4544–4549 (2017)
52. Arvand, M., Latify, L., Tajmehri, H., Yagubov, A.I., Nuriyev, A.N., Pourhabib, A., Mousavi, S.J., Abolhassani, M.R.: Comparative study for the removal of oxadiazon from aqueous solutions by adsorption on chitosan and activated carbon. Anal. Lett. **42**, 856–869 (2009)
53. Ignatowicz, K.: A mass transfer model for the adsorption of pesticide on coconut shell based activated carbon. Int. J. Heat Mass Transf. **54**, 4931–4938 (2011)
54. So, J., Pang, C., Dong, H., Jang, P., Juhyok, U., Ri, K., Yun, C.: Adsorption of 1-naphthyl methyl carbamate in water by utilizing a surface molecularly imprinted polymer. Chem. Phys. Lett. **699**, 199–207 (2018)
55. Shankar, A., Kongot, M., Saini, V.K., Kumar, A.: Removal of pentachlorophenol pesticide from aqueous solutions using modified chitosan. Arab. J. Chem. **13**, 1821–1830 (2020)
56. de Freitas, D.A., Barbosa, J.A., Labuto, G., Nocelli, R.C.F., Carrilho, E.N.V.M.: Removal of the pesticide thiamethoxam from sugarcane juice by magnetic nanomodified activated carbon. Environ. Sci. Pollut. Res. **29**, 79855–79865 (2022)
57. Bayat, M., Alighardashi, A., Sadeghasadi, A.: Fixed-bed column and batch reactors performance in removal of diazin on pesticide from aqueous solutions by using walnut shell-modified activated carbon. Environ. Technol. Innov. **12**, 148–159 (2018)
58. Chen, Q., Zheng, J., Yang, Q., Dang, Z., Zhang, L.: Insights into the glyphosate adsorption behavior and mechanism by a $MnFe_2O_4$@cellulose-activated carbon magnetic hybrid. ACS Appl. Mater. Interfaces **11**, 15478–15488 (2019)
59. Dehaghi, S.M., Rahmanifar, B., Moradi, A.M., Azar, P.A.: Removal of permethrin pesticide from water by chitosan–zinc oxide nanoparticles composite as an adsorbent. J. Saudi Chem. Soc. **18**, 348–355 (2014)
60. Ali, U., Syed, J.H., Malik, R.N., Katsoyiannis, A., Li, J., Zhang, G., Jones, K.C.: Organochlorine pesticides (OCPs) in South Asian region: a review. Sci. Total. Environ. **476**, 705–717 (2014)
61. Jaiswal, M., Chauhan, D., Sankararamakrishnan, N.: Copper chitosan nanocomposite: synthesis, characterization, and application in removal of organophosphorous pesticide from agricultural runoff. Environ. Sci. Pollut. Res. **19**, 2055–2062 (2012)
62. Celis, R., Adelino, M.A., Hermosín, M.C., Cornejo, J.: Montmorillonite-chitosan bionanocomposites as adsorbents of the herbicide clopyralid in aqueous solution and soil/water suspensions. J. Hazard. Mater. **209**, 209–210 (2012)

63. Hnana, K., Barhoun, A., Draoui, K.: Improvement of glyphosate adsorption using new composites based on ghassoul and chitosan: kinetics and equilibrium study. Mediterr. J. Chem. **9**, 266–279 (2019)
64. Huang, X., Feng, S., Zhu, G., Zheng, W., Shao, C., Zhou, N., Meng, Q.: Removal of organic herbicides from aqueous solution by ionic liquid modified chitosan/metal-organic framework composite. Int. J. Biol. Macromol. **149**, 882–892 (2020)
65. Motaghi, H., Arabkhani, P., Parvinnia, M., Asfaram, A.: Simultaneous adsorption of cobalt ions, azo dye, and imidacloprid pesticide on the magnetic chitosan/activated carbon@ UiO-66 bio-nanocomposite: optimization, mechanisms, regeneration, and application. Sep. Purif. Technol. **284**, 120258 (2022)
66. Mostafa, M., Bin Jumah, M.N., Othman, S.I., Alruhaimi, R.S., Salama, Y.F., Allam, A.A., Abukhadra, M.R.: Effective removal of different species of organophosphorus pesticides (acephate, omthosate, and methyl parathion) using chitosan/Zeolite-A as multifunctional adsorbent. Environ. Technol. Innov. **24**, 101875 (2021)
67. Carneiro, R.T.A., Taketa, T.B., Gomes Neto, R.J., Oliveira, J.L., Campos, E.V.R., de Moraes, M.A., da Silva, C.M.G., Beppu, M.M., Fraceto, L.F.: Removal of glyphosate herbicide from water using biopolymer membranes. J. Environ. Manage. **151**, 101875 (2015)
68. Dinu, I.A., Ghimici, L., Raschip, I.E.: Macroporous 3D chitosan cryogels for fastac 10EC pesticide adsorption and antibacterial applications. Polymers (Basel) **14**, 3145 (2022)
69. Sabbagh, N., Tahvildari, K., Mehrdad Sharif, A.A.: Application of chitosan-alginate bio composite for adsorption of malathion from wastewater: characterization and response surface methodology. J. Contam. Hydrol. **242**, 103868 (2021)
70. Attallah, O.A., Wafa, M.M.A., Al-Ghobashy, M.A., Nebsen, M., Monir, H.H.: Adsorptive removal of pesticides from aqueous solutions using chitosan/gelatin polymeric composite: process monitoring and optimization. Int. J. Environ. Sci. Technol. (Tehran) **19**, 1–12 (2022)
71. Lu, L.C., Wang, C.I., Sye, W.F.: Applications of chitosan beads and porous crab shell powder for the removal of 17 organochlorine pesticides (OCPs) in water solution. Carbohydr. Polym. **83**, 1984–1989 (2011)
72. Rangumagar, A.B., Chhetri, B.P., Parameswaran-Thankam, A., Watanabe, F., Sinha, A., Kim, J.W., Saini, V., Biris, A.S., Ghosh, A.: Nanocrystalline cellulose-derived doped carbonaceous material for rapid mineralization of nitrophenols under visible light. ACS Omega **3**, 8111–8121 (2018)
73. Khalaf, B., Hamed, O., Jodeh, S., Hanbali, G., Bol, R., Dagdag, O., Samhan, S.: Novel, environment-friendly cellulose-based derivatives for tetraconazole removal from aqueous solution. Polymers (Basel) **13**, 450 (2021)
74. Khalaf, B., Hamed, O., Jodeh, S., Bol, R., Hanbali, G., Safi, Z., Dagdag, O., Berisha, A., Samhan, S.: Cellulose-based hectocycle nanopolymers: synthesis, molecular docking and adsorption of difenoconazole from aqueous medium. Int. J. Mol. Sci. **22**, 6090 (2021)
75. Kodali, J., Arunraj, B., Sathvika, T., Krishna Kumar, A.S., Nagarathnam, R.: Prospective application of diethylaminoethyl cellulose (DEAE-cellulose) with a high adsorption capacity toward the detoxification of 2,4-dichlorophenoxyacetic acid (2,4-D) from water. RSC Adv. **11**, 22640–22651 (2021)
76. Shad, S., Lynch, I., Shah, S.W.H., Bashir, N.: Remediation of water using a nanofabricated cellulose membrane embedded with silver nanoparticles. Membranes **12**, 1035 (2022)
77. Tüysüz, M., Köse, K., Aksüt, D., Uzun, L., Evci, M., Köse, D.A., Youngblood, J.P.: Removal of atrazine using polymeric cryogels modified with cellulose nanomaterials. Water Air Soil Pollut. **233**, 472 (2022)
78. Lin, C., Qiu, Y., Fan, J., Wang, M., Ye, L., Liu, Y., Ye, X., Huang, X., Lv, Y., Liu, M.: Fabrication of photo-responsive cellulose based intelligent imprinted material and selective adsorption on typical pesticide residue. J. Chem. Eng. **394**, 124841 (2020)
79. Xu, C., Xia, T., Li, P., Yin, J., Chen, L., Chen, S., Chi, R., Han, Q., Luo, X., Lyu, R.: Treatment of glyphosate wastewater by Zr-amino bi-functionalized worm-like mesoporous silica absorbents. New J. Chem. **47**, 4288–4298 (2023)

80. Takács, E., Wojnárovits, L., KoczogHorváth, É., Fekete, T., Borsa, J.: Improvement of pesticide adsorption capacity of cellulose fibre by high-energy irradiation-initiated grafting of glycidyl methacrylate. Radiat. Phys. Chem. **81**, 1389–1392 (2012)
81. Desmet, G., Takács, E., Wojnárovits, L., Borsa, J.: Cellulose functionalization via high-energy irradiation-initiated grafting of glycidyl methacrylate and cyclodextrin immobilization. Radiat. Phys. Chem. **80**, 1358–1362 (2011)
82. Yang, J., Ma, C., Tao, J., Li, J., Du, K., Wei, Z., Chen, C., Wang, Z., Zhao, C., Ma, M.G.: Optimization of polyvinyl amine-modified nanocellulose for chlorpyrifos adsorption by central composite design. Carbohydr. Polym. **245**, 116542 (2020)
83. Gupta, K., Kumar, V., Tikoo, K.B., Kaushik, A., Singhal, S.: Encrustation of cadmium sulfide nanoparticles into the matrix of biomass derived silanized cellulose nanofibers for adsorptive detoxification of pesticide and textile waste. Chem. Eng. J. **385**, 123700 (2020)
84. Abdelhameed, R.M., Abdel-Gawad, H., Elshahat, M., Emam, H.E.: Cu-BTC@cotton composite: design and removal of ethion insecticide from water. RSC Adv. **6**, 42324–42333 (2016)
85. Abdelhameed, R.M., Abdel-Gawad, H., Emam, H.E.: MacroporousCu-MOF@cellulose acetate membrane serviceable in selective removal of dimethoate pesticide from wastewater. J. Environ. Chem. Eng. **9**, 105121 (2021)
86. Zhu, J., Yu, J., Zhang, B., Li, C., Wang, J., Ji, J., Liu, D., Gao, R., Li, J., Ma, Y.: Hydrophobic-action-driven removal of six organophosphorus pesticides from tea infusion by modified carbonized bacterial cellulose. Food Chem. **412**, 135546 (2023)
87. Cao, S., Low, J., Yu, J., Jaroniec, M.: Polymeric photocatalysts based on graphitic carbon nitride. Adv. Mater. **27**, 2150–2176 (2015)
88. Mamba, G., Mishra, A.K.: Graphitic carbon nitride (g-C_3N_4) nanocomposites: a new and exciting generation of visible light driven photocatalysts for environmental pollution remediation. Appl. Catal. B. **198**, 347–377 (2016)
89. Safaei, J., Mohamed, N.A., Noh, M.F.M., Soh, M.F., Ludin, N.A., Ibrahim, M.A., Isahak, W.N.R.W., Teridi, M.A.M.: Graphitic carbon nitride (g-C_3N_4) electrodes for energy conversion and storage: a review on photoelectrochemical water splitting, solar cells and supercapacitors. J. Mater. Chem. A. Mater. **6**, 22346–22380 (2018)
90. Chen, X., Liu, Q., Wu, Q., Du, P., Zhu, J., Dai, S., Yang, S.: Incorporating graphitic carbon nitride (g-C_3N_4) quantum dots into bulk-heterojunction polymer solar cells leads to efficiency enhancement. Adv. Funct. Mater. **26**, 1719–1728 (2016)
91. Ilager, D., Shetti, N.P., Foucaud, Y., Badawi, M., Aminabhavi, T.M.: Graphene/g-carbon nitride (GO/g-C_3N_4) nanohybrids as a sensor material for the detection of methyl parathion and carbendazim. Chemosphere **292**, 133450 (2022)
92. Gusain, R., Kumar, N., Ray, S.S.: Recent advances in carbon nanomaterial-based adsorbents for water purification. Coord. Chem. Rev. **405**, 213111 (2020)
93. Watcharenwong, A., Kaeokan, A., Rammaroeng, R., Upama, P., Kajitvichyanukul, P.: Adsorption of paraquat dichloride by graphitic carbon nitride synthesized from melamine scraps. IOP Conf. Ser. Earth Environ. Sci. **78**, 012012 (2017)
94. Liang, X., Fan, J., Liang, D., Xu, Y., Zhi, Y., Hu, H., Qiu, X.: Surface hydroxyl groups functionalized graphite carbon nitride for high efficient removal of diquat dibromide from water. J. Colloid Interface Sci. **582**, 70–80 (2021)
95. Lü, J., Liu, J., Wei, Y., Jiang, K., Fan, S., Liu, J., Jiang, G.: Preparation of single-walled carbon nanotube fiber coating for solid-phase microextraction of organochlorine pesticides in lake water and wastewater. J. Sep. Sci. **30**, 2138–2143 (2007)
96. Memon, G.Z., Moghal, M., Memon, J.R., Memon, N.N., Bhanger, M.I.: Adsorption of selected pesticides from aqueous solutions using cost effective walnut shells. Adsorption **4**, 43–56 (2014)

Adsorption and Desorption of Dyes

S. Gopika Devi, A. Joel Amalan, S. Subasini, and Anitha Pius

Abstract The color of the water is the most important and evident indicator of water quality parameter. It may not only be unsettling visually, but also the sign of contamination. Adsorption is the most widely used dye removal technology since it is straightforward, economical and very successful method. This chapter give current information on the most extensively research for the dyes adsorption and desorption methods. Illustrations are provided for the impacts of the initial dye concentration, pH, adsorbent dose, particle size and temperature. Likewise, potential eluents might be used to wash out contaminants from adsorbent materials. In this chapter, the adsorption and desorption investigations of dyes Methylene Blue, Congo Red, Malachite Green and Crystal Violet are discussed.

Keywords Adsorption · Efficient adsorbents · Classification of dyes · Adsorption influencing parameters · Desorption eluents

1 Introduction

A hygienic living environment and safe drinking water are the basic requirements to support healthy living. For residential use as well as industrial and agricultural purposes, clean water is an absolute need. Larger wastewater effluents would be produced as a result of the increased water demanding units. Aquifers have been constantly depleted as the result of the excessive use of water and the release of hazardous chemicals during industrial and agricultural operations and placing the constant demand on the life-sustaining resource of water. Water pollution can be caused by the variety of activity including mining, industrial waste, sewage, pesticides and agricultural fertilizers etc. Major pollutants in the wastewater effluents are heavy metals, organic chemicals, dyes and halogenated hydrocarbons. Adsorption, membrane filtration, coagulation/flocculation, oxidation, biological treatment and

S. G. Devi · A. J. Amalan · S. Subasini · A. Pius (✉)
Department of Chemistry, The Gandhigram Rural Institute—Deemed to be University, Gandhigram, Dindigul 624 302, Tamil Nadu, India
e-mail: dranithapius@gmail.com

J. Tharini and S. Thomas (eds.), *Carbon Nanomaterials and their Composites as Adsorbents*, Carbon Nanostructures, https://doi.org/10.1007/978-3-031-48719-4_13

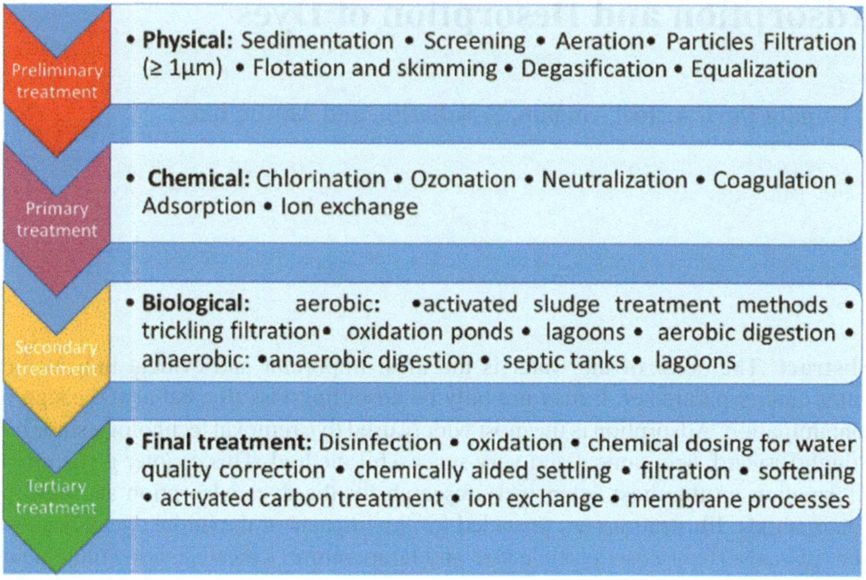

Fig. 1 Modern and traditional methods of wastewater treatment

other techniques have all been developed to date to reduce the risks of contaminants [1] (Fig. 1).

However, these procedures have lot of drawbacks. Chemical treatments, for instance, are quick, easy operations with the variety of oxidant approaches. In addition to all of these characteristics, chemical procedures are typically laboratory size and economically unviable for small enterprises to meet their energy needs. Even at high concentrations, physical methods like membrane filtration are quick and simple procedures. They are thought to be useful for practically all forms of contaminants, including suspended particles, mineral ions and dyes [2]. However, they are not practical for small companies due to their expensive energy consumption, maintenance and high operating costs, which result in quick membrane blockage at high concentrations. Coagulation and flocculation are straightforward physiochemical processes, however managing sludge volume creation (big-size flocs) is difficult and raises the cost of operation. The public wells adopt the biological treatment since it is an easy and appealing procedure from an economic standpoint. However, it takes a long time to degrade slowly and need the maintenance of healthy microbes in an ideal habitat [3]. Following an examination of several wastewater treatment methods, it is concluded that adsorption is the surface phenomena. Unbalanced residual force is present at the surface of a liquid or solid when it is under strain or unsaturation. Higher surface energy is the outcome of these uneven residual forces. As the result, the molecular species that come into touch with the surface of liquids or solids always have the tendency to attract and hold them [4]. It is the non-destructive, technologically straightforward, effective, and economical method. It is very effective in removing the different pollutants from the wastewater, including colors, metal ions, minerals and other impurities from textile, pharmaceutical, food, leather, paint and varnish industries [5].

After textiles are dyed, the leftover dye-concentrated effluent is frequently released into the environment at high pH and temperatures without being treated. This phenomenon will affect the oxygen transfer mechanism and the self-purification procedure of environmental water bodies, because of the high biological, chemical oxygen demand, high suspended particle content and other dangerous compounds in the wastewater. These compounds are difficult to remove from the water bodies because they are frequently synthetic in origin and have a complex aromatic molecular structure that enhances their chemical stability and also adding dyes to the water supply system.

- Dye make the water more turbid.
- The majority of dyes are carcinogenic, (bladder, liver, kidney) mutagenic and toxic to living organisms.
- They can cause allergic reactions to skin, eye, mucous membrane irritation, dermatitis, and respiratory problems.
- They have a significant impact on the photosynthetic activity of the aquatic environment because they block light from penetrating the water, inhibiting the growth of algae, which are crucial for the production of oxygen as well as being a foundation of the food chain.

In the last 30 years, the issue of dyes in natural waterways has not received much attention, and it has only lately been included in the environmental regulations [6].

2 Classification of Dyes

Dye can be classified based on the structure, source, color, solubility, and application techniques. In essence, the most typical classification was based on their chemical structure and it uses. The grouping by ionic nature (particle charge upon dissolving in aqueous media) and the application are combined in Fig. 2 [7]. Let us talk about the cationic and anionic dyes. Dyes are classified as the reactive, direct and acid (anionic dyes), basic (cationic dyes), or dispersion and vat (non-ionic dyes) depending on the application technique [1].

2.1 Reactive Dyes

Reactive dyes (Fig. 3) enable the attainment of high wet strength. Reactive groups includes vinyl-sulfone, chlorotriazine, trichloropyrimidine and difluorochloropyrimidine which form the covalent link with the fibre [8]. The findings of the adsorption method indicate that the anionic dyes were attracted to the positively

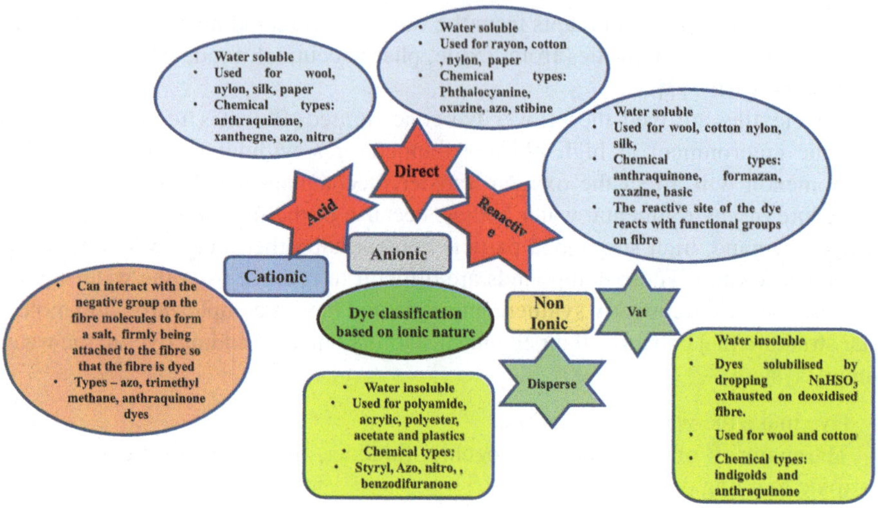

Fig. 2 Ionic-based categorization of dyes

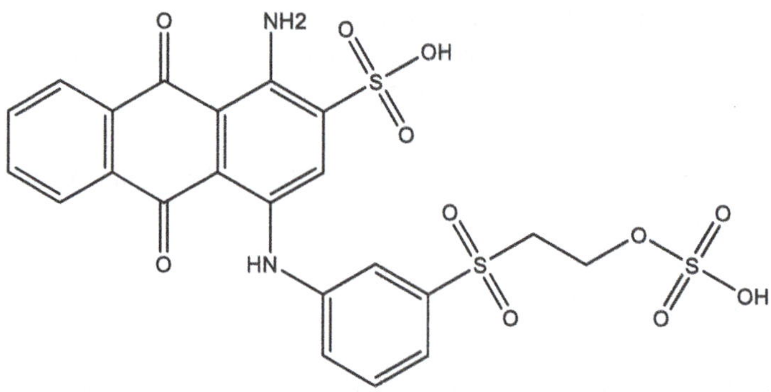

Fig. 3 Struture of reactive blue 19

charged surfaces of the adsorbent via electrical attraction because reactive dyes are soluble in aqueous media and have higher negative charge density. Originally intended for cellulose fibers, these dyes are now used on cotton, wool and polyamide textiles. Additionally, several fiber-reactive dyes for protein and polyamide fibers are also readily accessible on the market. It is impossible to overstate the significance of reactive dyes in the worldwide coloring industry. Reactive dyes are thought to be the most difficult to remove of all dyes since they frequently slip through traditional treatment techniques unscathed [9].

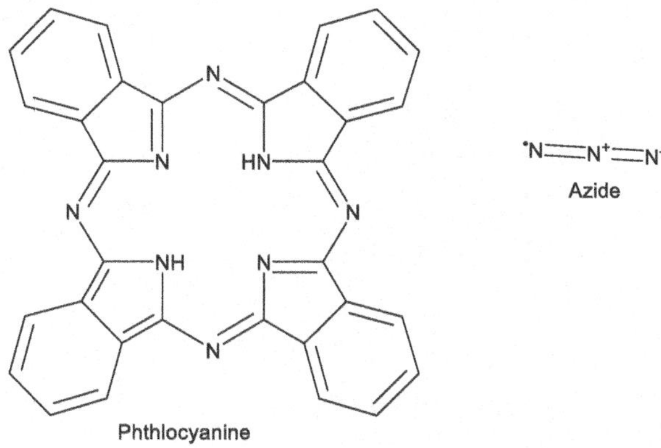

Fig. 4 Struture of phthlocyanine

2.2 Direct Dyes

The textile industry still uses direct dye most frequently for printing and dying purposes. It find significant contribution to the textile industry due their extensive usage in printing and dying cotton, viscose, silk, wool, and leather, direct dyes. Although these dyes are anionic and water-soluble, they cannot be categorized as acid dyes since they do not connect to the fiber via acid groups. The following are the main categories of chromophore: azo, stilbene, phthalocyanine, Di oxazine, formazan, anthraquinone, quinoloneand thiazole. In contrast to reactive dyes, direct dyes are less resistant to washing while having a large range of hues and tones and being well-known for being simple to use [9]. Reactive Blue 19 (Fig. 4) is the typical example of direct dye.

2.3 Acid Dyes

As implied by the name, the molecules of acid dyes include one or more acidic functionalities (SO_3H and COOH). They have exceptional chemical and photochemical stability, which accounts for the complex composition, low biodegradability and high tinctorial value. They are difficult to eliminate using standard techniques. Aquatic ecosystems can be harmed by their breakdown products or metabolites, which have the potential to be mutagenic or carcinogenic. Due to their vivid color and excellent solubility, water-soluble acid dyes, particularly sulphonic acid dyes, are used extensively. The dye molecules have highly distinct structural differences and frequently include certain metal complexes. The presence of sulfonated groups (Fig. 5), which enable water solubility and azo (the most significant group), anthraquinone, triphenylmethane or copper phthalocyanine are the group's distinguishing characteristics [10].

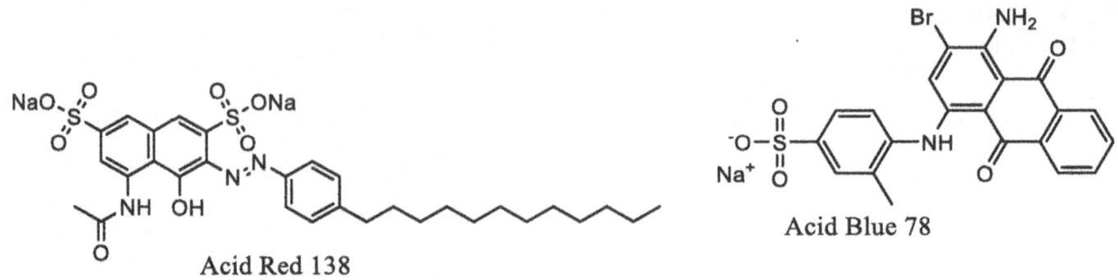

Acid Blue 78

Acid Red 138

Fig. 5 Struture of acid red 138

2.4 Cationic-Basic Dyes

Basic dyes fall under the category of cationic dyes due to the coloured cationic salt in an aqueous solution. Later, the anionic surface of the substrate (acrylic, paper, and nylon) reacts with these cationic salts to form electrostatic attraction between the resultant cations and the negatively charged substrates. The cationic functional groups ($-NR^{3+}$ or $= NR^{2+}$) are often substituted amino compounds and amino acids that are acid-soluble compound [11]. By creating ionic connections with the fiber's anionic groups, they would adhere to it. Basic dyes have a significant number of aromatic rings that contribute to their resonance capacity, a complex and massive structure that makes them robust and stable in the environment, and they are resistant to degradation. This feature prevents sunlight from penetrating, which lowers photosynthetic activity and lowers the effectiveness of natural biological self-cleansing [12].

2.5 Disperse Dyes

Disperse dyes are water-insoluble dyes having tiny, non-ionic structures that are connected with polar functional groups like $-NO_2$ and $-CN$. They are dispensed from an aqueous dispersion onto hydrophobic fibers. Disperse dyes are used on the surface of nylon, acrylic, cellulose, and cellulose acetate fibers [13]. Chemically speaking, simple azo compounds make up more than 50% of dispersion colors, anthraquinones make up around 25% and the remaining dyes are methine, nitro or naphthoquinone dyes. Disperse dyes are sometimes known as "sublimation" inks because, when heated, the ink molecules "sublimate" or shift instantly from a solid to a gas without ever passing through a liquid stage. The bulk of dispersion dyes are based on azo compounds, however, anthraquinone derivatives are frequently used to produce violet and blue hues. Due to their nano size, dispersed dye particles can maintain higher stability, particularly in high-temperature dyeing procedures [14]. Figure 6 shows the structure of Disperse Red 90.

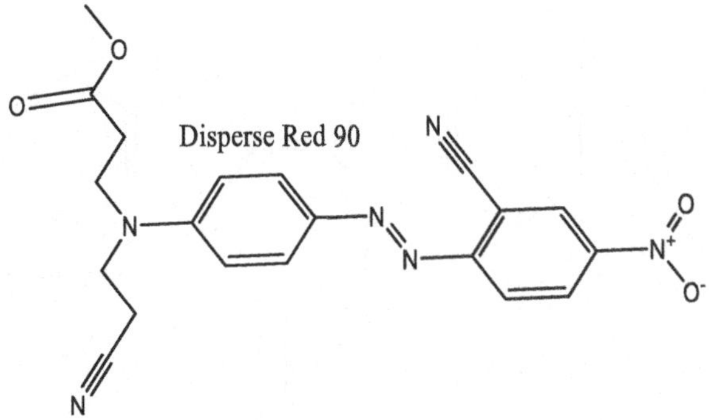

Fig. 6 Structure of disperse red 90

2.6 *Vat Dyes*

Vat dyes are frequently used to color cellulosic cotton garments. This is the principal causes of pollution in wastewater from textile and other industrial effluents. These kinds of colors are insoluble in water. Excellent color fastness, washability and chlorine-bleachable colored fibers are the some of the hallmarks of vat dyes. They are practically insoluble in water and therefore have no affinity for cellulosic fibers. In traditional tank dyeing processes, the dye is reduced in an alkaline medium in the presence of strong reducing agents such as sodium dithionite ($Na_2S_2O_4$) [15]. Figure 7 shows the structure of indigo dye in its oxidation and reduction form.

3 Factors Affecting the Adsorption Process

Several factors affect the effectiveness of liquid phase adsorption. The adsorbent/ adsorptive interaction, the adsorbent's surface chemistry, pore structure, particle size, the adsorbent's nature, the presence of other ions in the aqueous solution, pH, temperature, pressure and contact time are the some examples of these physicochemical factors affecting the adsorption process. It is also vital to consider the adsorbate's polarity, size, molecular weight, and molecular structure (Fig. 8) [16].

4 Types of Adsorbents

It is critical to develop novel adsorbents with superior qualities, i.e., cheap and conveniently accessible, high removal efficiency even at trace levels. The adsorbents may be extracted from industrial waste, animal waste or agricultural waste or nanoadorbent (Fig. 9). This kind of adsorbent is the crucial component in the process of getting rid of waste since all adsorbents include functional groups that are essential to the adsorption. The porosity, pore structure and adsorbent surface area are the unique

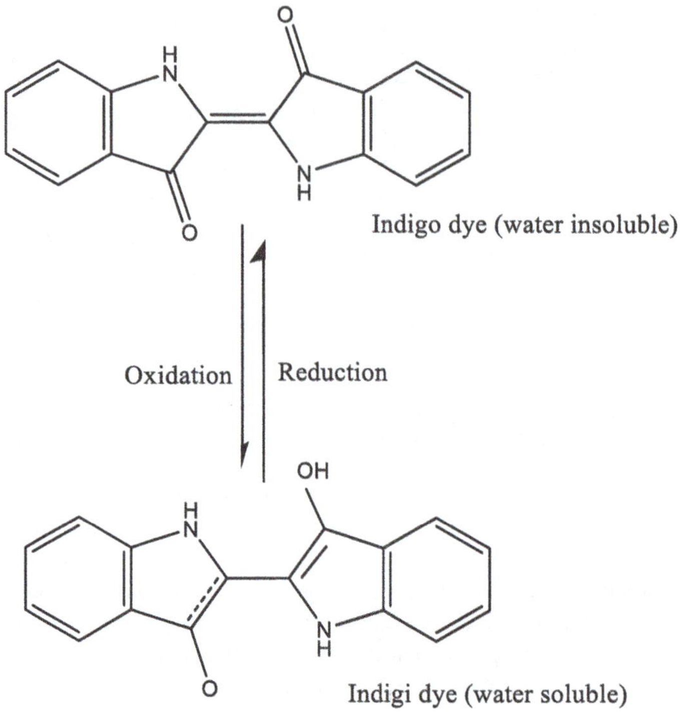

Fig. 7 Structure of indigo dye both in oxidized and reduced form

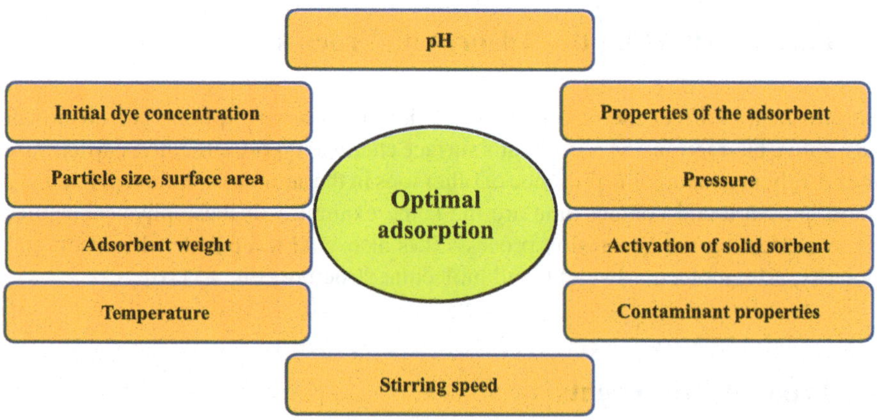

Fig. 8 Factors influencing the adsorption process

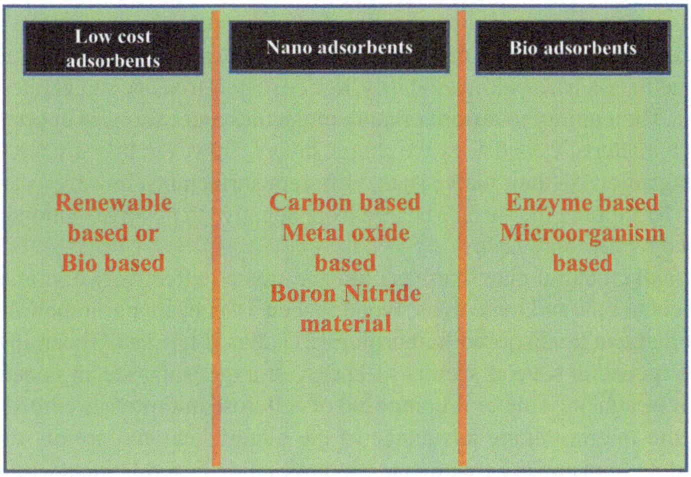

Fig. 9 Various adsorbents for the wastewater treatment

properties of adsorbent. Their classification and sorting have become crucial due to the rise in the usage of adsorbents [17].

4.1 Low-Cost Adsorbents

Wastewater can be purified using the variety of in expensive adsorbents. They are most favoured material because of their economic viability and decreased waste and less by-product production. For example—red mud, coalfly ash, coffee residue, waste tea, sugarcane bagasse, eggshell waste [18].

4.2 Natural Adsorbents

Several substances in nature can be used as adsorbents for the removal of contaminants, heavy metal ions, organic compounds and colors from the wastewater. Natural adsorbent includes chitin, clay, zeolite, peat moss, coal and wood. Zeolites are alumina silicates that resemble crystals and are formed naturally. They have tetrahedral framework structure which are connected by the oxygen atoms. Zeolites that have been surfactant-modified could absorb various organic contaminants. Natural zeolites may exhibit efficient adsorption ranging from 45 to 64% in 4 h [19].

The biopolymers chitin and chitosan is made of fiber which have special qualities and biodegradability. These fibers have garnered a lot of interest due to their superior removal efficiency and strength compared to the activated carbon. Chitin is a naturally occurring, renewable material that has potential usage as the adsorbent in wastewater treatment. Alternative inorganic adsorbents like clay minerals provide several benefits, such as wide availability, low cost, non-toxicity and greater pollutant adsorption. Their ability to absorb organic molecules and cations is improved by the colloidal characteristics and negative charge layers. However, they are not very good at adsorbing anions in their native state. Different surfactants, however, can enhance adsorption by flipping the surface charge from negative to positive. Anionic dyes can be absorbed by raw clay [20].

Additionally, natural clay that has not been treated effectively eliminates colors of several chemicals and ionic dyes. Wood is seen as a cheap, non-renewable natural adsorbent that can be disposed by burning [21]. Wood has lately been investigated to develop its use in several sectors of energy storage, solar steam generation and wastewater treatment. Timber is composed of cellulose microfibrils embedded in the porous lignin matrix, taking advantage of the hierarchical and porous structure. It can be used without any pre-treatment to remove color. Several plants with wood and bark, such as Eucalyptus bark, have been employed for pollutant adsorption from wastewater [22].

4.3 Agricultural Wastes

Agricultural waste products include vegetable and fruit peels which are used with pre-treatment and they are easily used as inexpensive adsorbents. Agricultural wastes are mostly made of cellulose and lignin, which gives them a flexible structure, a range of chemical characteristics and the ability to serve as the appealing alternatives to the traditional adsorbents. They are found in polymer chains with certain functional groups, such as carboxyl, alcohol, aldehyde, phenol and ketone, which aid in the removal of different pollutants from water. A range of agricultural wastes, like pulse seed coat, wheat bran, rice husk, coconut, orange, lemon and banana peels are used as adsorbents [23].

Both natural and modified forms of agricultural waste are beneficial. The waste product is properly cleansed, crushed and sieved until it reaches the necessary particle size in the natural system. Finally it is employed as an adsorbent for the water purification. filter Granular activated carbon (GAC) is created by pre-treating the material namely hazelnut, almond, walnut and apricot shells into a modified form. The rice husk was converted into the urea activated carbon for the removal of nitrate ions from wastewater. In order to remove malachite green, a variety of plant materials, including plant bark, coconut fiber, coconut shell, pine needles, neem leaf powder and cactus leaves have been used with the removal efficiencies of more than 90% at the ideal pH [24].

Citric acid is added to rice bran, rice straws and soybean hulls to make them more effective in removing various dyes from the wastewater. Methylene blue, a frequently dye could be removed by the mean of lotus leaves and pumpkin seeds [25].

4.4 Industrial Wastes

Solid waste was created due to the industrial activity; certain by-products are recycled, while others are dumped in the landfills. Heavy metals, organic compounds and colors are removed from the wastewater using inexpensive adsorbents. These materials include fly ash, palm oil ash, red mud, bagasse ash, coffee waste, etc., are affordable and widely accessible locally [26].

If these wastes are not handled, they may result in significant environmental issues. Many different treatment techniques are used, but they are difficult, expensive, and produce secondary contaminants. Pollution will be greatly reduced if these wastes are converted into useful materials.

Fly ash has been found to have a porous surface with tiny, irregularly shaped, glassy spherical particles. Fly ash's surface has been altered to make it more adsorption-friendly. It is the by-product of combustion that is primarily employed in the manufacture of cement, bricks and other building materials. Due to its high alumina and silica content and suitability for adsorption, fly ash is the cheap adsorbent materials. Fly ash from bagasse, a waste product of the sugar industry is difficult to dispose it from the environment and hence utilized as the filler in building products; however, it has been transformed into an effective adsorbent for the removal of harmful substances [27].

Red mud, an effluent from the aluminum industry, has drawn attention as an active adsorbent to reduce pollution by effectively absorbing a variety of aquatic toxins Due to the presence of sodium hydroxide solution (used in the refining process), red mud has an alkaline pH range of 10–12. Due to its alkaline makeup, it may constitute hazardous waste and must be neutralized before being used as an adsorbent. Red mud after neutralisation has a pH between 8 and 8.5. Red mud adsorption has been used extensively to remove colors and arsenic from wastewater [28].

4.5 Nano-adsorbents

Although nanotechnology has many effective uses, wastewater treatment has emerged as a particularly intriguing topic of study. The production and investigation of nanoscale elements are fundamental to nanotechnology. In several areas of interest recently, nanomaterials, nano-adsorbents or nanoparticles have demonstrated promising uses, primarily in the removal of metallic contaminants from industrial effluent. Various analytes have been separated from one another and purified using nanomaterials, particularly in the form of nano-adsorbents and their variations. At the nanoscale, materials are categorized according to different biological, physical and chemical characteristics than according to their conventional or real size. As a result, materials with a high surface-to-volume ratio at the nanoscale include metal oxides, ceramics, polymers and carbon derivatives. The surface area of materials quickly rises with decreasing particle size and alternatively, materials' macroscopic-level properties are altered at the nanoscale [29].

4.6 Carbon-Based Nanomaterials

Numerous carbon-based nanomaterials have been generated due to the rising need for nanotechnology. Due to its wide surface area and adjustable porosity, activated carbon has the great potential for adsorption studies. Nanomaterials made of carbon and graphene, like nanotubes, come in functionalized and non-functionalized varieties. Higher adsorption activities are possible by the functionalization of carbon-based nanomaterials on the surfaces. Carbon based nanomaterials has two major advantages namely (1) improved hydrophilicity which boosts the dispersion in an aqueous media (2) improves the electrostatic interactions between adsorbent and adsorbate to maximize adsorption capacity and provides greater surface area for the exposure to the contaminants. If the surface is functionalized by a negative charge, the electrostatic interactions are the main driving mechanism for the adsorption of especially positively charged species such as heavy metals. By effectively eliminating contaminants from the water, carbon nanotubes (CNTs) could help to maintain the healthy environment.

Iijima developed CNTs in 1991, and since then, several research have done to exploit the CNT for the water purification. CNTs have a huge surface area that makes them as the ideal materials for the adsorption of pollutants. Three-dimensional (3D) graphene is a viable choice for the adsorption process due to its uniform structure, high surface area, chemical stability, structural sustainability and highly oleophilic and hydrophobic surfaces. In comparison to current adsorbents, graphene has made substantial advancements in its capacity to absorb organic liquids and oils. Heavy metal ions and organic pollutants have been successfully removed using nano-adsorbents made of graphene and CNTs [30].

4.7 Metal Oxide-Based Nanomaterials

The removal of harmful metal ions, inorganic and organic contaminants from wastewater and raw water has previously been accomplished using a sizable variety of metal oxide nanomaterials, such as Fe_3O_4, CuO, ZnO, and their composites. Due to their enormous surface area and effective catalytic activity, metal nanoparticles are regarded as the ideal material for eliminating harmful dyes, particularly azo dyes. The best materials for the adsorption of arsenic and other ions are iron and its compounds, goethite, hematite, granular ferric hydroxide, and iron oxide-coated surfaces. CuO is the good nano adsorbent for the removal of arsenic from water due to easily regenerable and reusable. ZnO is a substance that is extremely stable and safe which function in the pH range between 5.8 and 6. Compared to individual metal oxides, zinc oxide nanocomposites exhibit a high adsorption capacity. Fe–Mn, Fe–Ce, Fe–Ti, Fe–Zr, Ce–Ti, Fe–Cu, Fe–Cr, and Mn–Co are some examples of these composites [31].

4.8 Nanocomposites

Several nano-sized adsorbents have been created and used in the water filtration systems. Therefore, there is still much work to be done in creating innovative nano-adsorbents for the waste water treatment. To achieve this goal, nanomaterials are being thoroughly investigated to the develop adsorbent with flexible separation properties, such as a large surface area and more binding sites. For the filtration of water, spinal ferrite nanoparticles (SFNPs) and their nanocomposites are used. Numerous studies have shown that nanocomposites based on SFNPs have improved photodegradation capabilities and may be used to entirely remove pollutants [32].

4.9 Boron Nitride Material

The chemical compound boron nitride (BN) is made up of an equal amount of boron (B) and nitrogen (N) atoms. In 1842, Balmain created BN for the first time through the interaction of molten boric acid (H_3BO_3) and potassium cyanide (KCN). Since then, several studies on the creation of various BN nanostructures, including nanotubes, nanofibers, nanoparticles, nanoflowers and nanosheets, have been developed. With its enormous surface area, oxidation resistance, and chemical durability, hexagonal boron nitride in a porous form exhibits special chemical and physical properties. Due to these properties, BN is the favorable material for the storage of hydrogen, adsorption of inorganic and organic contaminants, and as the catalysis [33].

4.10 Bio-sorbents

An adsorptive technique called bio-sorption can remove contaminants, metal ions, and organic dyes from aqueous solutions. As suggested by the name, bio-sorbents are made from biological sources or by using biological methods. These bio-sorbents contain certain functional groups on their surfaces that tend to attract or bind to contaminants. Depending on the functional group polarity, this technique successfully eliminates the variety of organic and inorganic substances by the passive attachment to the bio-sorbents. Algal charcoal was used as the bio-sorbent to remove the hazardous pigments, heavy metals, different inorganic, organic contaminants from the waste effluents.

The alteration of walnut shells has shown to be the potential technique for producing bio-sorbents for the practical sorption capabilities. The sorption capacity was further enhanced by the modifications with inorganic acid [1].

5 Desorption Studies

Desorption may arise from heat (thermal energy), incoming light such as infrared, visible, or ultraviolet photons, or an incident beam of energetic particles such as electrons. Additionally, it take place after chemical processes like oxidation or reduction in an electrochemical cell or following a chemical process involving adsorbed substances in which the surface might function as the catalyst.

Desorption experiments were conducted to regenerate the used adsorbent and hence making the procedure more cost-effective and cheaper. There are significant amount of frequently dangerous by-products and trashes. These solids may be recycled and hence allowing the the recovery of the adsorbent and the contaminant.

The desorption depends on the nature of adsorbent, adsorbate (various types of dyes with an ionic nature) and the adsorption process. Examining the adsorbent's reusability is the crucial factor in the adsorption–desorption investigations. The adsorbent has to be cleaned and regenerated in between the dye removals so that it can be reused again. Various desorption techniques and a wide variety of eluents were utilized to renew the used adsorbents [34].

Siroos S. et al. investigated the adsorption of the dyes malachite green, and auramine-O, on the NaX nano zeolites. The used NaX nano zeolites were cleaned with a little quantity of methanol before dried in the vacuum oven. The outcomes show that the adsorption efficiency declines after five cycles [35].

Malachite green (MG) dye desorption on cellulose nano fibril aerogels was studied by Feng J. et al. The first rounds of utilized aerogels were dissolved in the deionized water and finally 16% of the MG was regenerated. The material was also subjected to desorption process using 50 mL of 0 mM, 50 mM, and 200-mM sodium chloride solutions [36].

The adsorption–desorption behavior of the dyes Direct Orange-26 (DO-26), Direct Red-31 (DR-31), Direct Blue-67 (DB-67) and Ever direct Orange-3GL (EDO-3) over native, modified rice husk was thoroughly studied by Bhatti et al. After the biosorbent was dried at 60 °C, the dye desorption was examined using distilled H_2O (pH 8, 10, and 12), NaOH, and Na_2CO_3 (0.1 M). Under standard circumstances, it was shown that rice husk biomass can adsorb EDO-3, DR-31, DO-26, and DB-67 dyes at rates of 75.32, 80.59, 62.88, and 53.97 (mg/g), respectively. After 10 sorption/desorption cycles, the rice husk biomass's adsorption capability was decreased by 17% [37].

Ilknur S. Entürk and Mazen Alzein evaluated the adsorption–desorption of Acid Violet 17 on a regenerated acid-activated pistachio shell. As per the protocol, 100 mL of HCl, NaCl, CH_3COOH, and NaOH desorption agents were prepared at various concentrations (0.1, 0.2, 0.4, and 0.8 M), along with solvents (ethanol and distilled water), and 1 g of the dye-loaded adsorbent was obtained (0.1, 0.2, 0.4, and 0.8 M). The desorption treatments was carried out separately with water and ethanol which shows poor desorption efficiency. After three rounds of desorption, the AV 17 dye's adsorption efficiency dropped from 94.76 to 75.84% [38].

In desorption investigations, Mohammad A. Al-Ghouti and Rana S. Al-Absi introduced black and green olive stones to the 50 mL of acidic solution of acetic acid and ethanol (%vol) (10:1, 5:1, and 1:1) that are loaded with 600 mg/L of methylene blue. The mixture was shaken for 24 h at 25 °C and 150 rpm. Finlly MB-loaded green and black olive stones show the desorption removal capabilities of 92.5 and 88.1%, respectively [39].

Momina et al. studied the chemical regeneration. on the bentonite surface for the adsorption of methylene blue dye. The following solvents were used: sodium chloride (NaCl), nitric acid (HNO_3), ethanol (C_2H_5OH), propanol (C_3H_7OH), acetone ($(CH_3)_2CO$) and distilled water (H_2O). Using an aqueous HCl solution, significant amout of desorption of MB (70%) was accomplished. Ainoa Murcia-Salvador et al. used NaOH as the effective solvent for the adsorption–desorption of Direct Blue 78 on the eggshell surface. Figure 10 shows potential eluents that might be used to rejuvenate adsorbent materials.

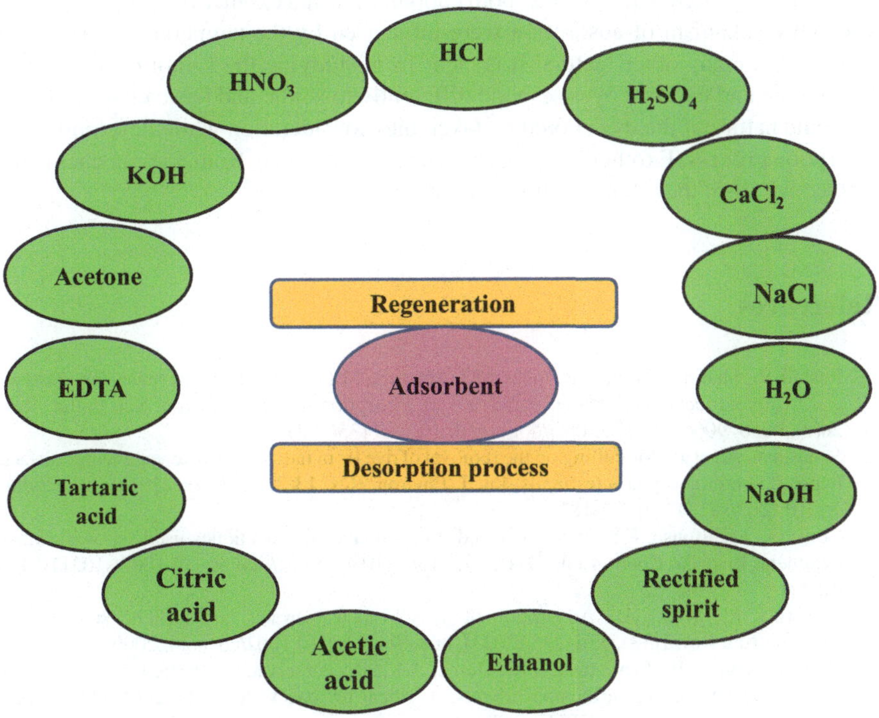

Fig. 10 Eluents that could be used to remove impurities from adsorbent substances

Desorption agents such as HCl, NaCl, CH_3COOH, and NaOH were synthesized with solvents such as ethanol and distilled water in an orbital mixer running at 125 rpm for 24 and 48 h. The desorption treatments was carried separately with water and ethanol which shows poor desorption efficiency. After three rounds of desorption, the AV 17 dye's adsorption efficiency dropped from 94.76 to 75.84% [40].

6 Conclusion

Due to the advancement of science and the chemical industry, researchers were able to develop several synthetic dyes. Many of the synthetic dyes are hazardous to the environment and human health. Therefore, we must address the problem of treating industrial wastewater (mostly textile and dye waste) and create effective and sustainable water treatment systems. Although there are many methods for the water treatment, adsorption is perhaps the most widely used commercial technology. Several external factors affect the remediation process, and it is crucial to optimize these factors so that the system can be used even at the lower concentrations with low costs, minimal by-products and high efficiency. With various adsorbents, Methylene Blue and Congo Red demonstrated both endothermic and exothermic properties. The type and mechanism of adsorption were influenced by the temperature, pH of the solution and many other factors. It is clear from studying the solution's chemistry that anionic and cationic dyes respond differently to acidic and basic environments. The ionic nature of the dye should be taken into consideration while developing the adsorption process. Exothermic processes are advantageous from a green chemistry perspective since they don't require extra energy.

References

1. Rashid, R., Shafiq, I., Akhter, P., Iqbal, M.J., Hussain, M.: A state-of-the-art review on wastewater treatment techniques: the effectiveness of adsorption method. Environ. Sci. Pollut. Res. **28**(8), 9050–9066 (2021). https://doi.org/10.1007/s11356-021-12395-x
2. Saravanan, A., et al.: Modelling on the removal of dye from industrial wastewater using surface improved enteromorpha intestinalis. Int. J. Environ. Res. **13**(2), 349–366 (2019). https://doi.org/10.1007/s41742-019-00181-0
3. Crini, G., Lichtfouse, E.: Advantages and disadvantages of techniques used for wastewater treatment. Environ. Chem. Lett. **17**(1), 145–155 (2019). https://doi.org/10.1007/s10311-018-0785-9
4. Agarwala, R., Mulky, L.: Adsorption of dyes from wastewater: a comprehensive review. ChemBioEng Rev. **10**(3), 326–335 (2023). https://doi.org/10.1002/cben.202200011
5. Dutta, S., Gupta, B., Srivastava, S.K., Gupta, A.K.: Recent advances on the removal of dyes from wastewater using various adsorbents: a critical review. Mater. Adv. **2**(14), 4497–4531 (2021). https://doi.org/10.1039/d1ma00354b
6. Jadhav, A.C., Jadhav, N.C.: Treatment of textile wastewater using adsorption and adsorbents (January) (2021). https://doi.org/10.1016/b978-0-323-85829-8.00008-0

7. Bhardwaj, D., Bharadvaja, N.: Phycoremediation of effluents containing dyes and its prospects for value-added products: a review of opportunities. J. Water Process Eng. **41**(February), 102080 (2021). https://doi.org/10.1016/j.jwpe.2021.102080
8. Abdoul, H.J., et al.: Efficient adsorption of bulky reactive dyes from water using sustainably-derived mesoporous carbons. Environ. Res.**221**, 115254 (2023). https://doi.org/10.1016/j.envres.2023.115254
9. Rapo, E., Tonk, S.: Factors affecting synthetic dye adsorption; desorption studies : a review of results from the last five years. Molecules **26**, 5419–5450 (2021)
10. Mo, J., Yang, Q., Zhang, N., Zhang, W., Zheng, Y., Zhang, Z.: A review on agro-industrial waste (AIW) derived adsorbents for water and wastewater treatment. J. Environ. Manage. **227**(August), 395–405 (2018). https://doi.org/10.1016/j.jenvman.2018.08.069
11. Benvenuti, J., Fisch, A., Dos Santos, J.H.Z., Gutterres, M.: Silica-based adsorbent material with grape bagasse encapsulated by the sol-gel method for the adsorption of basic blue 41 dye. J. Environ. Chem. Eng. **7**(5), 103342 (2019). https://doi.org/10.1016/j.jece.2019.103342
12. Ahmed, M.A., Ahmed, M.A., Mohamed, A.A.: Synthesis, characterization and application of chitosan/graphene oxide/copper ferrite nanocomposite for the adsorptive removal of anionic and cationic dyes from wastewater. RSC Adv. **13**(8), 5337–5352 (2023). https://doi.org/10.1039/d2ra07883j
13. Qin, Y., et al.: Preparation and interaction mechanism of Nano disperse dye using hydroxypropyl sulfonated lignin. Int. J. Biol. Macromol. **152**, 280–287 (2020). https://doi.org/10.1016/j.ijbiomac.2020.02.261
14. López-Rodríguez, D., Micó-Vicent, B., Jordán-Núñez, J., Montava-Seguí, I., Bou-Belda, E.: Complete desorption of hybrid nanoclays composed of hydrotalcite and disperse Dye. Int. J. Mol. Sci. **24**(13), 10950 (2023). https://doi.org/10.3390/ijms241310950
15. Chaari, I., Medhioub, M., Jamoussi, F., Hamzaoui, A.H.: Acid-treated clay materials (Southwestern Tunisia) for removing sodium leuco-vat dye: characterization, adsorption study and activation mechanism. J. Mol. Struct. **1223**, 128944 (2020). https://doi.org/10.1016/j.molstruc.2020.128944
16. Mohammad Razi, M.A., Mohd Hishammudin, M.N.A., Hamdan, R.: Factor affecting textile dye removal using adsorbent from activated carbon: a review. MATEC Web Conf. **103**, 1–17 (2017). https://doi.org/10.1051/matecconf/201710306015
17. Chakraborty, R., Asthana, A., Singh, A.K., Jain, B., Susan, A.B.H.: Adsorption of heavy metal ions by various low-cost adsorbents: a review. Int. J. Environ. Anal. Chem. **102**(2), 342–379 (2022). https://doi.org/10.1080/03067319.2020.1722811
18. De Gisi, S., Lofrano, G., Grassi, M., Notarnicola, M.: Characteristics and adsorption capacities of low-cost sorbents for wastewater treatment: a review. Sustain. Mater. Technol. **9**, 10–40 (2016). https://doi.org/10.1016/j.susmat.2016.06.002
19. Kuleyin, A.: Removal of phenol and 4-chlorophenol by surfactant-modified natural zeolite. J. Hazard. Mater. **144**(1–2), 307–315 (2007). https://doi.org/10.1016/j.jhazmat.2006.10.036
20. Wei, Y.C., Hudson, S.M., Mayer, J.M., Kaplan, D.L.: The crosslinking of chitosan fibers. J. Polym. Sci. Part A Polym. Chem. **30**(10), 2187–2193 (1992)
21. Errais, E., Duplay, J., Darragi, F.: Textile dye removal by natural clay—case study of Fouchana Tunisian clay. Environ. Technol. **31**(4), 373–380 (2010). https://doi.org/10.1080/09593330903480080
22. Morais, L.C., Freitas, O.M., Gonçalves, E.P., Vasconcelos, L.T., González Beça, C.G.: Reactive dyes removal from wastewaters by adsorption on eucalyptus bark: Variables that define the process. Water Res. **33**(4), 979–988 (1999). https://doi.org/10.1016/S0043-1354(98)00294-2
23. Tasaso, P.: Adsorption of copper using pomelo peel and depectinated pomelo peel. J. Clean Energy Technol. **2**(2), 154–157 (2014). https://doi.org/10.7763/jocet.2014.v2.112
24. Dakiky, M., Khamis, M., Manassra, A., Mer'eb, M.: Selective adsorption of chromium(VI) in industrial wastewater using low-cost abundantly available adsorbents. Adv. Environ. Res. **6**(4), 533–540 (2002). https://doi.org/10.1016/S1093-0191(01)00079-X
25. Rangabhashiyam, S., Anu, N., Selvaraju, N.: Sequestration of dye from textile industry wastewater using agricultural waste products as adsorbents. J. Environ. Chem. Eng. **1**(4), 629–641 (2013). https://doi.org/10.1016/j.jece.2013.07.014

26. Benkhaya, S., M' rabet, S., El Harfi, A.: A review on classifications, recent synthesis and applications of textile dyes. Inorg. Chem. Commun. **115**, 107891 (2020). https://doi.org/10.1016/j.inoche.2020.107891
27. Bhatnagar, A., Sillanpää, M.: Utilization of agro-industrial and municipal waste materials as potential adsorbents for water treatment-a review. Chem. Eng. J. **157**(2–3), 277–296 (2010). https://doi.org/10.1016/j.cej.2010.01.007
28. Kumar, S., Kumar, R., Bandopadhyay, A.: Innovative methodologies for the utilisation of wastes from metallurgical and allied industries. Resour. Conserv. Recycl. **48**(4), 301–314 (2006). https://doi.org/10.1016/j.resconrec.2006.03.003
29. Kyzas, G.Z., Matis, K.A.: Nanoadsorbents for pollutants removal: a review. J. Mol. Liq. **203**(2014), 159–168 (2015). https://doi.org/10.1016/j.molliq.2015.01.004
30. Wan Ngah, W.S., Teong, L.C., Hanafiah, M.A.K.M.: Adsorption of dyes and heavy metal ions by chitosan composites: a review. Carbohydr. Polym. **83**(4), 1446–1456 (2011). https://doi.org/10.1016/j.carbpol.2010.11.004
31. Lata, S., Samadder, S.R.: Removal of arsenic from water using nano adsorbents and challenges: a review. J. Environ. Manage. **166**, 387–406 (2016). https://doi.org/10.1016/j.jenvman.2015.10.039
32. Kefeni, K.K., Mamba, B.B.: Photocatalytic application of spinel ferrite nanoparticles and nanocomposites in wastewater treatment: review. Sustain. Mater. Technol. **23**, e00140 (2020). https://doi.org/10.1016/j.susmat.2019.e00140
33. Meng, X.L., et al.: Simple synthesis of mesoporous boron nitride with strong cathodoluminescence emission. J. Solid State Chem. **184**(4), 859–862 (2011). https://doi.org/10.1016/j.jssc.2011.01.040
34. Crini, G.: Non-conventional low-cost adsorbents for dye removal: a review. Bioresour. Technol. **97**(9), 1061–1085 (2006). https://doi.org/10.1016/j.biortech.2005.05.001
35. Shojaei, S., Shojaei, S., Band, S.S.: Application of chemometrics into removal of dyes by NaX nanozeolites: simultaneous model. SSRN Electron. J. (2021). https://doi.org/10.2139/ssrn.3809612
36. Jiang, F., Dinh, D.M., Lo Hsieh, Y.: Adsorption and desorption of cationic malachite green dye on cellulose nanofibril aerogels. Carbohydr. Polym. **173**, 286–294 (2017). https://doi.org/10.1016/j.carbpol.2017.05.097
37. Bhatti, H.N., Safa, Y., Yakout, S.M., Shair, O.H., Iqbal, M., Nazir, A.: Efficient removal of dyes using carboxymethyl cellulose/alginate/polyvinyl alcohol/rice husk composite: adsorption/desorption, kinetics and recycling studies. Int. J. Biol. Macromol. **150**, 861–870 (2020). https://doi.org/10.1016/j.ijbiomac.2020.02.093
38. Şentürk, İ, Alzein, M.: Adsorption of acid violet 17 onto acid-activated pistachio shell: isotherm, kinetic and thermodynamic studies. Acta Chim. Slov. **67**(1), 55–69 (2020). https://doi.org/10.17344/acsi.2019.5195
39. Al-Ghouti, M.A., Al-Absi, R.S.: Mechanistic understanding of the adsorption and thermodynamic aspects of cationic methylene blue dye onto cellulosic olive stones biomass from wastewater. Sci. Rep. **10**(1), 1–18 (2020). https://doi.org/10.1038/s41598-020-72996-3
40. Momina, M.S., Suzylawati, I.: Study of the adsorption/desorption of MB dye solution using bentonite adsorbent coating. J. Water Process Eng. **34**(July 2019), 2020. https://doi.org/10.1016/j.jwpe.2020.101155

Two-Dimensional Carbon-Based Materials for Sorption of Selected Aromatic Compounds in Water

Adedapo O. Adeola and Patricia B. C. Forbes

Abstract The availability of clean water is of pressing concern in developing countries and has been a key area of focus for research and development worldwide. The sixth Sustainable Development Goal of the United Nations emphasizes the need for clean water. Aromatic hydrocarbons are emerging organic contaminants that are being found frequently in drinking water, municipal wastewater, and surface water. Conventional wastewater treatment plants have been shown to have limited efficiency in removing these trace pollutants from water. Due to advantages including cost, effectiveness, simplicity of use, and reusability, the adsorption process is recognized as a promising water remediation technology for aromatic compound removal. Advanced carbon-based materials discussed in this chapter are two-dimensional materials, such as graphene and carbon nanotubes, and their composites. Their performance towards remediating monocyclic aromatic hydrocarbons (MAHs) and polycyclic aromatic hydrocarbons (PAHs) was explored using recent published results between 2007 and 2022 from reputable sources. The concluding section of the chapter presents recommendations for bridging knowledge gaps, as well as suggestions for future research direction.

Keywords Aromatic compounds · Adsorbents · Graphene · Carbon nanotubes · Water treatment

---

A. O. Adeola
Department of Chemical Sciences, Adekunle Ajasin University, Ondo State, Akungba-Akoko, Nigeria
e-mail: adedapo.adeola@aaua.edu.ng

P. B. C. Forbes (✉)
Department of Chemistry, Faculty of Natural and Agricultural Sciences, University of Pretoria, Lynnwood Road, Hatfield, Pretoria 0002, South Africa
e-mail: patricia.forbes@up.ac.za

J. Tharini and S. Thomas (eds.), *Carbon Nanomaterials and their Composites as Adsorbents*, Carbon Nanostructures, https://doi.org/10.1007/978-3-031-48719-4_14

1 Introduction

The cyclic ring structure of aromatic compounds alternates between double and single bonds. The presence of at least one planar carbon ring and the alternation of double and single bonds in the carbon ring are the two main requirements for classifying a compound as aromatic [1, 2]. In industries, aromatic compounds are widely employed in the manufacture of pharmaceuticals, plastics, insecticides, etc. These are chemicals that can have a negative impact on all forms of life, particularly in terms of human health [3, 4].

Adsorption methods using various carbon-based materials are often used to decontaminate water containing aromatic chemicals. As a result, numerous research studies have been conducted to better understand the adsorption mechanisms involved in these processes [5–10]. Carbon-based adsorbents such as graphene, carbon nanotubes, and their derivatives, are frequently utilized as the effective adsorbents for aromatic pollutants in water due to their tunable surface functional groups and relatively large surface areas [11]. The adsorbent type, sorbate properties and hydrochemistry (such as pH, temperature, ionic strength) may influence the adsorption capacity of carbonaceous materials [6, 9]. The four primary components of the surface of carbon materials are the basal planes, crystal edges and defects, ash impurities, and surface groups containing oxygen. These surface groups are often found around the graphitic basal planes' edges. There are two types of surface functional groups: basic groups (pyrone- and chromene-like structures) and acidic groups (carbonyl, carboxyl, phenolic, lactone, hydroxyl and anhydride) [12].

Researchers have attempted to comprehend how the various aromatic ring substituents influence the sorption and desorption processes, since aromatic molecules are ubiquitous water pollutants [13]. The solubility of the molecule, the density of electrons in the aromatic ring, and pKa values have all been reported to impact the adsorption process [7, 14]. There is strong evidence supporting the importance of $\pi - \pi$ interaction mechanisms, although the nature of the driving factors for the adsorption of aromatic compounds is still being elucidated. Through $\pi - \pi$ interactions between the electrons of the aromatic ring and the electrons of the graphene layers, aromatic compounds (as neutral molecules) are adsorbed on the carbon surface [15]. The charge of phenolic aromatic compounds can be impacted by the pH of the solution, which in turn affects the electrostatic interactions between the adsorbent and the adsorbate [7].

This chapter addresses the application of various 2D carbon-based adsorbents for the sorption of aromatic compounds in water, highlighting challenges, merits and prospects.

2 Monocyclic and Polycyclic Aromatic Hydrocarbons (MAHs and PAHs)

Monocyclic aromatic hydrocarbons (MAHs) are made up of one aromatic ring, while polycyclic aromatic hydrocarbons (PAHs) are a fusion of two or more aromatic rings. Benzene, toluene, ethylbenzene, and xylenes (BTEX) are the most water-soluble aromatic hydrocarbons and are well-known environmental contaminants [16]. On the other hand, there are hundreds of distinct PAHs and their derivatives, but the United States Environmental Protection Agency (US EPA) has designated 16 of them as priority pollutants [6, 17]. They are frequently discovered in water bodies where there are nearby wood/coal processing, gas production, and crude oil exploration activities. Humans may be exposed to MAHs and PAHs through a variety of pathways, including air, water, food, and work-related exposures. However, contaminated water is one of the main routes of exposure [18, 19]. Due to the toxicological implications of PAH-contaminated water, several studies have been carried out towards the remediation of PAH-polluted water globally based on adsorptive technologies.

3 Adsorption of Selected MAHs and PAHs Using Graphene-Based Materials

In 1995, the International Union of Pure and Applied Chemistry (IUPAC) adopted the term "graphene," which had already been proposed in 1986. Two-dimensional (2D) graphene has hexagonally organized sp^2 hybridized carbon atoms in a close-packed crystal lattice structure that contains σ- and π-bonds [20]. Graphene has been synthesized in a variety of forms, including pristine graphene, graphene nanoshells, graphene oxide (GO), reduced graphene oxide (RGO), graphene quantum dots (GQDs), graphene wool (GW), graphene-based composites, and functionalized graphene [5, 21, 22]. Figure 1 presents a schematic illustration of the preparation of graphene-based materials with graphite as a precursor. Numerous scientific fields use graphene and its composites for a myriad of applications due to their thermal stability, large specific surface area, high tensile strength, thermal conductivity, chemical resistance, flexibility, tunability and charge mobility [6, 23]. For water remediation objectives, such as the removal of MAHs and PAHs from water, graphene also serves as an effective adsorbent.

In general, the interaction mechanisms between MAHs, PAHs, and various forms of graphene in water follows second-order reaction kinetics (chemisorption) and are predominantly defined by partitioning and adsorption [6]. It is clear from the investigations listed in Table 1 that the dosage, PAH concentration, reaction temperature, and solution ionic strength all have a significant influence on the efficiency of adsorption by various types of graphene. The absence of specific moieties in PAHs is responsible for the minimal effect of pH variation [25, 26]. A study carried out on MAHs by Zhou and Zhang [27], employed a density-functional tight-binding approach

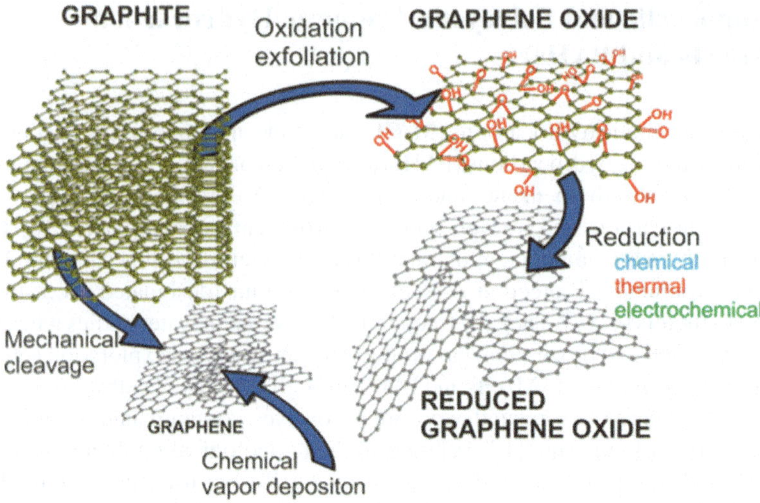

Fig. 1 Representation of potential methods for synthesizing graphene and reduced graphene oxide. Adapted with modification from Rowley-Neale et al. [24]

to examine the adsorption of benzene derivatives containing various substituents onto graphene. The aromatic compounds with either an electrophile or an electron-donating substituent showed stronger adsorption than benzene. Additionally, the strength of the adsorption depends on the magnitude of steric effects, and the electron transfer from graphene is influenced by the direction and quantity of H-atoms in the substituent. The stereo-electronic impact of the combination of substituent and the ringed structure of benzene has a significant impact on the region where the molecule interacts with graphene [27].

Due to the elimination of PAHs with a removal efficiency > 98%, as well as the flexibility and reusability associated with graphene wool (GW), it has a competitive edge over other graphene forms reported in Table 1. GW may be an excellent filtration medium and a useful polishing tool for water treatment because of its low density and high porosity.

The removal of benzene, toluene, ethylbenzene, and xylene using adsorption methods was assessed in two studies (Table 1). Pourmand et al. [28] studied the adsorption of BTX by nanoporous graphene, while Azizi et al. [29]evaluated the adsorption of TEX by graphene oxide. Graphene was not as effective as nanoporous graphene at eliminating these monocyclic aromatic compounds. The surface area of graphene, which was reported to be 410.99 m^2/g, was nearly 30 times larger than that of GO (12.41 m^2/g), while its total pore volume, which was reported to be 1.07 cm^3/g, was roughly eight times larger than that of GO (0.13 cm^3/g). However, the size of graphene pores (45–63 nm) was comparable to that of GO pores (43 nm). According to the analysis of the material, the adsorption maxima (q_{max}) of graphene for xylene and toluene was more than the adsorption maxima (q_{max}) of graphene oxide for these compounds. Graphene-based materials are emerging nanomaterials,

Table 1 Comparison of different forms of graphene used for the removal of aromatic compounds in contaminated water

Adsorbent	Removal efficiency (%)	Adsorption capacity	References
Graphene	NR	1460 mg/g (1-napthalenesulfonic acid)	Wu et al. [26]
Graphene	NR	206.00 mg/g (1-naphthylamine), 167 mg/g (naphthalene), 152.82 mg/g (1-naphthol)	Ji et al. [30]
Graphene nanosheets Graphene oxide Graphene SWNT Graphene MWNT	NR	150.2 mg/g (phenanthrene), 139.0 mg/g (biphenyl) 127.4 mg/g (phenanthrene), 92.9 mg/g (biphenyl) 202.0 mg/g (phenanthrene), 104.9 mg/g (biphenyl) 37.1 mg/g (phenanthrene), 915.5 mg/g (biphenyl)	Apul et al. [31]
Reduced graphene oxide (RGO)	NR	5912 mg/g (naphthalene), 183 mg/g (anthracene), 979 mg/g (pyrene)	Sun et al. [32]
RGO/FeO.Fe_3O_4 composite	NR	337.1 mg/g (naphthalene), 408.1 mg/g (1-naphthylamine), 389.3 mg/g (1-naphthol)	Yang et al. [33]
Graphene oxide/ brilliant blue (BBGO)	72.7 – 93.2	349.0 mg/g (anthracenemethanol), 447.4 mmol/g (fluoranthene)	Zhang et al. [34]
Graphene nanosheets	–	126.0 mg/g (naphthalene), 116 mg/g (phenanthrene), 123 mg/g (pyrene)	Wang et al. [35]
Graphene oxide (GO)	NR	3.67 mg/g (naphthalene), 5.9 mg/g (phenanthrene), 6.12 mg/g (pyrene)	
Graphene	NR	24.1 mg/g (phenanthrene)	Zhao et al. [36]
Graphene-coated materials (GCMs)	80	1.74 mg/g (phenanthrene)	Yang et al. [37]
Graphene wool	98.5–99.9	5 mg/g (phenanthrene), 20 mg/g (pyrene)	Adeola and Forbes [15]
GO-iron oxide	NR	285.7 mg/g (1-naphthylamine), 228.4 mg/g (1-naphthol)	Yang et al. [38]

(continued)

Table 1 (continued)

Adsorbent	Removal efficiency (%)	Adsorption capacity	References
RGO-iron oxide	NR	303.0 mg/g (1-naphthylamine), 243.2 mg/g (1-naphthol)	
Graphene oxide	NR	11.3 mg/g (toluene), 13.6 mg/g (ethylbenzene), 14.0 mg/g (xylene)	Azizi et al. [29]
Nanoporous graphene	NR	118.8 mg/g (benzene), 123.5 mg/g (toluene), 125.4 mg/g (xylene)	Pourmand et al. [28]
N-Doped RGO	40–90	5.77 mg/g (anthracene), 9.29 mg/g (2-methylanthraquinone)	Song et al. [23]

NR: Not Reported

as a result, a methodical and scientific approach must be used to better understand their stability, durability, mechanical qualities, and toxicity.

4 Adsorption of Selected MAHs and PAHs Using Carbon Nanotubes

Carbon nanotubes (CNTs) are a 2D nanomaterial with cylindrically-shaped hollow graphitic layers. Multi-walled and single-walled carbon nanotubes (MWCNTs and SWCNTs, repectively) are the two forms of carbon nanotubes that are widely discussed (Fig. 2) [39]. Numerous techniques for creating carbon nanotubes have been investigated since their discovery in 1991 (MWCNTs) and 1993 (SWCNTs), respectively [40]. SWCNTs, double-walled carbon nanotubes (DWCNTs), and MWCNTs have been made using arc-evaporation of graphite electrodes, pyrolysis of organic molecular precursors, laser ablation, electrochemical methods, vapor phase decomposition and chemical vapor deposition (CVD) of carbon-containing molecules [41, 42]. CNTs contain contaminants and may cause harm when utilized in their pristine form (after preparation). However, their toxicity may be significantly reduced when they are purified and surface-functionalized, making them suitable substrates for various applications. Additionally, it has been demonstrated that continuous CNT fibers offer a suitable means of avoiding the potential issue of CNT leaching [40].

The physicochemical properties and structures of selected monocyclic aromatic compounds (MAH) are presented in Table 2. Lu et al. [43] reported the fabrication of CNTs using the catalytic chemical vapor deposition technique and its application as an adsorbent for the removal of selected monocyclic aromatic compounds

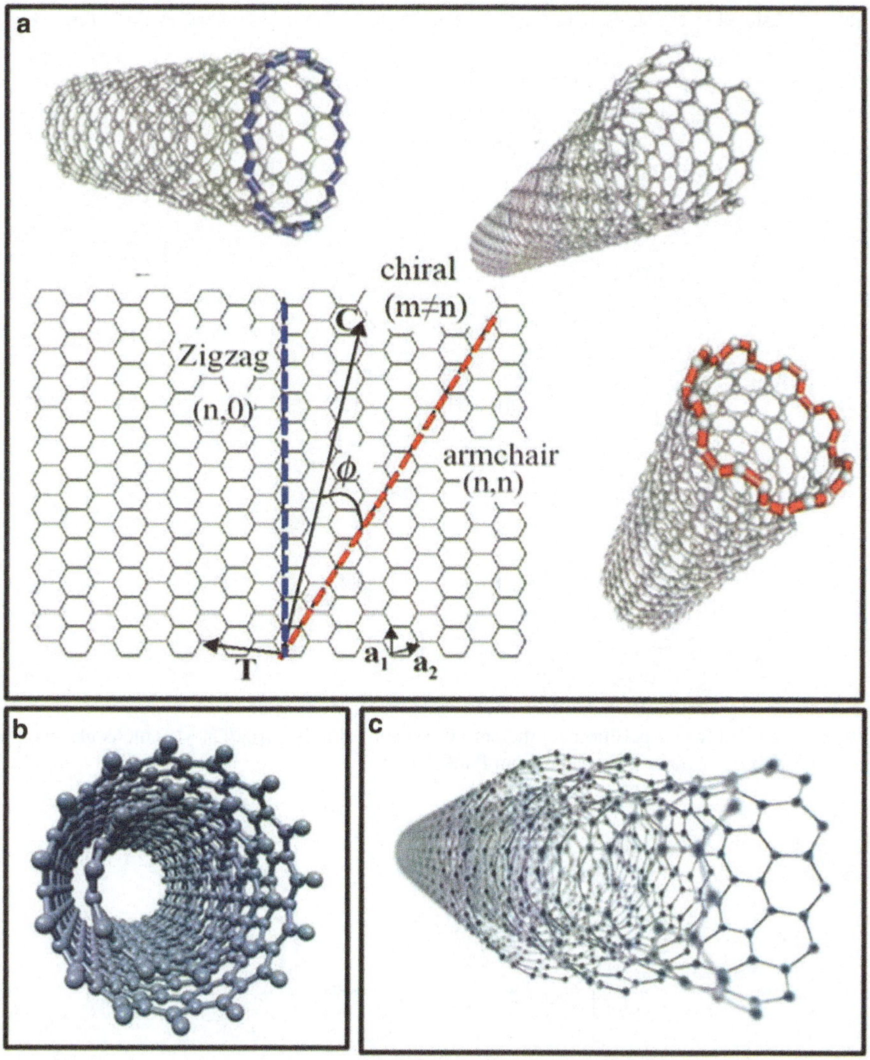

Fig. 2 Structure of carbon nanotubes according to the number of walls. (A) SWCNTs according to their chirality (zigzag, armchair, and chiral), (B) DWCNTs and (C) MWCNTs with multiple overlapping shells. Adapted from Tilmaciu and Morris [40]

(benzene, toluene, ethylbenzene, and p-xylene) in contaminated water (Fig. 3). Pristine CNTs were modified using oxidation with H_2SO_4, HNO_3, and NaOCl, and the surface composition of CNTs was altered, resulting in CNTs that adsorb more BTEX. Following the nitric acid-oxidized CNTs and the sulfuric acid-oxidized CNTs, the NaOCl-oxidized CNTs exhibited the largest improvement in BTEX adsorption. The interactivity between the surface carboxylic groups of CNTs and the aromatic ring

Table 2 Basic physicochemical properties of the selected monocyclic aromatic hydrocarbons (MAHs)

Hydrocarbon	Molecular structure	Molecular formula	$LogK_{ow}$	S_w	M_w	B_p
Benzene		C_6H_6	2.17	1790	78.11	80
Ethylbenzene	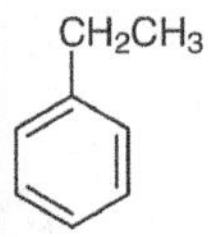	$C_6H_5C_2H_5$	3.15	170	106.16	136
p-Xylene	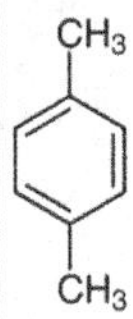	$C_6H_4(CH_3)_2$	3.15	162	106.16	138.3
Toluene	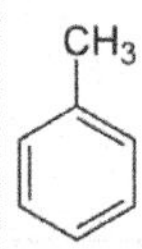	$C_6H_5CH_3$	2.69	526	92.14	110.6

Log K_{ow}: octanol–water partition coefficient, S_w: water solubility (mg/L), M_{w}: molecular weight (g/mol), B_p: boiling point (°C). Cited from PubChem

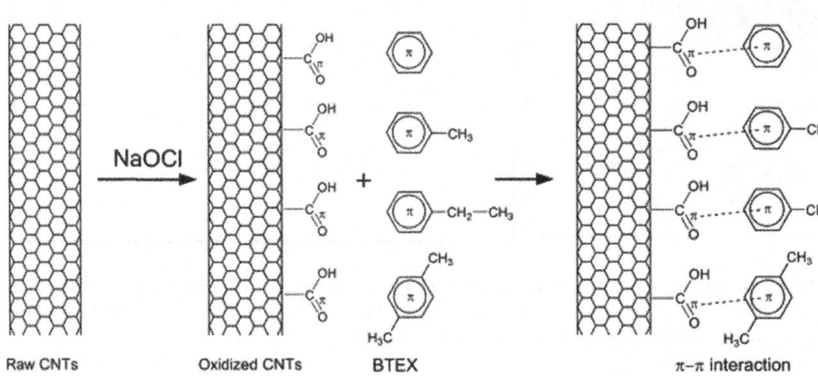

Fig. 3 Illustration of adsorption of BTEX onto oxidized CNTs. Adapted with permission Lu et al. [43], Copyright 2008, Elsevier

of BTEX is considered to be primarily responsible for the adsorption mechanism of BTEX via CNTs. The NaOCl-oxidized CNTs out performed several different types of carbon and silica based adsorbents [43].

In a recent study, the role of humic acid (used as a model for natural organic matter) on the ability of MWCNTs to adsorb selected aromatic compounds was investigated [44]. In batch adsorption tests, toluene and ethylbenzene adsorption from aqueous solution onto multi-walled carbon nanotubes in the presence of various humic acid concentrations were evaluated. According to the findings, ethylbenzene and toluene had the highest MWCNT adsorption rates of 72 mg/g and 35 mg/g, respectively, in an aqueous solution devoid of humic acid. The two main models of the adsorption process were the pseudo-second-order kinetic model and the Langmuir isotherm model. When studies were conducted using MWCNT pre-loaded with humic acid from 0 to 30%, the adsorption capacity of MWCNTs decreased from 14 mg/g to 8 mg/g for toluene and from 25 mg/g to 10 mg/g for ethyl benzene [44]. This decrease in adsorption capacity as a result of the presence of NOM is similar to what was reported by Adeola and Forbes [13], for pyrene interaction with graphene wool in the presence of NOM. Thus the presence of particulate or dissolved NOM in water often negatively impacts the adsorption capacity of carbon-based materials. Table 3 presents a summary of various CNTs utilized for the removal of MAHs.

As a result of superior sorption potential and ease of retrieval, magnetic carbon nanotubes have attracted significant interest as adsorbents for the treatment of water containing organic contaminants [50]. Three main groups of CNT-based magnetic composites have been identified: (i) CNT-metal nanocomposites, (ii) CNT-metal oxide nanocomposites, and (iii) CNT-mixed metal oxide nanocomposites. Iron and

Table 3 Comparison of different forms of carbon nanotubes used for remediation of selected aromatic compounds in contaminated water

Carbon nanotubes	Maximum adsorption capacity, q_e (mg/g)				References
	Benzene	Toluene	Ethylbenzene	p-Xylene	
MWCNT (NaOCl)	36.2				Chen et al. [45]
SWCNT (NaOCl)	60.1	103.2			Wibowo et al. [46]
SWCNT (HNO_3)				85.5	Chin et al. [47]
SWCNT				77.5	Chin et al. [47]
MWCNT		31.15	54.94		Abedi et al. [44]
MWCNT (H_2SO_4)	42.6	36.7		78.4	Lu et al. [43]
MWCNT (HNO_3)	105.7	160.8		108.9	Lu et al. [43]
MWCNT (HCl)	45.0	60.0		148.0	Lu et al. [43]
MWCNT (NaOCl)	247.87	279.81		413.8	Lu et al. [43]
MWCNT (KOH)	99.52	339.37		278.28	Yu et al. [48]
MWCNT (NaOCl)		99.47		112.19	Yu et al. [48]
MWCNT		23.28	92.71	44.15	Yu et al. [48]
MWCNT (O_3 and NaOCl)	16.6				Anjum et al. [49]
MWCNT (HNO_3 and NaOCl)		59.48		103.40	Anjum et al. [49]

iron oxide nanoparticles are the main contributors to magnetism in these nanocomposites. Even CNT-based composites without any additional metal or metal oxide components show magnetic properties [51], as a result of the inclusion of metal catalysts during CNT production (i.e., iron) to catalyse the chemical vapour deposition (CVD) process used to synthesize CNTs [52].

The adsorption of low molecular weight aromatic compounds from water has been accomplished using iron-impregnated MWCNTs with Fe_3O_4 adsorbed on the exterior walls of the MWCNTs [51]. Due to the inclusion of Fe_3O_4 nanoparticles, the composite has greater adsorption effectiveness than pure MWCNTs. Polycyclic aromatic hydrocarbons (PAHs) have been sorbed from aqueous solution using a magnetic N-doped CNT nanocomposite. To develop undoped carbon nanotubes, the nanomaterial was made utilizing (i) ethylene and iron catalyst, and (ii) acetonitrile and nitrogen. These procedures result in CNTs that contain both N-doped and undoped components [51, 53]. These N-doped amphiphilic CNTs exhibit great effectiveness for the sorption of various PAH pollutants due to their extremely hydrophobic surface and they can easily diffuse in any aqueous matrix because of the N-containing hydrophilic regions (Fig. 4). After adsorption, a straightforward magnetic separation can be used to separate the CNTs from the medium with ease. These magnetic CNTs demonstrated greater extraction efficiency for the ten low molecular weight PAHs tested, as compared to conventional PDMS fibers [54]. The potential to adjust the surface characteristics of the CNTs by regulating the doping process allows for the development of specialized materials for the adsorption of numerous additional organic contaminants. This is arguably the most interesting aspect of this field of research and opens the door to promising applications.

5 Conclusion and Recommendations

Graphene- and carbon nanotube-based materials are useful adsorbents for the elimination of aromatic organic compounds in polluted water bodies. To effectively deploy these materials on a full scale for environmental remediation, there is still much work that has to be done. Many of the reported studies did not account for naturally occurring substances found in water bodies, such as natural organic matter (NOM) and metals in surface waters, while assessing sorption processes. Rather, they were restricted to studying how adsorbent physicochemical properties influence the sorption performance. According to a few studies, the presence of NOM and metals dramatically affects the adsorption efficiency obtained for targeted aromatic organic compounds due to competitive sorption onto the sorbent's active sites. Therefore, future studies should take into account a deeper investigation as to how these competitive processes may affect adsorption efficiencies.

Physicochemical aspects and solution chemistry of aqueous media, which includes pH, adsorbent dosage, initial contaminant concentration, and temperature, among others, require in-depth analysis. Statistical analysis and the design of trials for process optimization are necessary for method validation. Future research should

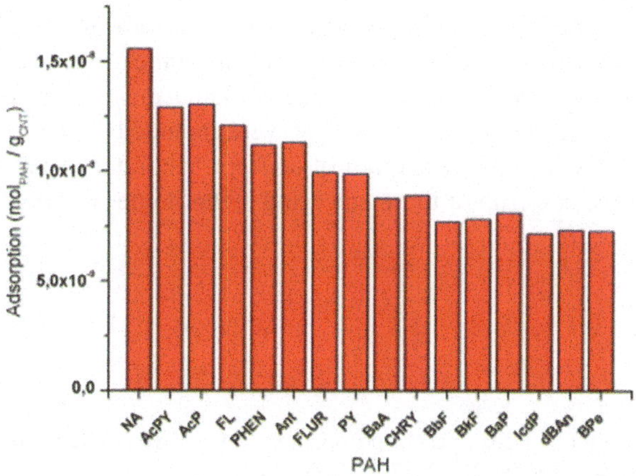

Fig. 4 Adsorption of 16 polycyclic aromatic hydrocarbons by carbon nanotubes. NA = Naphthalene, AcPY = Acenaphthene, AcP = Acenaphthylene, FL = Fluorene, PHEN = Phenanthrene, Ant = Anthracene, FLUR = Fluoranthene, PY = Pyrene, BaA = Benzo[a]anthracene, CHRY = Chrysene, BbF = Benzo[b]fluoranthene, BkF = Benzo[k]fluoranthene, BaP = Benzo[a]pyrene, IcdP = Indeno[1,2,3-cd]pyrene, dBAn = Dibenz[a,h]anthracene, BPe = Benzo[g,h,i]perylene. Adapted with permission from Menezes et al. [54], copyright 2015 Elsevier

use statistical methods for process optimization to assess the relative significance of specific variables and both cooperative and competitive interactions. More in-depth research is required in connection with aromatic compounds to comprehend how variation in molecular structure may influence the sorption processes.

In some cases, harmful chemicals are used in the production of adsorbents when it comes to the synthesis of graphene, carbon nanotubes, and their derivatives. Therefore, it is necessary to create eco-friendly and sustainable processes for making such carbon-based adsorbents. The development of commercially viable industrial processes for material preparation with specific features, such as a defined number of layers and structural quality, remains one of the primary challenges in the synthesis of graphene and carbon nanotubes.

To assess the toxicity of synthesized adsorbents, it is essential to be aware of the toxicity of the precursors for the synthesis of carbon-based materials. Without this understanding, procedures could be used that increase the potential threat to the environment and human health. Furthermore, there is currently limited knowledge on the biocompatibility and ecotoxicity of graphene and carbon nanotubes, and none of the research reported investigated the toxicity of adsorbents with respect to biological organisms. Therefore, in order to fill these knowledge gaps, future research should concentrate on evaluating the ecotoxicity of graphene and carbon nanotubes, and the potential impact of their application in water treatment on the larger ecosystem.

One major consideration when deciding on the technology to be utilized is the cost of carrying out environmental restoration operations. Sorbents produced from cheap materials, but with high removal efficiency are generally always chosen because they improve the cost and efficiency of the water treatment process. To ensure the economic sustainability of using these adsorbents in wastewater treatment plants or household water purification, a cost–benefit analysis of the usage of advanced carbon nanomaterials for the removal of aromatic contaminants must be carried out. It should be emphasized that none of the studies that were included in this chapter reported on operational costs, as they were proof of concept, laboratory-scale investigations. The viability of the scale-up of these emerging technologies has thus not been fully elucidated.

References

1. Honda, M., Suzuki, N.: Toxicities of polycyclic aromatic hydrocarbons for aquatic animals. Int. J. Environ. Res. Public Health **17**(4), 1363 (2020)
2. Furukawa, S., et al.: Double aromaticity arising from σ- and π-rings. Commun. Chem. **1**(1), 60 (2018)
3. Ololade, I.A., et al.: Bioavailability of polycyclic aromatic hydrocarbons (PAHs) and Environmental Risk (ER) Assessment: the case of the Ogbese river, Nigeria. Reg. Stud. Mar. Sci. **9**, 9–16 (2017)
4. Adeola, A.O., Forbes, P.B.C.: Antiretroviral drugs in african surface waters: prevalence, analysis, and potential remediation. Environ. Toxicol. Chem. **42**(2), 247–262 (2021)
5. Adeola, A.O., et al.: Facile synthesis of graphene wool doped with oleylamine-capped silver nanoparticles (GW-αAgNPs) for water treatment applications. Appl. Water Sci. **11**(11), 172 (2021)
6. Adeola, A.O., Forbes, P.B.C.: Advances in water treatment technologies for removal of polycyclic aromatic hydrocarbons: existing concepts, emerging trends, and future prospects. Water Environ. Res. **93**(3), 343–395 (2021)
7. Adeola, A.O., de Lange, J., Forbes, P.B.C.: Adsorption of antiretroviral drugs, efavirenz and nevirapine from aqueous solution by graphene wool: kinetic, equilibrium, thermodynamic and computational studies. Appl. Surf. Sci. Adv. **6**, 100157 (2021)
8. Adeola, A.O., Forbes, P.B.C.: Assessment of reusable graphene wool adsorbent for the simultaneous removal of selected 2–6 ringed polycyclic aromatic hydrocarbons from aqueous solution. Environ. Technol. **43**(8), 1255–1268 (2022)
9. Pathak, S., et al.: A state-of-the-art review of various adsorption media employed for the removal of toxic Polycyclic aromatic hydrocarbons (PAHs): An approach towards a cleaner environment. J. Water Proc. Eng. **47**, 102674 (2022)
10. Ilyas, M., Ahmad, W., Khan, H.: Utilization of activated carbon derived from waste plastic for decontamination of polycyclic aromatic hydrocarbons laden wastewater. Water Sci. Technol. **84**(3), 609–631 (2021)
11. Sabzehmeidani, M.M., et al.: Carbon based materials: a review of adsorbents for inorganic and organic compounds. Mater. Adv. **2**(2), 598–627 (2021)
12. Rehman, A., et al.: Current progress on the surface chemical modification of carbonaceous materials. Coatings **9**, 103 (2019)
13. Adeola, A.O., Forbes, P.B.C.: Influence of natural organic matter fractions on PAH sorption by stream sediments and a synthetic graphene wool adsorbent. Environ. Technol. Innov. **21**, 101202 (2021)
14. Ololade, I.A., et al.: In-situ modification of soil organic matter towards adsorption and desorption of phenol and its chlorinated derivatives. J. Environ. Chem. Eng. **6**(2), 3485–3494 (2018)
15. Adeola, A.O., Forbes, P.B.C.: Optimization of the sorption of selected polycyclic aromatic hydrocarbons by regenerable graphene wool. Water Sci. Technol. **80**(10), 1931–1943 (2018)

16. Astrahan, P.: Monocyclic aromatic hydrocarbons (phthalates and BTEX) and aliphatic components in the SE Mediterranean costal Sea-surface microlayer (SML): Origins and phase distribution analysis. Mar. Chem. **205**, 56–69 (2018)
17. USEPA.: U.S. Environmental Protection Agency Office of the Science Advisor Risk Assessment Forum. Framework for Human Health Risk Assessment to Inform Decision Making, April 5, 2012 (2014)
18. Abdel-Shafy, H.I., Mansour, M.S.M.: A review on polycyclic aromatic hydrocarbons: source, environmental impact, effect on human health and remediation. Egypt. J. Pet. **25**(1), 107–123 (2016)
19. Maletic, S.P., et al.: State of the art and future challenges for polycyclic aromatic hydrocarbons is sediments: sources, fate, bioavailability and remediation techniques. J. Hazard. Mater. **365**, 67–482 (2019)
20. Ali, I., et al.: Graphene based adsorbents for remediation of noxious pollutants from wastewater. Environ. Int. **127**, 160–180 (2019)
21. Schoonraad, G., et al.: Synthesis and optimisation of a novel graphene wool material by atmospheric pressure chemical vapour deposition. J. Mater. Sci. **55**, 545–564 (2020)
22. Kubheka, G., Adeola, A.O., Forbes, P.B.C.: Hexadecylamine functionalised graphene quantum dots as suitable nano-adsorbents for phenanthrene removal from aqueous solution. RSC Adv. **12**(37), 23922–23936 (2022)
23. Song, T., Tian, W., Qiao, K., Zhao, J., Chu, M., Du, Z., et al.: Adsorption behaviors of polycyclic aromatic hydrocarbons and oxygen derivatives in wastewater on N-doped reduced graphene oxide. Sep. Purif. Technol. **254**, 117565 (2021)
24. Rowley-Neale, S.J., et al.: An overview of recent applications of reduced graphene oxide as a basis of electroanalytical sensing platforms. Appl. Mater. Today **10**, 218–226 (2018)
25. Lamichhane, S., Bal Krishna, K.C., Sarukkalige, R.: Polycyclic aromatic hydrocarbons (PAHs) removal by sorption: a review. Chemosphere **148**, 336–353 (2016)
26. Wu, T., et al.: Adsorption characteristics of acrylonitrile, p-toluenesulfonic acid, 1-naphthalenesulfonic acid and methyl blue on graphene in aqueous solutions. Chem. Eng. J. **173**(1), 144–149 (2011)
27. Zhou, P.P., Zhang, R.Q.: Physisorption of benzene derivatives on graphene: critical roles of steric and stereoelectronic effects of the substituent. Phys. Chem. Chem. Phys. **17**(18), 12185–12193 (2015)
28. Pourmand, S., Abdouss, M., Rashidi, A.M.: Preparation of nanoporous graphene via nanoporous zinc oxide and its application as a nanoadsorbent for benzene, toluene and xylenes removal. Int. J. Environ. Res. **9**(4), 1269–1276 (2015)
29. Azizi, A., et al.: Adsorption performance of modified graphene oxide nanoparticles for the removal of toluene, ethylbenzene, and xylenes from aqueous solution. Desalin. Water Treat. **57**(59), 28806–28821 (2016)
30. Ji, L., et al.: Graphene nanosheets and graphite oxide as promising adsorbents for removal of organic contaminants from aqueous solution. J. Environ. Qual. **42**(1), 191–198 (2013)
31. Apul, O.G., et al.: Adsorption of aromatic organic contaminants by graphene nanosheets: comparison with carbon nanotubes and activated carbon. Water Res. **47**(4), 1648–1654 (2013)
32. Sun, Y., et al.: Adsorption of polycyclic aromatic hydrocarbons on graphene oxides and reduced graphene oxides. Chem. Asian J. **8**(11), 2755–2761 (2013)
33. Yang, X., Li, J., et al.: Adsorption of naphthalene and its derivatives on magnetic graphene composites and the mechanism investigation. Colloids Surf. A Physicochem. Eng. Asp. **422**, 118–125 (2013)
34. Zhang, C., et al.: Adsorption of polycyclic aromatic hydrocarbons (Fluoranthene and Anthracenemethanol) by functional graphene oxide and removal by pH and temperature-sensitive coagulation. ACS Appl. Mater. Interf. **5**(11), 4783–4790 (2013)
35. Wang, J., Chen, Z., Chen, B.: Adsorption of polycyclic aromatic hydrocarbons by graphene and graphene oxide nanosheets. Environ. Sci. Technol. **48**(9), 4817–4825 (2014)
36. Zhao, J., et al.: Adsorption of phenanthrene on multilayer graphene as affected by surfactant and exfoliation. Environ. Sci. Technol. **48**(1), 331–339 (2014)

37. Yang, K., Chen, B., Zhu, L.: Graphene-coated materials using silica particles as a framework for highly efficient removal of aromatic pollutants in water. Sci. Rep. **5**, 11641 (2015)
38. Yang, X., et al.: Graphene oxide-iron oxide and reduced graphene oxide-iron oxide hybrid materials for the removal of organic and inorganic pollutants. RSC Adv. **2**(23), 8821–8826 (2012)
39. Xu, L., et al.: Efficient synthesis of high quality double walled carbon nanotubes and their cost-effective properties. New J. Chem. **46**(39), 18724–18731 (2022)
40. Tilmaciu, C.M., Morris, M.C.: Carbon nanotube biosensors. Front. Chem. **3**, 59 (2015)
41. Golnabi, H.: Carbon nanotube research developments in terms of published papers and patents, synthesis and production. Sci. Iran. **19**(6), 2012–2022 (2012)
42. Poudel, Y.R., Li, W.: Synthesis, properties, and applications of carbon nanotubes filled with foreign materials: a review. Mater. Today Phys. **7**, 7–34 (2018)
43. Lu, C., Su, F., Hu, S.: Surface modification of carbon nanotubes for enhancing BTEX adsorption from aqueous solutions. Appl. Surf. Sci. **254**(21), 7035–7041 (2008)
44. Abedi, Z., et al.: The effect of natural organic compounds on the adsorption of toluene and ethylene benzene on MWCNT. J. Environ. Health Sci. Eng. **17**(2), 1055–1065 (2019)
45. Chen, W., Duan, L., Zhu, D.: Adsorption of polar and nonpolar organic chemicals to carbon nanotubes. Environ. Sci. Technol. **41**(24), 8295–8300 (2007)
46. Wibowo, N., et al.: Adsorption of benzene and toluene from aqueous solutions onto activated carbon and its acid and heat treated forms: Influence of surface chemistry on adsorption. J. Hazard. Mater. **146**(1), 237–242 (2007)
47. Chin, C.J.M., et al.: Adsorption of o-xylene and p-xylene from water by SWCNTs. Carbon **45**(6), 1254–1260 (2007)
48. Yu, F., et al.: Magnetic iron oxide nanoparticles functionalized multi-walled carbon nanotubes for toluene, ethylbenzene and xylene removal from aqueous solution. Chemosphere **146**, 162–172 (2016)
49. Anjum, H., et al.: Investigation of green functionalization of multiwall carbon nanotubes and its application in adsorption of benzene, toluene & p-xylene from aqueous solution. J. Clean. Prod. **221**, 323–338 (2019)
50. Alves, D.C.D.S., et al.: Chapter 18—carbon nanotube-based materials for environmental remediation processes. In: Giannakoudakis, D., Meili, L., Anastopoulos, I. (eds.) Advanced Materials for Sustainable Environmental Remediation. Elsevier, pp. 475–513 (2022)
51. Agasti, N., et al.: Carbon nanotube based magnetic composites for decontamination of organic chemical pollutants in water: a review. Appl. Surf. Sci. Adv. **10**, 100270 (2022)
52. Yahyazadeh, A., Khoshandam, B.: Carbon nanotube synthesis via the catalytic chemical vapor deposition of methane in the presence of iron, molybdenum, and iron–molybdenum alloy thin layer catalysts. Results Phys. **7**, 3826–3837 (2017)
53. Mahgoub, H.A.: Nanoparticles used for extraction of polycyclic aromatic hydrocarbons. J. Chem. **2019**(4), 816–849 (2019)
54. Menezes, H.C., et al.: Magnetic N-doped carbon nanotubes: a versatile and efficient material for the determination of polycyclic aromatic hydrocarbons in environmental water samples. Anal. Chim. Acta **873**, 51–56 (2015)

Purification of Water Using Carbon Nanomaterials

S. Subasini, M. Kanagalakshmi, and Anitha Pius

Abstract Water deterioration and the scarcity of pure drinking water are the long-haul global problems faced by humanity. To keep the aquatic environment clean, toxic organic and inorganic contaminants must be removed from the water bodies. Even in micromolar concentrations, these contaminants can harm the environment and human health. The salinity levels of seawater are steadily rising due to global warming and climate change, which reduce the amount of freshwater available for domestic and industrial purposes. The water treatment technique not only has to eliminate the pollutants but also significantly desalinate the water. Carbon nanomaterials such as graphene, carbon nanotubes, fullerenes, graphene, graphitic carbon nitride and nanodiamonds are excellent materials for water purification due to their antifouling property, self-cleaning properties, easy modification, large specific surface area, high chemical stability, porosity and simplicity of regeneration and reusability. This chapter presents a comprehensive overview of the state-of-the-art carbon nanomaterials, including significant recent and past advancements and plans for their application in water treatment.

S. Subasini · M. Kanagalakshmi · A. Pius (✉)
Department of Chemistry, The Gandhigram Rural Institute—Deemed to be University, Gandhigram, Dindigul, Tamil Nadu 624 302, India
e-mail: dranithapius@gmail.com

J. Tharini and S. Thomas (eds.), *Carbon Nanomaterials and their Composites as Adsorbents*, Carbon Nanostructures, https://doi.org/10.1007/978-3-031-48719-4_15

Graphical Abstract

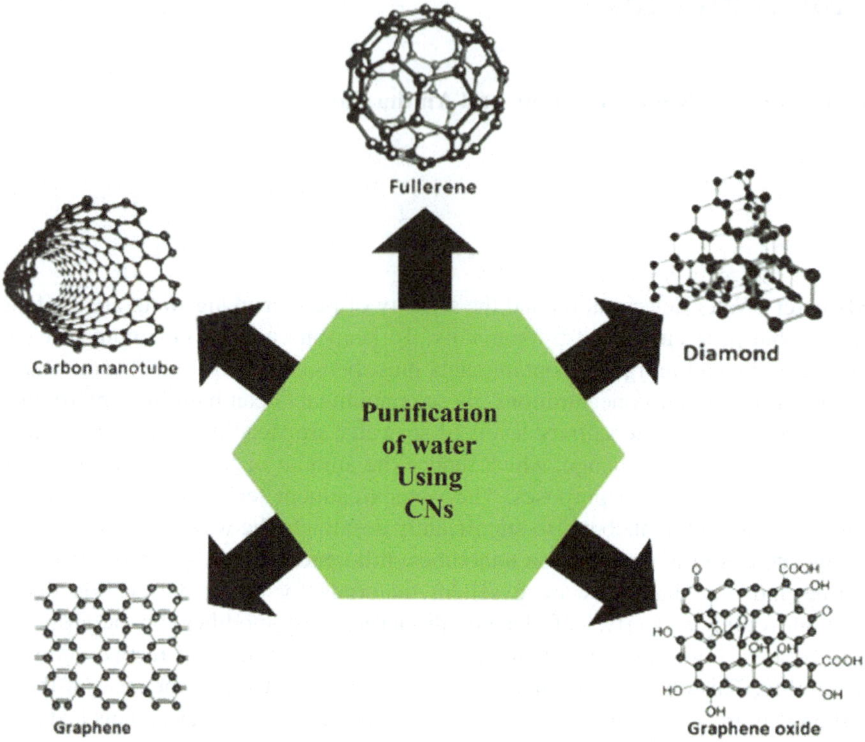

Keywords Carbon nanotubes · Fullerenes · Nanodiamonds · Graphene · Graphene oxides

1 Introduction

An exponential increase in the global population has increased the demand for pure water. Water pollution is a global crisis since it has a significant influence on aquatic ecosystems, food quality parameters and human health [1]. Over the past few decades, concentration of pollutants such as drugs, dyes, heavy metals and pesticides has increased considerably in the water resources. Contaminants often originate from industries, domestic activities, hospitals and farms [2]. The significant difference in freshwater availability and usage has led to significant growth in water purification. The development of cutting-edge water filtration technology and public awareness campaigns would help in overcoming the obstacles in highly populated countries like China and India [1].

Material science plays a vital role in cleaning environmental pollutants and finding suitable solutions for water purification methods. Much attention has been paid to nanotechnology methods to address water purification around the globe [3]. Nanomaterials are excellent materials for water purification due to their unique physical and chemical properties namely small size, large surface area, high porosity, high catalytic activity and reactivity, hydrophobicity, ease of isolation and regeneration [4]. Nanotechnology has significantly improved conventional wastewater treatment methods namely filtration, coagulation, flocculation, reverse osmosis, distillation and sorption.

The special catenation feature of carbon allowed it to establish covalent bonds with other carbon atoms in the sp, sp^2, and sp^3 hybridization states, resulting in a variety of carbon structures [5]. Carbonous materials are abundantly available in nature in the form of coal, graphite and diamond [6]. Countless new varieties of carbon nanomaterials (CNs) have been invented and are currently used in the purification of water. Examples of CNs include carbon nanotubes (CNTs), graphene and its variations, carbon nanofibers, nanodiamonds (NDs), fullerenes and nanoporous carbons [7]. Their distinctive layered and hollow architectures make it possible to interact with a variety of organic molecules. These include stacking interactions, hydrogen bonds, Van der Waals forces, hydrophobic interactions and electrostatic forces [8].

Fouling is the fundamental issue with conventional membranes. CN membranes, which exhibit adequate rejection, high permeability and low operating pressure, have been attempted to replace conventional membranes. For instance, by incorporating oxidized multi-walled carbon nanotubes (O-MWCNT) and graphene oxide (GO) into a polyether sulfone substrate via a phase inversion technique, Behdarvand et al. (2021) created a novel class of thin-film nanocomposite (TFN) membrane. Additionally, thin-film composite (TFC) membranes without CNs were created for better comparison. The TFN/O-MWCNTs and TFN/GO membranes showed 54% and 35% improvement in water permeability and exhibited better filtration behaviour than the TFC membrane [9].

This chapter will focus on the application of advanced, green CNs for the cleansing of aqueous contaminants and impurities from aquatic environments.

2 Purification of Water Using Carbon Nanotubes

An integrated electrochemical filtering system for the degradation of aqueous microcontaminants was developed by Zheng et al. (2012) employing composites of CNT and nanoscale zero-valence copper (nZVC). The nZVC-CNT filter eliminated all traces of congo red when the pH was neutral. They noticed a combination of radical and nonradical reaction pathways that together contributed to the decomposition of congo red. According to the mechanism, the CNT's carbonyl group and its electrophilic oxygen functioned as electron donors and acceptors to produce $^{\bullet}OH$ and $^1O_2$, as illustrated in Fig. 1. As a result, an integrated electrochemical filtering system having CNT offers the technology that can effectively remove pollutants [10].

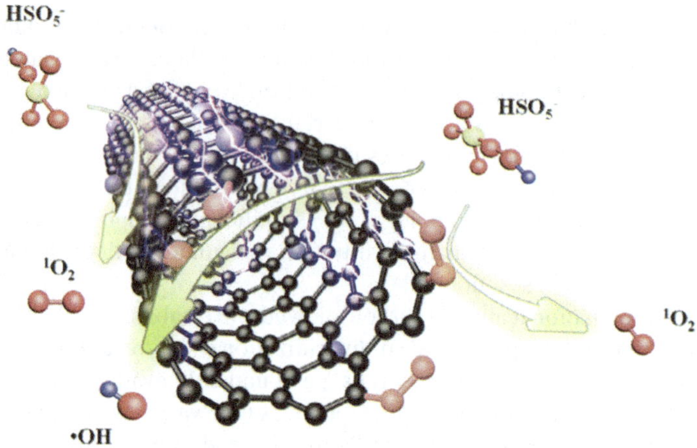

Fig. 1 Radical and nonradical pathway of electrochemical nZVC-CNT filter [10]

For the on-site, real-time, quantitative detection of p-cresol in the wastewater, Zhao et al. (2021) created a new and stable biosensor. A screen-printed carbon electrode was modified using CNTs made from debris that had been trapped with laccase to create the biosensor which displayed great levels of repeatability, stability and reusability. Even with the interference of metal ions and organic compounds, the designed biosensor was effectively used to detect p-cresol in wastewater. Figure 2 shows the detection of p-cresol before and after Fenton degradation by biosensor [11].

A lot of effort has been paid to the development of super hydrophilic, underwater superoleophobic oil/water separation material because of their effective antifouling qualities. By using a one-step thermal chemical vapour deposition technique, Yin et al. (2021) created inorganic carbon nanotube stainless steel meshes (CNT@SSMs), which exhibit super hydrophilic-underwater superoleophobic qualities as illustrated in Fig. 3. Small pore sizes in CNT@SSMs enabled a high-water flux of 10,639 $Lm^{-2}h^{-1}$ that successfully separated oily wastewater with a rejection ratio of > 98.89%. It exhibits exceptional flame resistance, strong chemical stability and outstanding flexibility. In addition, these super-wetting CNT@SSMs membranes are potential candidates for the treatment of risky oily wastes as well as for the actual use of water purification [12].

Oztekin and Sponza (2023) developed CNT/TiO_2 hybrid fibres and investigated the photocatalytic removal of 4-chloro-2-methylphenoxyacetic acid (MCPA), an endocrine disruptor, from several water bodies and effluent from the herbicide production line. Different CNT/TiO_2 volume ratios like 8%, 10%, 15% and 20% were used for the photocatalytic removal of MCPA. The elimination effectiveness of MCPA was reported to be 100% for 10 vol% CNT/TiO_2, 55% for TiO_2 fibre and 15% for TiO_2 nanoparticles which shows the addition of CNT improves the removal efficacy [13].

A catalytic CNT membrane for $KMnO_4$ activation towards efficient micropollutant degradation was thoughtfully created by Wang X et al. (2023). Under ideal

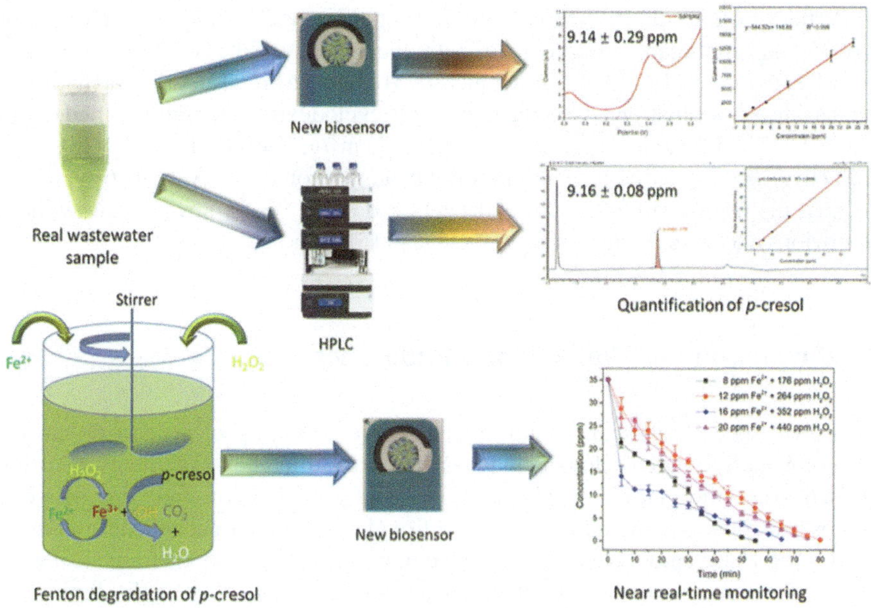

Fig. 2 Real-time quantification and monitoring of p-cresol by biosensor [11]

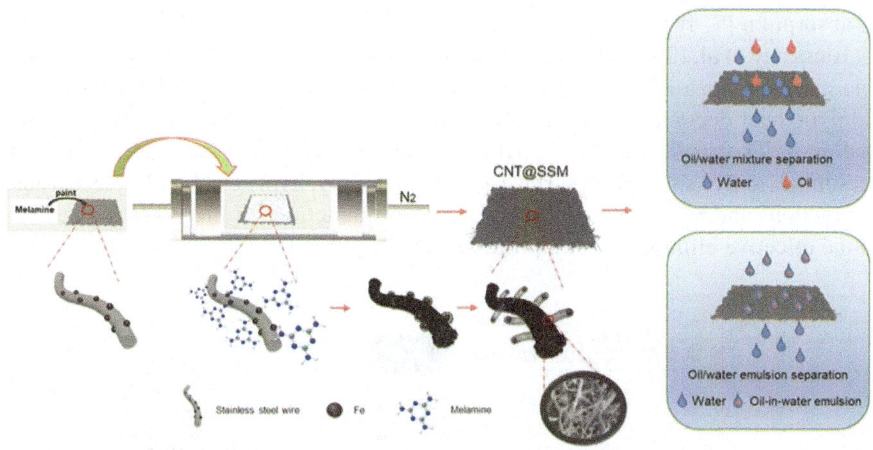

Fig. 3 Preparation of CNT@SSMs and application over oil/water separation [12]

conditions, a single-pass mode may remove >70% of sulfamethoxazole (SMX). The toxicity evaluation's findings indicated that the intermediates from SMX breakdown had lower toxicity. Additionally, the $KMnO_4$/CNT system had excellent reusability and CNT could maintain a long-lasting reactivity, serving as a green method for the remediation of micropollutants in an environmentally friendly way. Hence, the $KMnO_4$/CNT system can serve as an efficient CN for water purification [14].

Perfluorooctanesulfonic acid (PFOS), is categorised as a chemically stable anthropogenic micropollutant, which contaminates the majority of the aquatic environment globally. To remove PEOS, Zakaria et al. (2023) created a brand-new hybrid hydrogel bead using a combination of sodium alginate, β-cyclodextrin and multi-walled CNTs (SA-β-CD/CNTs) having a surface area of 193.73 m^2/g. The addition of CNTs to SA-β-CD hydrogel increased its maximum PFOS adsorption capacity by 91.6%. Subsequently, the manufactured hybrid hydrogel bead could be employed again without an obvious decrease in the removal efficiency [15].

3 Purification of Water Using Fullerenes

A discrete soccer ball-shaped molecular structure made up of at least 60 carbon atoms joined by single or double bonds is known as fullerenes. Fullerenes are being utilised in a wide range of applications, including medication and gene delivery, photosensitizers, organic photovoltaics, antioxidants, biopharmaceuticals and diagnostics [6]. Fullerenes have been employed in water purification to eliminate a wide range of pollutants, because of their greater surface area, sleek look, diminutive size and hierarchical structures. The high prevalence of adsorption of contaminants through π-π interactions is caused by the numerous π electrons in the inner and outer layers of spheres. Surface functionalization can boost a fullerene's capability of absorption [7, 16].

Elessawy et al. (2019) developed an innovative systematic way that only requires one step to create functionalized magnetic fullerene nanocomposite (FMFN) through catalytic heat decomposition of polyethene terephthalate bottle waste as illustrated in Fig. 4. The surface area of FMFN is relatively high and has both micro- and mesopores active sites. High surface area and porosity, hydrogen bonds, electrostatic interactions and π- π electron donor–acceptor interactions of FMFN play a key role in the effective elimination of ciprofloxacin (CIP) [17, 18].

Fig. 4 Preparation of FMFN and its interactions with pollutants [17, 18]

4 Purification of Water Using Nanodiamonds

NDs are the carbon allotropic residue produced as a result of explosions of trinitrotoluene. NDs have excellent levels of biocompatibility, a compact, narrow size distribution and ease in surface modification and bioconjugation. NDs are comparable to bulk diamonds in terms of optical and mechanical properties, they are also stable in severe environmental conditions and have high thermal conductivity, rigidity, Young's modulus and refractive index. The diamond core's chemical inertness makes it feasible for many functional groups to be bounded readily, either by covalent or non-covalent attachment. NDs could be used to produce therapeutic agents, antiviral and antibacterial medications, gene therapy, innovative medical equipment and tissue scaffolds because of their unique features [6, 7].

The adsorption of the azo dyes sulfasalazine (SSZ) and acid orange 7 (AO7) onto the surface of the NDs was investigated by Wang et al. (2012). For AO7 and SSZ, NDs showed corresponding adsorption capabilities of 1288 and 925 mmol/kg in wastewater. The adsorption was significantly influenced by hydrogen bonding as well as electrostatic interactions between the NDs and the azo molecules [19].

On thermally oxidised NDs, Molavi et al. (2018) reported the selective adsorption of the organic dyes methylene blue (MB) and methyl orange (MO). Due to the NDs' reduced zeta potential caused by thermal oxidation, MB was successfully and preferentially adsorbed to the oxidised NDs rather than MO. A blue aqueous solution containing MO was left behind after 16 min when untreated NDs were utilised as an adsorbent for a combination of MB and MO in polluted water as shown in Fig. 5 [20].

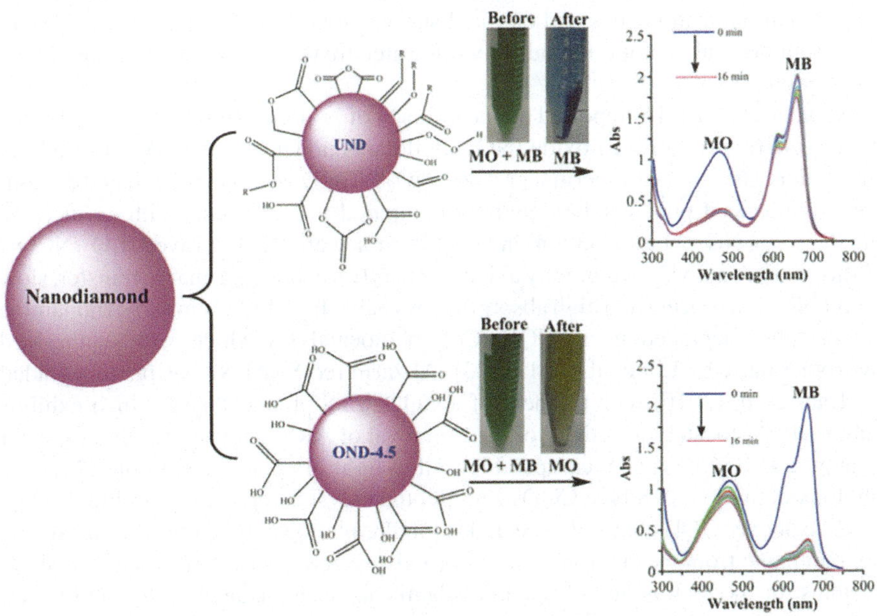

Fig. 5 Adsorption difference between oxidized NDs and unoxidized NDs [20]

5 Purification of Water Using Carbon Dots

The newest member of the nanocarbon family, carbon dots (CDs), has attracted a lot of interest in a variety of applications, including fluorescent probes, biosensing, drug delivery, bioimaging and photovoltaic devices, because of their exceptional optical properties, high chemical stability, high dispersibility and low toxicity. Different methods have been devised to add functional groups to the surface of CDs to enhance the adsorption capabilities [7].

Yao et al. (2017) examined the removal efficiency of U(VI) ions using layered double oxides (LDO) and LDO/CDs nanocomposite, which were prepared using a conventional calcination method in an inert atmosphere. In comparison to LDO (237.6 mg/g), LDO/CDs nanocomposites showed improved U(VI) uptake adsorption capability (354.2 mg/g). Since there are oxygen functions and a lot of surface area, this is the latter case [21].

6 Purification of Water Using Carbon Quantum Dots

Small fluorescent carbon nanoparticles having a diameter of less than 10 nm are known as carbon quantum dots (CQDs). A subtype of carbon quantum dots called graphene quantum dots (GQDs) is often generated from graphene or graphene oxide. Quantum dots made of carbon and graphene exhibit outstanding physical–chemical characteristics, including sturdy chemical stability, low cytotoxicity, a multitude of edge functionalization sites and elevated surface areas, which have attracted attention from several sectors, including wastewater treatment, sensor and membrane fabrication [2].

Mendes et al. (2016) reported on the efficacy of N-doped CQDs and TiO_2 for the photooxidation of nitrogen oxide (NO_x) pollutants under UV and UV–visible light irradiation. Using a hydrothermal process, TiO_2/NCQD was synthesized and demonstrated enhanced photocatalytic activity compared to pure TiO_2, with an increase of 36.3% in NO photooxidation and 16.8% in selectivity. Improved TiO_2/NCQD photocatalytic activity was greatly aided by NCQD's enhancing charge transfer, slow recombination process and high absorption of visible light [22]. Similarly, rhodamine B was photodegraded using CQDs/CdS photocatalysts which was synthesized hydrothermally by Liu et al. (2013) [23]. Alizarin red S (ARS) was photodegraded by Kaur et al. (2016) with the help of a CQDs/ZnS photocatalyst, which exhibits enhanced photocatalytic activity of 89% removal of ARS, which is 1.4 times greater than pure ZnS (63%) under comparable visible-light irradiation conditions [24]. The photooxidation of ARS by a CQDs/ZnS photocatalyst is represented in Fig. 6.

Like above, GQDs have also been used in the photocatalytic removal of organic contaminants from the environment. As promising peroxidase-mimic nano catalytic systems for the elimination of phenolic chemicals, CNs made of GQDs and Fe_3O_4 NPs were prepared by Wu et al. (2014). In comparison to individual Fe_3O_4

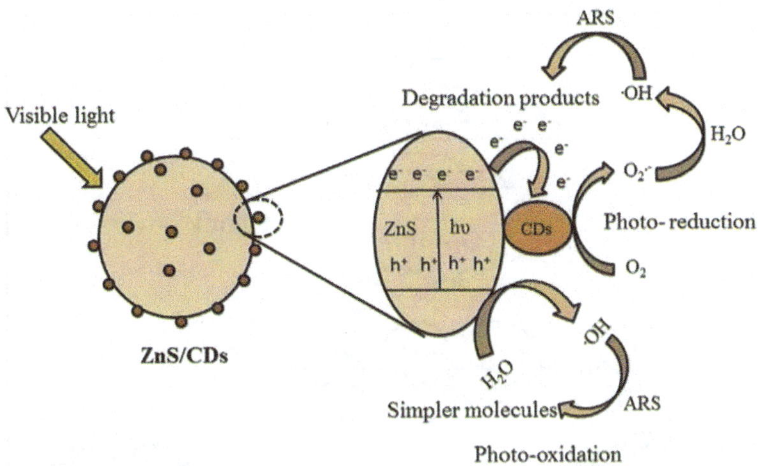

Fig. 6 Photooxidation mechanism of ARS by a CQDs/ZnS [24]

NPs (22-fold higher) and GQDs (25-fold higher), the as-prepared GQDs/Fe_3O_4 nanocomposite showed comparable or superior catalytic capabilities [25].

According to a study by Yan et al. (2016), N-GQDs-$BiVO_4$/g-C_3N_4-Z scheme heterojunctions have higher photocatalytic activity than $BiVO_4$/g-C_3N_4 for degrading antibiotic pollutants like oxytetracycline, tetracycline (TC) and CIP. The as-prepared heterojunction showed impressive photocatalytic efficiency, degrading 91.5% of TC in 30 min [26].

7 Purification of Water Using Graphene

Graphene is indeed the key structural component of various carbon allotropes, including zero-dimensional fullerenes, one-dimensional CNTs and three-dimensional graphite as shown in Fig. 7 [6]. Due to graphene's hydrophobicity, it was converted into reduced graphene oxide (rGO) and GO via a variety of chemical oxidation and reduction processes for the water purification process. Graphene in its highly oxidised form, referred to as GO, contains functional groups like hydroxyl, carboxyl, carbonyl and epoxy, while rGO is usually generated by chemical reduction of GO with a suitable reducing agent and exfoliating it. Since most of the functional groups on the surface of GO and GO-based materials comprise oxygen-containing functional groups, these functional groups can be exchanged for other functional groups using simple and direct chemical processes. This alteration enhances their competence to remove pollutants from water [27].

For the adsorption of phenanthrene and 1-naphthol in the water sample, Wang et al. (2019) synthesised rGO which displays a 99% removal capacity. The hydrophobic attraction, π-π interaction and H-bonding between graphene sheets and water pollutants could be to attribute for this enhanced removal efficacy [28].

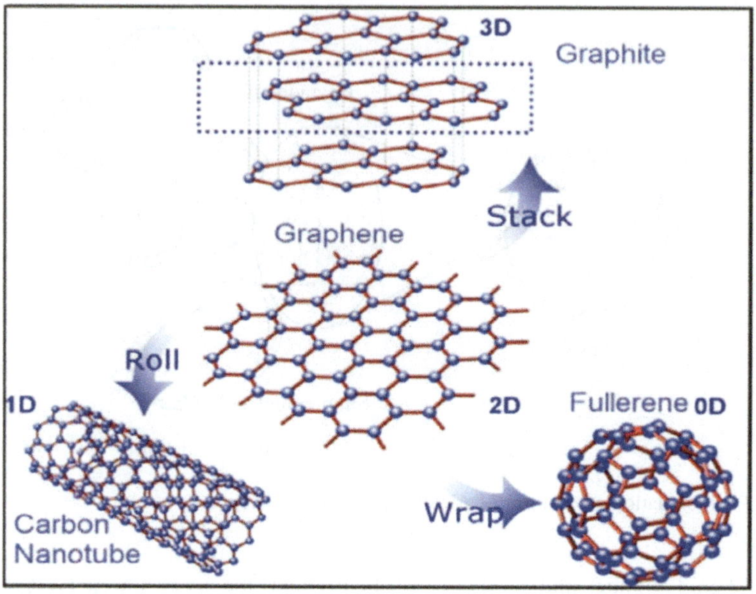

Fig. 7 Different constitutes of graphene [6]

Pharmaceutical contaminants are effectively treated with graphene-based adsorbents. They can also be used to remove radionuclides and heavy metals from wastewater. Jones et al. (2017) synthesized a multipurpose chitosan-attached plasmonic Au nanoparticle conjugated GO architecture-based 3D porous membrane for the supply of fresh water. To effectively separate pharmaceutical contaminants from methicillin-resistant *Staphylococcus aureus* (MRSA) in contaminated water and to kill MRSA, the 3D porous membrane was used as a novel CN. A highly porous chitosan-Au NPs GO membrane is easily able to trap MRSA bacteria, confirming that the MRSA superbug was destroyed utilising a produced porous membrane [29].

To remove target pharmaceutical pollutants from municipal wastewater being treated by an anaerobic membrane bioreactor (AnMBR), Casabella-Font et al. (2023) evaluated the effects of adding various GO concentrations. The target antibiotics diclofenac (DCF), naproxen (NPX), roxithromycin (ROX), sulfamethoxazole (SMX), triclosan (TCL), trimethoprim (TMP) and metronidazole (MTR) were removed more effectively with the addition of GO. As a result, antibiotic elimination might be improved and their environmental release could be decreased by adding GO to AnMBR [30].

Kumar et al. (2023) purpose of the study was to prepare a new boron oxide decorated nitrogen-rich reduced graphene oxide (B_2O_3/N-rGO) using a solvothermal method for photocatalytic CIP degradation. According to the findings, the B_2O_3/N-rGO significantly improved photocatalytic performance by 98% towards the degradation of CIP when exposed to visible light. The developed B_2O_3/N-rGO photocatalyst seems to have the potential to be used for environmental applications [31].

8 Purification of Water Using Graphitic Carbon Nitride

Graphite-like planes of C and N atoms make up the fundamental structure of graphitic carbon nitride (g-C_3N_4). The presence of nitrogen atoms in graphite makes it more attractive for various applications which include solar cells, sensors, energy storage and photocatalysis. The reduced band gap of g-C_3N_4 makes it a popular choice for solar radiation photocatalysis. The surface of g-C_3N_4 has many functional groups and electron-rich N atoms, which creates active sites for the adsorption of hazardous contaminants [16].

Yang et al. [32] successfully developed biochar-coupled Fe_3O_4@SiO_2/TiO_2/g-C_3N_4 using conventional sol–gel and calcination techniques. The composites effectively eliminated 91.88% hazardous high concentration TC using $^{\bullet}OH$ because of the large specific surface area, improved visible light responsiveness and the addition of magnetic nanoparticles in coupled Fe_3O_4@SiO_2/TiO_2/g-C_3N_4. Intriguingly, in the combined TC/Cr (VI) polluted water, a synergistic photocatalytic effect was found, leading to massively increased removal of Cr (VI). All of these analysis shows that the catalyst was an effective photocatalyst for the removal of TC/Cr (VI) contaminated water [32].

To adsorb a variety of cationic (MB, azure B, acriflavine and safranin O) and anionic (rhodamine B, eosin Y and MO) dyes, Zhang et al. (2016) developed a g-C_3N_4 hydrogel (h-CN). With an astounding 99% adsorption efficiency, h-CN preferentially adsorbed the cationic dyes due to the surface charge. The removal of cationic dyes from combinations of cationic and anionic contaminants might be made easier by h-CN [33].

Zhang et al. [34] examined the photocatalytic remediation capabilities of g-C_3N_4, TiO_2-graphene aerogel (TiO_2-GA) and g-C_3N_4-TiO_2-GA nanohybrid. In comparison to TiO_2-GA and g-C_3N_4, g-C_3N_4-TiO_2-GA successfully absorbed 96.5% of MB dye from an aqueous medium due to impregnation of g-C_3N_4 between TiO_2 and GA which increases the material's porosity and adsorption. g-C_3N_4 is an environmentally beneficial adsorbent for removing water contaminants because of its graphitic properties, low cost and good mechanical, chemical and thermal stabilities [34].

9 Conclusion

The unexpected, unprecedented discoveries have led to rapid progress in technological applications. The unique structure of carbon-based nanomaterials shows superior electronic, optoelectronic, mechanical and chemical performance. The carbon nanomaterials including fullerenes, NDs and CDs have significantly contributed to water purification. Various properties such as high surface area, presence of a large number of functional groups and simple surface modification of CNM have improved the efficiency of water purification.

References

1. Aslam, A.A., Ishtaiq, M., Faryal, S., Nazir, M.S., Akhtar, M.N., Ali, Z., Abdullah, M.A.: Application of biosurfactants and nanomaterials in the treatment of polluted water. In: Green Sustainable Process for Chemical and Environmental Engineering and Science: Microbially-Derived Biosurfactants for Improving Sustainability in Industry. Elsevier, pp. 203–234 (2021)
2. Fallah, Z., Zare, E.N., Ghomi, M., Ahmadijokani, F., Amini, M., Tajbakhsh, M., Arjmand, M., Sharma, G., Ali, H., Ahmad, A., Makvandi, P., Lichtfouse, E., Sillanpä, M., Varma, R.S.: Toxicity and remediation of pharmaceuticals and pesticides using metal oxides and carbon nanomaterials. Chemosphere **275**, 130055 (2021). https://doi.org/10.1016/j.chemosphere.2021.130055
3. Abd Elkodous, M., El-Sayyad, G.S., Abdel Maksoud, M.I.A., Kumar, R., Maegawa, K., Kawamura, G., Tan, W.K., Matsuda, A.: Nanocomposite matrix conjugated with carbon nanomaterials for photocatalytic wastewater treatment. J. Hazard. Mater. **410**, 124657 (2021). https://doi.org/10.1016/j.jhazmat.2020.124657
4. Zahid, M., Nadeem, N., Tahir, N., Farid-Un-Nisa, Majeed, M.I., Mansha, A., Naqvi, S.A.R., Hussain, T.: Hybrid nanomaterials for water purification. In: Multifunctional Hybrid Nanomaterials for Sustainable Agri-food and Ecosystems. Elsevier, pp. 155–188 (2020)
5. Sonawane, S., Thakur, P., Sonawane, S.H., Bhanvase, B.A.: Nanomaterials for membrane synthesis: introduction, mechanism, and challenges for wastewater treatment. In: Handbook of Nanomaterials for Wastewater Treatment: Fundamentals and Scale up Issues. Elsevier, pp. 537–553 (2021)
6. Datta, D., Das, K.P., Deepak, K.S., Das, B.: Candidates of functionalized nanomaterial-based membranes. In: Membranes with Functionalized Nanomaterials: Current and Emerging Research Trends in Membrane Technology. Elsevier Inc, pp. 81–127 (2022)
7. Ray, S.S., Gusain, R., Kumar, N.: Zero-dimensional carbon nanomaterials-based adsorbents. In: Carbon Nanomaterial-Based Adsorbents for Water Purification. Elsevier, pp. 181–193 (2020)
8. Fathy, N.A.: Carbon-based nanomaterials for wastewater treatment. In: Aquananotechnology. Elsevier, pp. 367–384 (2021)
9. Behdarvand, F., Valamohammadi, E., Tofighy, M.A., Mohammadi, T.: Polyvinyl alcohol/polyethersulfone thin-film nanocomposite membranes with carbon nanomaterials incorporated in substrate for water treatment. J. Environ. Chem. Eng. **9**, 104650 (2021). https://doi.org/10.1016/j.jece.2020.104650
10. Zheng, W., Liu, Y., Liu, W., Ji, H., Li, F., Shen, C., Fang, X., Li, X., Duan, X.: A novel electrocatalytic filtration system with carbon nanotube supported nanoscale zerovalent copper toward ultrafast oxidation of organic pollutants. Water Res. **194**, 116961 (2021). https://doi.org/10.1016/j.watres.2021.116961
11. Zhao, K., Veksha, A., Ge, L., Lisak, G.: Near real-time analysis of para-cresol in wastewater with a laccase-carbon nanotube-based biosensor. Chemosphere **269**, 128699 (2021). https://doi.org/10.1016/j.chemosphere.2020.128699
12. Yin, X., He, Y., Li, H., Ma, X., Zhou, L., He, T., Li, S.: One-step in-situ fabrication of carbon nanotube/stainless steel mesh membrane with excellent anti-fouling properties for effective gravity-driven filtration of oil-in-water emulsions. J. Colloid Interface Sci. **592**, 87–94 (2021). https://doi.org/10.1016/j.jcis.2021.02.043
13. Öztekin, R.R., Sponza, D.T.: The effect of dissolved natural organic matter (NOM) on the photocatalytic removal of 4-Chloro-2-Methylphenoxyacetic Acid (MCPA) endocrine disrupting compound from the surface water using carbon nanotubes/titanium dioxide (CNT-TiO_2) nanocatalysts. J. Biosci. Bioeng. (2023). ISSN 2836–4740
14. Wang, X., Guo, D., Zhang, J., Yao, Y., Liu, Y.: Insights into the electron transfer mechanisms of permanganate activation by carbon nanotube membrane for enhanced micropollutants degradation. Front. Environ. Sci. Eng. **17**(9), 106 (2023). https://doi.org/10.1007/s11783-023-1706-0

15. Zakaria, A.F., Kamaruzaman, S., Abdul Rahman, N., Yahaya, N.: Sodium alginate immobilized β-Cyclodextrin/Multi-walled carbon nanotubes as hybrid hydrogel adsorbent for perfluorinated compounds removal. J. Polym. Environ. **31**, 1895–1914 (2023). https://doi.org/10.1007/s10924-022-02737-2
16. Gusain, R., Kumar, N., Ray, S.S.: Recent advances in carbon nanomaterial-based adsorbents for water purification. Coord. Chem. Rev. 405 (2020)
17. Elessawy, N.A., El-Sayed, E.M., Ali, S., Elkady, M.F., Elnouby, M., Hamad, H.A.: One-pot green synthesis of magnetic fullerene nanocomposite for adsorption characteristics. J. Water Process Eng. **34** (2020). https://doi.org/10.1016/j.jwpe.2019.101047
18. Elessawy, N.A., Elnouby, M., Gouda, M.H., Hamad, H.A., Taha, N.A., Gouda, M., Mohy Eldin, M.S.: Ciprofloxacin removal using magnetic fullerene nanocomposite obtained from sustainable PET bottle wastes: adsorption process optimization, kinetics, isotherm, regeneration and recycling studies. Chemosphere **239**, 124728 (2020). https://doi.org/10.1016/j.chemosphere.2019.124728
19. Wang, H.D., Yang, Q., Hui Niu, C., Badea, I.: Adsorption of azo dye onto nanodiamond surface. Diam. Relat. Mater. **26**, 1–6 (2012). https://doi.org/10.1016/j.diamond.2012.03.006
20. Molavi, H., Shojaei, A., Pourghaderi, A.: Rapid and tunable selective adsorption of dyes using thermally oxidized nanodiamond. J. Colloid Interface Sci. **524**, 52–64 (2018). https://doi.org/10.1016/j.jcis.2018.03.088
21. Yao, W., Wang, X., Liang, Y., Yu, S., Gu, P., Sun, Y., Xu, C., Chen, J., Hayat, T., Alsaedi, A., Wang, X.: Synthesis of novel flower-like layered double oxides/carbon dots nanocomposites for U(VI) and ^{241}Am(III) efficient removal: Batch and EXAFS studies. Chem. Eng. J. **332**, 775–786 (2018). https://doi.org/10.1016/j.cej.2017.09.011
22. Martins, N.C.T., Ângelo, J., Girão, A.V., Trindade, T., Andrade, L., Mendes, A.: N doped carbon quantum dots/TiO_2 composite with improved photocatalytic activity. Appl. Catal. B **193**, 67–74 (2016). https://doi.org/10.1016/j.apcatb.2016.04.016
23. Liu, Y., Yu, Y.X., De, Z.W.: Carbon quantum dots-doped CdS microspheres with enhanced photocatalytic performance. J. Alloys Compd. **569**, 102–110 (2013). https://doi.org/10.1016/j.jallcom.2013.03.202
24. Kaur, S., Sharma, S., Kansal, S.K.: Synthesis of ZnS/CQDs nanocomposite and its application as a photocatalyst for the degradation of an anionic dye, ARS. Superlattices Microstruct. **98**, 86–95 (2016). https://doi.org/10.1016/j.spmi.2016.08.011
25. Wu, X., Zhang, Y., Han, T., Wu, H., Guo, S., Zhang, J.: Composite of graphene quantum dots and Fe_3O_4 nanoparticles: peroxidase activity and application in phenolic compound removal. RSC Adv. **4**, 3299–3305 (2014). https://doi.org/10.1039/c3ra44709j
26. Yan, M., Zhu, F., Gu, W., Sun, L., Shi, W., Hua, Y.: Construction of nitrogen-doped graphene quantum dots-$BiVO_4$/g-C_3N_4 Z-scheme photocatalyst and enhanced photocatalytic degradation of antibiotics under visible light. RSC Adv. **6**, 61162–61174 (2016). https://doi.org/10.1039/c6ra07589d
27. Adewuyi, A., Lau, W.J.: Nanomaterial development and its applications for emerging pollutant removal in water. In: Handbook of Nanotechnology Applications: Environment, Energy, Agriculture and Medicine. Elsevier, pp. 67–97 (2020)
28. Wang, F., Jia, Z., Su, W., Shang, Y., Wang, Z.L.: Adsorption of phenanthrene and 1-naphthol to graphene oxide and L-ascorbic-acid-reduced graphene oxide: effects of pH and surfactants. Environ. Sci. Pollut. Res. **26**, 11062–11073 (2019). https://doi.org/10.1007/s11356-019-04549-9
29. Jones, S., Pramanik, A., Kanchanapally, R., Priya, B., Nellore, V., Begum, S., Sweet, C., Ray, P.C.: Multifunctional three dimensional chitosan/gold nanoparticle/graphene oxide architecture for separation, label-free SERS identification of pharmaceutical contaminants and effective killing of superbugs. ACS Sustain. Chem. Eng. **5**(8), 7175–7187 (2017). https://doi.org/10.1021/acssuschemeng.7b01351
30. Casabella-Font, O., Ponzelli, M., Papapanou, M., Balcazar, J.L., Pijuan, M., Radjenovic, J.: Impact of graphene oxide addition on pharmaceuticals removal in anaerobic membrane bioreactor. Bioresour. Technol. **383**, 129252 (2023). https://doi.org/10.1016/j.biortech.2023.129252

31. Kumar, N., Gusain, R., Masukume, M., Ray, S.S.: B_2O_3 nanoparticles decorated N-rich reduced graphene oxide composites for the enhanced visible-light assisted photocatalytic degradation of ciprofloxacin. Solar RRL (2023). https://doi.org/10.1002/solr.202300475
32. Yang, B., Dai, J., Zhao, Y., Wang, Z., Wu, J., Ji, C., Zhang, Y., Pu, X.: Synergy effect between tetracycline and Cr(VI) on combined pollution systems driving biochar-templated Fe_3O_4@SiO_2/TiO_2/g-C_3N_4 composites for enhanced removal of pollutants. Biochar **5**, 1 (2023). https://doi.org/10.1007/s42773-022-00197-4
33. Zhang, Y., Zhou, Z., Shen, Y., Zhou, Q., Wang, J., Liu, A., Liu, S., Zhang, Y.: Reversible assembly of graphitic carbon nitride 3D network for highly selective dyes absorption and regeneration. ACS Nano **10**, 9036–9043 (2016). https://doi.org/10.1021/acsnano.6b05488
34. Zhang, J.J., Fang, S.S., Mei, J.Y., Zheng, G.P., Zheng, X.C., Guan, X.X.: High-efficiency removal of rhodamine B dye in water using g-C_3N_4 and TiO_2 co-hybridized 3D graphene aerogel composites. Sep. Purif. Technol. **194**, 96–103 (2018). https://doi.org/10.1016/j.seppur.2017.11.035

Mathematical Modeling of Nonlinear Vibrations of Single-Walled Carbon Nanotubes

Sekar Rekha and Lakshmanan Rajendran

Abstract In this article, the nonlinear vibrations of single-walled carbon nanotubes were discussed. The governing equations of each layer are connected with those of its neighboring layers via the Van der Waals interlayer forces. By comparing the new results to those from earlier research, the effects of changes in the geometrical parameters of the nanotubes and the material constants of the elastic media around them on the vibration characteristics are studied. The variational iteration technique (VIM), the Adomian decomposition method (ADM), and the new homotopy perturbation method are compared. These strategies are solid and effective for resolving various linear and nonlinear differential equations that can appear in various branches of research and engineering. Also contrasting with numerical simulations are our results. The answers derived using these three methods and a numerical solution agrees extremely well. Please be aware that your name and affiliation and if applicable those of your co-author(s) will be published as presented in this proof. If you want to make any changes, please correct the details now. Please note that after publication corrections won't be possible. Due to data protection we standardly publish professional email addresses, but not private ones, even if they have been provided in the manuscript and are visible in this proof. If you or your co-author(s) have a different preference regarding the publication of your mail address(s) please indicate this clearly. If no changes are required for your name(s) and affiliation(s) and if you agree with the handling of email addresses, please respond with "Ok". The author and co-author names and affiliations are correct.

Keywords Single-walled carbon nanotubes · Adomian decomposition method · Variational iteration method · New homotopy perturbation method · Nonlinear vibration · Carbon nanotube

---

S. Rekha
Department of Mathematics, St. Joseph College of Engineering, Sriperumbudur, Chennai 602117, India

L. Rajendran (✉)
Department of Mathematics, AMET Deemed to be University, Chennai 603112, India
e-mail: raj_sms@rediffmail.com

J. Tharini and S. Thomas (eds.), *Carbon Nanomaterials and their Composites as Adsorbents*, Carbon Nanostructures, https://doi.org/10.1007/978-3-031-48719-4_16

1 Introduction

The rapid development of nanotechnology has increased interest in this field of study among scientists and researchers. Nanomaterials have dominated numerous scholarly studies due to their exceptional mechanical, physical, and chemical properties. Modern nanoelectronics, nanodevices, and nanocomposites utilise them as the essential components. Carbon nanotubes (CNT) are materials whose remarkable mechanical strength has gathered much attention. In 1991, Iijima discovered CNTs which emerged as the new branch of knowledge in the materials science. Even though they are minimal and light, they have very high Young's modulus in the axial direction. CNTs are unquestionably eligible to be the novel materials in the twenty-first-century. The vibration of CNTs is significant because of its various nanomechanical devices, including oscillators, charge detectors, field emission devices, and sensors. The issue of these nanomaterials' vibration has received much attention from researchers. However, most studies on the vibration of multi-walled carbon nanotubes (MWNTs) have been limited to the linear regime, and there have been fewer works on the nonlinear vibration of these materials.

Nonlinear differential equations are used to model many scientific issues in carbon nanotubes. It is generally known that when non-Newtonian fluids are considered, finding exact solutions to these nonlinear boundary values are challenging. Numerical and analytical techniques are applied to solve this nonlinear problem. However, numerical approaches are comparatively time-consuming and challenging because of stability and convergence issues. The homotopy analysis method (HAM) [1], the variational iteration method (VIM) [2–7], the new homotopy perturbation method (NHPM) [8, 9], and the Adomian decomposition method (ADM) [10–15] are the some of the numerous analytical techniques developed in the last ten years to solve nonlinear problems. Numerous numerical works have been compared with various analytical methodologies (Table 1).

The nonlinear vibrations of single-walled carbon nanotubes embedded in an elastic medium have been estimated in this study using VIM, NHPM, and ADM. Siddiqui and Farooq recently compared the approaches of variational iteration and Adomian decomposition for resolving nonlinear thin film flow problems [16]. These techniques produce the answer in a convergent series with beautifully computed components. These analytical techniques also avoid the complications offered by other pure numerical techniques [17–20]. These findings show that the suggested analytical techniques offer the powerful mathematical tool for handling the sizable class of linear and nonlinear differential equations in engineering and chemical sciences.

Table 1 Recent contributions to the theoretical modeling of the comparison of analytical and Numerical solutions of two or three methods

Publication		Name of equation	Modeling method (HPM, HAM, ADM, etc.)
Author(s)	Reference		
A. M. Wazwaz	Journal of Computational and Applied Mathematics 207 (2007) 129–136	Homogeneous and nonhomogeneous advection problem	ADM and VIM
D. D. Ganjiet al.	International Journal of Science & Technology 2 (2) (2007) 179–188	Generalized Hirota–Satsuma coupled Kdv equation	VIM and ADM
M. S. H. Chowdhury et al.	Communications in Nonlinear Science and Numerical Simulation 14 (2009) 371–378	The temperature distribution along a straight Fin equation	HPM and HAM
D. D. Ganji et al.	Walailak Journal of Science &Technology 11 (7) (2014) 593–609	Three-dimensional flow of a Walter's B fluid in vertical channel	ADM, HPM and VIM
Elaf Jaafar Ali et al.	Journal of Basrah Researches (Sciences) 37(2011) 1817–2695	Extraordinary differential equation	HPM and VIM
E. Babolian et al.	Journal of Applied Sciences, (2012) 1812–5654	Nonlinear differential equations	HPM and ADM
Jamshad Ahmad et al.	International Journal of Basic Sciences & Applied Research.3 (3) (2014) 173–179,	Cubic nonlinear schrödinger equation	LDM and ADM
K. R. Raslanl et al.	General Mathematics Notes 20 (2) (2014) 125–135 2219–7184	Improved modified Kortweg-de varies equation	ADM and VIM
Malihe Bagheri et al.	The Journal of Mathematics and Computer Science 5 (4) (2012) 288–296	Volterra and Feredholm integral equations	DTM and HPM
Muhammad Shakil et al.	International Journal of Research in Applied, Natural and Social Sciences (Impact: Ijranss) 1 (3) (2013) 37–48	Nonlinear wave equation, klein–gordon equation	ADM and HPM
A. M. Siddiqui et al.	Applied Mathematical Sciences, 6 (99) (2012) 4911–4919	Thin film flow	ADM and VIM

(continued)

Table 1 (continued)

Publication		Name of equation	Modeling method (HPM, HAM, ADM, etc.)
Author(s)	Reference		
N. Shawagfeh et al.	Applied Mathematics Letters 17 (2004) 323–328	Ordinary differential equations	ADM and RKM
S. Abbasbandy et al.	Applied Mathematics and Computation173 (1) (2006) 493–500	Non-singular integral equations	ADM and HPM
S. Abbasbandy et al.	Chaos, Solitons & Fractals 31 (1) (2007) 257–260	Blasius equation	ADM and HPM
S. Abbasbandy et al.	Applied Mathematics and Computation 172 (1) (2006) 431–438	Nonlinear equations	MHPM and ADM
A. Sadighi and D. D. Ganji	International Journal of Nonlinear Sciences and Numerical Simulation 7 (4) (2006) 411–418 2191–0294	Generalized nonlinear boussinesq equation	HPM and VIM
A. M. Wazwaz et al.	Kybernetes, 40 (9) (2010) 1305–1318	Lane-emden equations	ADM and VIM
Sunil Kumar et al.	Applied Mathematics and Computation 293 (2016) 508–522	Parabolic singularly perturbed differential equation	FDM, ADM, MSI And TGM
Z. A. Firoozjae et al.	American Journal of Applied Mathematics 3 (3) (2015) 90–94	Nonlinear partial differential equations	DQM and ADM

2 Theory

Consider the SWNT of length l, Young's modulus E, density, cross-sectional area A, and cross-sectional inertia moment I, embedded in an elastic medium with material constant k. The nonlinear vibration equation for this carbon nanotube (CNT) is in the following form [1]:

$$\frac{d^2W}{dt^2} + \left(\frac{\pi^4 EI}{\rho A l^4} + \frac{k}{\rho A}\right)W + \frac{\pi^4 E}{4\rho l^4}W^3 = 0, \tag{1}$$

Using the following dimensionless variables

$$r = \sqrt{\frac{I}{A}},\ x = \frac{W}{r},\ W_l = \frac{\pi^2}{l^2}\sqrt{\frac{EI}{\rho A}},\ w_k = \sqrt{\frac{k}{\rho A}} \tag{2}$$

The Eq. (1) can be transformed to the following dimensionless nonlinear vibration equation.

$$\frac{d^2x}{d\tau^2} + ax + bx^3 = 0, \tag{3}$$

where

$$a = \frac{w_b^2}{\omega^2},\ b = \frac{aw_1^2}{\omega^2} \tag{4}$$

in which $\alpha = 0.25$ and $w_b = \sqrt{w_l^2 + w_k^2}$, is the linear free vibration frequency.

The initial and boundary conditions are

$$x(0) = X, \dot{x}(0) = 0. \tag{5}$$

3 Analytical Expression of Nonlinear Vibration Equation

The dimensionless non-linear Eq. (3) defines the initial value problem. The variational iteration method [2–7], the new homotopy perturbation method [8, 9], and the Adomian decomposition method [10–15] give the approximate solutions of the non-linear Eq. (3). Using these three methods, we can obtain the analytical expressions of the vibration of a string as follows:

$$\begin{aligned} x(\tau) =& X - (4aX + b^3X^3)\frac{\tau^2}{2} - \frac{(a + bX^2)}{24}(aX - 3b^3X^3)\tau^4 - \left(\frac{(b^3X^3)(a + bX^2)^2}{40}\right)\tau^6 \\ &+ \left(\frac{(b^3X^3)(a + bX^2)^3}{448}\right)\tau^8 \end{aligned} \tag{6}$$

The analytical expression of vibration using NHPM is,

$$x(\tau) = -\frac{b}{a} + \left(\left(\frac{b}{a} + X\right)\cos\left(\sqrt{a}\tau\right)\right) \tag{7}$$

The analytical expression of vibration using ADM is,

Table 2 Comparison of various analytical results with numerical results for vibration $x(\tau)$ when $a = 1, b = 0.5$ and $x(0) = 0.5$

$a = 1, b = 0.1$ Initial value X = 0.5							
τ	Numerical simulation	VIM Equation (6)	Error (%)	NHPM Equation (7)	Error (%)	ADM Equation (8)	Error (%)
0	0.5000	0.5000	0.00	0.5000	0.00	0.5000	0.00
0.1	0.4974	0.4975	0.02	0.4974	0.00	0.4974	0.00
0.2	0.4878	0.4898	0.41	0.4896	0.37	0.4896	0.37
0.3	0.4771	0.4772	0.02	0.4766	0.1	0.4767	0.08
0.4	0.4595	0.4597	0.04	0.4587	0.17	0.4588	0.15
0.5	0.4373	0.4376	0.07	0.436	0.3	0.4361	0.27
0.6	0.4106	0.411	0.1	0.4087	0.46	0.409	0.39
0.7	0.3798	0.3803	0.13	0.3771	0.71	0.3776	0.58
0.8	0.3451	0.3458	0.2	0.3416	1.01	0.3425	0.75
0.9	0.3069	0.3079	0.33	0.3024	1.47	0.3039	0.98
1	0.2656	0.2713	2.15	0.2644	0.45	0.2667	0.41
		Average error (%) 0.32		Average error (%) 0.46		Average error (%) 0.36	

$$x(\tau) = X - \frac{\tau^2}{2}(aX + bX^3) + \frac{\tau^4}{24}(aX + bX^3)(a + 3bX^2) \tag{8}$$

4 Numerical Simulation

The non-linear differential Eq. (3) for the given initial and boundary conditions is being solved numerically. The function pdex, which is a function of solving the initial-boundary value problems, is used to solve this equation in the two-dimensional differential equation solver and Grapher V 1.0 software. These analytical expressions of vibration are compared with simulation results for various kinetic parameters in Tables 1, 2, 3, 4, 5, 6 and 7. A satisfactory agreement is noted.

5 Discussion and Conclusions

The new and straightforward mathematical formulas for dimensionless SWNT vibration for all parameter values are given in Eqs. 6–8. Figures 1, 2, 3 and 4 shows how the vibration of an SWNT varies in dimensionless time for different parameter values. For all parameter values, it can be seen from the figures that the dimensionless vibration of SWNT decreases from its initial value. Using a nonlinear initial value in single-walled carbon nanotubes, we have shown how VIM, NHPM, and ADM could be used to achieve approximative analytical solutions. As a result, it is found

Table 3 Comparison of various analytical results with numerical results for vibration $x(\tau)$ when $a = 0.01,\ b = 1$ and $x(0) = 0.5$

$a = 0.01, b = 1$ Initial value X = 0.5							
τ	Numerical simulation	VIM Equation (6)	Error (%)	NHPM Equation (7)	Error (%)	ADM Equation (8)	Error (%)
0	0.5000	0.5000	0.00	0.5000	0.00	0.5000	0.00
0.1	0.4994	0.4993	0.02	0.4993	0.02	0.4993	0.02
0.2	0.4974	0.4974	0.00	0.4973	0.02	0.4974	0.00
0.3	0.4941	0.4941	0.00	0.494	0.02	0.4941	0.00
0.4	0.4897	0.4897	0.00	0.4894	0.06	0.4895	0.04
0.5	0.484	0.4843	0.06	0.4841	0.02	0.4837	0.06
0.6	0.4771	0.4786	0.31	0.4761	0.21	0.4767	0.08
0.7	0.4691	0.4708	0.36	0.4675	0.34	0.4685	0.13
0.8	0.46	0.4631	0.67	0.4576	0.52	0.4593	0.15
0.9	0.4496	0.4559	1.40	0.4463	0.73	0.4502	0.13
1	0.4388	0.4479	2.07	0.4351	0.84	0.4391	0.07
		Average error (%) 0.45		Average error (%) 0.25		Average error (%) 0.06	

Table 4 Comparison of various analytical results with numerical results for vibration $x(\tau)$ when $a = 1,\ b = 0.1$ and $x(0) = 0.8$

$a = 1, b = 0.1$ Initial value $X = 0.8$							
τ	Numerical simulation	VIM	Error (%)	NHPM	Error	ADM	Error (%)
0	0.8000	0.8000	0.00	0.8000	0.00	0.8000	0.00
0.1	0.7957	0.7959	0.03	0.7959	0.03	0.7959	0.03
0.2	0.783	0.7837	0.09	0.7836	0.08	0.7836	0.08
0.3	0.762	0.7635	0.20	0.7633	0.17	0.7633	0.17
0.4	0.7329	0.7356	0.37	0.7352	0.31	0.7352	0.31
0.5	0.6962	0.7002	0.57	0.6995	0.47	0.6995	0.47
0.6	0.6521	0.6578	0.87	0.6566	0.69	0.6567	0.71
0.7	0.6012	0.6088	1.26	0.607	0.96	0.6073	1.01
0.8	0.5441	0.5538	1.78	0.5511	1.29	0.5517	1.40
0.9	0.4815	0.4895	1.66	0.4896	1.68	0.4907	1.91
1	0.4139	0.4353	5.17	0.4299	3.87	0.4316	4.28
		Average error (%) 1.09		Average error (%) 0.87		Average error (%) 0.88	

Table 5 Comparison of various analytical results with numerical results for vibration $x(\tau)$ when $a = 1,\ b = 0.1$ and $x(0) = 1$

$a = 1, b = 0.1$ Initial value $X = 1$							
τ	Numerical simulation	VIM Equation (6)	Error (%)	NHPM Equation (7)	Error (%)	ADM Equation (8)	Error (%)
0	0.1000	0.1000	0.00	0.1000	0.00	0.1000	0.00
0.1	0.9945	0.9949	0.04	0.9944	0.01	0.9944	0.01
0.2	0.9781	0.9797	0.16	0.9776	0.05	0.9777	0.04
0.3	0.9509	0.9544	0.37	0.9499	0.11	0.95	0.09
0.4	0.9135	0.9195	0.66	0.9114	0.23	0.9118	0.19
0.5	0.8661	0.8801	1.62	0.8627	0.39	0.8636	0.29
0.6	0.8094	0.828	2.30	0.8041	0.65	0.8122	0.35
0.7	0.7442	0.7677	3.16	0.7435	0.09	0.7469	0.36
0.8	0.671	0.6999	4.31	0.66	1.64	0.6739	0.43
0.9	0.5909	0.6255	5.86	0.5846	1.07	0.5944	0.59
1	0.5047	0.5454	8.06	0.4943	2.06	0.5096	0.97
		Average error (%) 2.41		Average error (%) 0.57		Average error (%) 0.31	

Table 6 Comparison of various analytical results with numerical results for vibration $x(\tau)$ when $a = 1,\ b = 0.01$ and $x(0) = 1$

$a = 1, b = 0.01$ Initial value $X = 1$							
τ	Numerical simulation	VIM Equation (6)	Error (%)	NHPM Equation (7)	Error (%)	ADM Equation (8)	Error (%)
0	0.1000	0.1000	0.00	0.1000	0.00	0.1000	0.00
0.1	0.9949	0.995	0.01	0.995	0.01	0.995	0.01
0.2	0.9798	0.9799	0.01	0.9799	0.01	0.9799	0.01
0.3	0.9549	0.9549	0.00	0.9549	0.00	0.9549	0.00
0.4	0.9203	0.9203	0.00	0.9203	0.00	0.9203	0.00
0.5	0.8764	0.8764	0.00	0.8764	0.00	0.8764	0.00
0.6	0.8237	0.8237	0.00	0.8237	0.00	0.8237	0.00
0.7	0.7675	0.7628	0.61	0.7628	0.61	0.7628	0.61
0.8	0.6941	0.6914	0.39	0.6941	0.00	0.6941	0.00
0.9	0.6184	0.6234	0.81	0.6185	0.02	0.6185	0.02
1	0.5366	0.5421	1.02	0.5367	0.02	0.5367	0.02
		Average error (%) 0.26		Average error (%) 0.06		Average error (%) 0.06	

Table 7 Comparison of various analytical results with numerical results for vibration $x(\tau)$ when $a = 0.4$, $b = 0.8$ and $x(0) = 1$

$a = 0.4, b = 0.8$ Initial value $X = 1$							
τ	Numerical simulation	VIM Equation (6)	Error (%)	NHPM Equation (7)	Error	ADM Equation (8)	Error (%)
0	0.1000	0.1000	0.00	0.1000	0.00	0.1000	0.00
0.1	0.994	0.9954	0.14	0.9939	0.01	0.9951	0.11
0.2	0.9762	0.9815	0.54	0.9757	0.05	0.9781	0.19
0.3	0.9471	0.9585	1.20	0.9461	0.11	0.9495	0.25
0.4	0.9074	0.9268	2.14	0.9058	0.18	0.9103	0.32
0.5	0.8581	0.8867	3.33	0.8561	0.23	0.8614	0.38
0.6	0.8003	0.8385	4.77	0.7985	0.22	0.8046	0.54
0.7	0.7353	0.7825	6.42	0.735	0.04	0.7416	0.86
0.8	0.6641	0.7191	8.28	0.6747	1.60	0.6679	0.57
0.9	0.5879	0.6483	10.27	0.6065	3.16	0.5998	2.02
1	0.5077	0.5783	13.91	0.54	6.36	0.5465	7.64
		Average error (%) 4.64		Average error (%) 1.09		Average error (%) 1.17	

that these methods are particularly effective in resolving numerous nonlinear issues that can arise in various scientific and engineering fields. The error rate is less than 4.66 for all approaches. However, in some locations, one method's error percentage is higher than the other two techniques due to the nonlinear differential equations' coefficients and initial conditions.

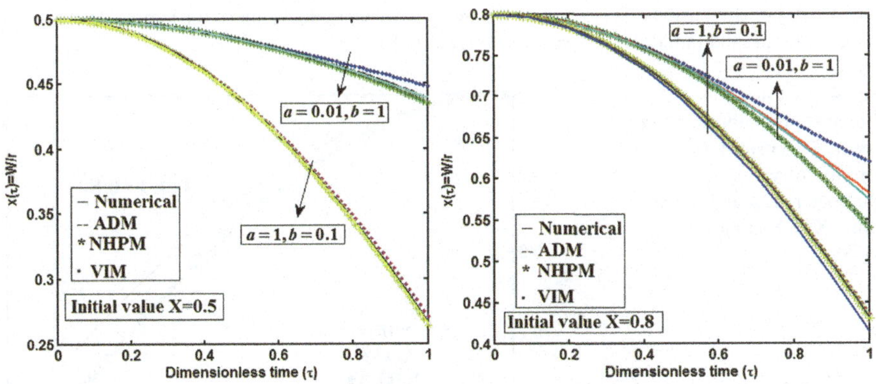

Fig. 1 Plot of dimensionless vibration $x(\tau)$ versus dimensionless time τ for various values of the parameter, a, b and initial values X = 0.5 and X = 0.8 using Eqs. (6–8)

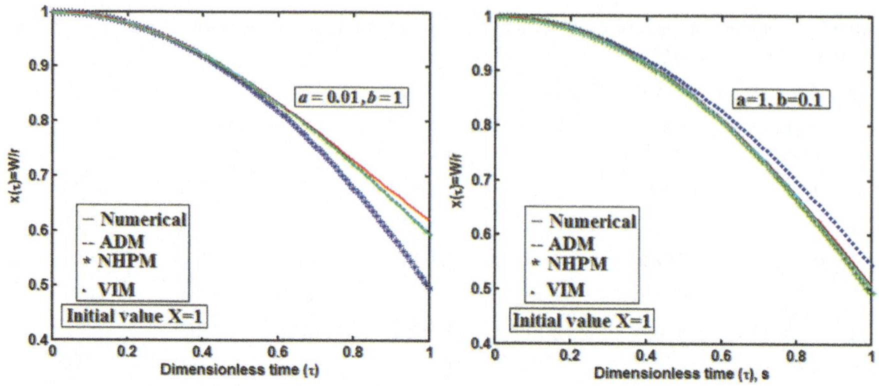

Fig. 2 Plot of dimensionlessvibration $x(\tau)$ versus dimensionless time τ for various values of the parameter, a, b and initial value X = 1 using Eqs. (6–8)

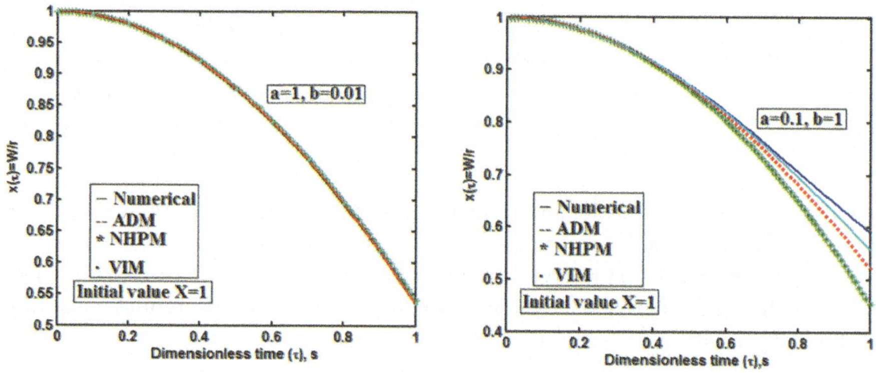

Fig. 3 Plot of dimensionless vibration $x(\tau)$ versus dimensionless time τ for various value of the parameter, a, b and initial value X = 1 using Eqs. (6–8)

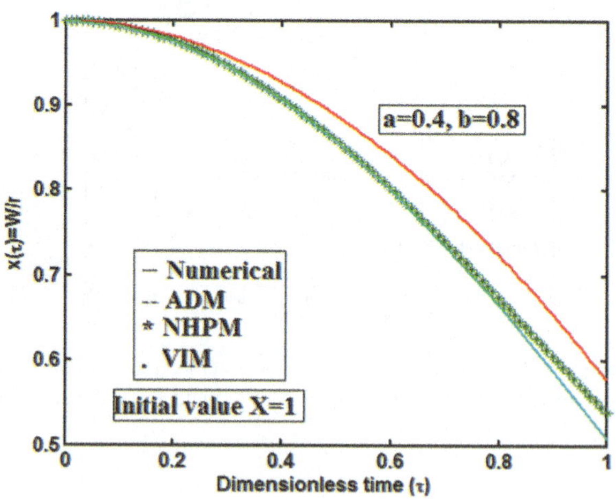

Fig. 4 Plot of dimensionlessvibration $x(\tau)$ versus dimensionless time τ for various value of the parameter, a, b and initial value X = 1 using Eqns. (6–8)

ADM requires the evaluation of the Adomian polynomials, which generally requires time-consuming algebraic operations, while VIM requires the evaluation of the Lagrangian multiplier. Adomian polynomials must be used with ADM for nonlinear terms, which requires additional research. The analytical results are obtained using ADM and VIM throughout the three iterations. However, NHPM got the analytical data after just one round. Therefore, the accuracy of the approaches depends on the parameters included in the differential equations and the beginning and boundary conditions of the presented problems, in addition to the iteration.

References

1. Khadar, M.M., Sweilam, N.H., EL-Sehrawy, Z.I., Ghwail, S.A.: Analytical study for the nonlinear vibration of multiwalled carbon nanotubes using homotopy analysis method. Appl. Math. Inf. Sci. **8**(4), 1675–1684 (2014)
2. He, J.H.: A variational iteration approach to nonlinear problems and its applications. Mech. Appl. **20**(1), 30–31 (1998)
3. He, J.H.: Variational iteration method—a kind of nonlinear analytical technique: some examples, Internat. J. Nonlinear Mech. **34**, 708–799 (1999)
4. He, J.H.: Variational iteration method for autonomous ordinary differential systems. Appl. Math. Comput. **114**(2/3), 115–123 (2000)
5. Meena, A., Eswari, A., Rajendran, L.: Mathematical modeling of enzyme kinetics reaction mechanisms and analytical solutions of non- linear reaction equations. J. Math. Chem. **48**, 179–186 (2010)
6. Rahamathunissa, G., Rajendran, L.: Application of He's variational iteration method in nonlinear boundary value problems in enzyme–substrate reaction diffusion processes: part 1: the steady-state amperometric response. J. Math. Chem. **44**, 849–861 (2008)
7. Pavithra, S., Saravanakumar, R., Rajendran, L.: Modeling of non linear enzyme reaction process using variational iteration method. Int. J. Comput. Eng. Res. (IJCER) **6**(08) (2016)
8. Biazar, J., Eslami, M.: A new homotopy perturbation method for solving systems of partial differential equations. Comput. Math. Appl. **62**(1), 225–323 (2011)
9. Rabbani, M.: New homotopy perturbation method to solve non-linear problems. J. Math. Comput. Sci. **7**, 272–275 (2013)
10. Wazwaz, A.: Adomian decomposition method for a reliable treatment of the Bratu-type equations. Appl. Math.Comput. **166**(3), 638–651 (2005)
11. Wazwaz, A.M.: The modified Adomian decomposition method for solving linear and nonlinear boundary value problems of tenthorder and 12th-order. Int. J. Nonlinear Sci. Numer. Simul. **1**, 17–24 (2000)
12. Wazwaz, A.M.: A reliable modification of Adomian decomposition method. Appl. Math.Comput. **102**, 77–86 (1999)
13. Anitha, A., Loghambal, S., Rajendran, L.: Analytical expressions for steady-state concentrations of substrate and product in an amperometric biosensor with the substrate inhibition—the Adomian decomposition method. Am. J. Anal. Chem. **3**, 495–502 (2012)
14. Muthukaruppan, S., Eswari, A., Rajendran, L.: Mathematical modeling of a biofilm: the Adomian decomposition method. Nat. Sci. **5**(4), 456–462 (2013)
15. Sivasankari, M.K., Rajendran, L.: Analytical expression of the concentration of species and effectiveness factors in porous catalysts using the Adomian decomposition method. Kinet. Catal. **54**(1), 95–105 (2013)
16. Siddiqui, M., Farooq, A.A.: A comparison of variational iteration and adomian decomposition methods in solving nonlinear thin film flow problems. Appl. Math. Sci. **6**(99), 4911–4919 (2012)

17. Cavlak, E.: Mustafa Inc, On numerical solutions of a new coupled system by using the Adomian decomposition method and He's Variational iteration method. Phys. Scr. **78**(045008), 1–7 (2008)
18. Wazwaz, A.M.: A reliable analytic tool for solving linear and nonlinear wave equations. Comput. Math. Apps. **54**, 926–932 (2007)
19. Bildik, N., Konuralp, A.: Two-dimensional differential transform method, Adomian decomposition method, and variational iteration method for Partial differential equations. Int. J. Comput. Math. **83**, 973–987 (2006)
20. Inokuti, M., Sekine, H., Mura, T.: General use of the lagrange multiplier in nonlinear mathematical physics. In: Nemat-Nasser, S. (ed.) Variational Method in the Mechanics of Solids, pp. 156–162. Pergamon Press, New York (1978)

SPRINGER NATURE

GPSR Compliance

The European Union's (EU) General Product Safety Regulation (GPSR) is a set of rules that requires consumer products to be safe and our obligations to ensure this.

If you have any concerns about our products, you can contact us on ProductSafety@springernature.com

In case Publisher is established outside the EU, the EU authorized representative is:

Springer Nature Customer Service Center GmbH
Europaplatz 3
69115 Heidelberg, Germany

The manufacturer's authorised representative in the EU is Springer Nature Customer Service Centre GmbH, Europaplatz 3, 69115 Heidelberg, Germany. If you have any concerns regarding our products, please contact ProductSafety@springernature.com

Printed and bound by CPI Group (UK) Ltd, Croydon, CR0 4YY
28/09/2026
02224086-0010